Prometheus Bedeviled

Prometheus Bedeviled

SCIENCE AND THE CONTRADICTIONS OF CONTEMPORARY CULTURE

Norman Levitt

RUTGERS UNIVERSITY PRESS

New Brunswick, New Jersey and London

Library of Congress Cataloging-in-Publication Data

Levitt, N. (Norman), 1943–
    Prometheus bedeviled : science and the contradictions of
contemporary culture / Norman Levitt.
        p.   cm.
    Includes bibliographical references and index.
    ISBN 0-8135-2652-3 (cloth : alk. paper)
    1. Science—Social aspects.   2. Science—Political aspects.
3. Science—Philosophy.   I. Title.
Q175.5.L475 1999
306.4'5—dc21   98-37628
                                                                                    CIP

British Cataloging-in-Publication data for this book is available from the British
Library

Manufactured in the United States of America

To Renee

Contents

Acknowledgments

First of all, I owe substantial thanks to the Open Society Institute, which supported this project generously through an Individual Project Fellowship. My colleagues in the Rutgers University Mathematics Department also provided invaluable help by giving me the time and freedom to write this book, notwithstanding its concern with matters far removed from mathematics. I am in their debt. Beyond that, thanks go to dozens of friends and associates who ungrudgingly gave me the benefit of their views and insights. I mention, in particular, Gerald Holton, who read early drafts, and whose encouragement was a most welcome sign that my efforts might be worthwhile. I am also particularly indebted to Mario Bunge, Ralph Raimi, and Rafael Pardo, important and active scholars in their fields who nonetheless took the time and trouble to read my manuscript and to offer useful comments and criticism. I thank Helen Hsu of Rutgers University Press for her enthusiasm. Deep thanks, too, go to my copy editor, Alice Calaprice, who valiantly put up with my execrable typing, my anarchic referencing system, and my occasional misadventures with the relative pronoun. Finally, the greatest thanks and gratitude go to my wife, Renee, who patiently tolerated my proclivity for working on the book at all sorts of ungodly hours.

Norman Levitt
New York, N.Y., October 1998

Prometheus Bedeviled

The Rule of Opinion and the Fate of Ideas

If, as the twentieth century staggers to its end, the world is less riven by national rivalry, ethnic hatred, political irrationality, and simple human bloody-mindedness than it has been in previous centuries, this information will come as a stunning surprise to those of us who tax ourselves with the disagreeable necessity of scanning the headlines each morning. The key to civic peace within nations and reasonably enduring concord among them remains obscure, if, indeed, it is even knowable. Nonetheless, if one insists, despite all, on finding grounds for a quantum of optimism, there is something to cheer in what seems an overwhelming agreement among the peoples of the world and those who speak for them to use a few particularly resonant words as symbols for what is presumably desirable in the foundations of civic and political life and efficacious in the promotion of humanity's control over its own fate. Two of these watchwords are "democracy" and "science."

Universal use of words is, of course, a very different thing from universal consensus as to what they mean, either in theory or in practice. Nor does their use exclude hypocrisy and outright fraud. Nonetheless, the terms "democracy" and "science" have not been drained of content quite yet, and if both are frequently abused by scoundrels, that abuse usually runs the risk of evoking countervailing resonances appropriate to the authentic meaning. These words are potent, and they always threaten to turn subversively upon their abusers. They are deeply unsettling to various forms of tyranny and imposture. "Democracy" cannot be invoked repeatedly without implanting the idea that there must be a limit to the abasement of even the humblest individuals by the forces of state and society.[1] "Science" suggests that even the most deeply entrenched authority is subject to scrutiny and reevaluation and must bend to the constraints of evidence and logic.[2] Social systems that prop themselves up upon the rhetoric of science and democracy without according them genuine respect are honing weapons that may well be turned against them.

Passionate admirers of either concept tend, as a general rule, to endorse the other, often with equal enthusiasm. They are ideas yoked together historically and, to a degree, by innate logic. Democracy, by definition, insists that everyone in a given culture must be allotted at least a minimal share of political and social authority. This doctrine is usually grounded on some notion of irreducible human dignity and hence is intertwined with theories of inherent human rights. Theories of this kind are as numerous as theorists, and there is, I think, no consensus on whether political rights are themselves fundamental, rather than being a necessary corollary, for practical reasons, of other rights considered to be more fundamental. In any case, the democratic thesis does not usually rely merely on "right" in the abstract for its justification.

Another argument, frequently held to be of equal force, is that only by giving some measure of political power to each and all can we assure that the final outcome of political processes tends, at least in the long run, to be maximally consistent with the general welfare. The image is that of an overall vector of social and political will arising as the sum of individual interests and preferences. The metapolitics behind this is that a political process so directed is, if not precisely optimal, at least unlikely to stray into unredeemed disaster. Clearly, democracy defined in such generality, not to say vagueness, does not mandate any specific political or constitutional system, nor any set of state institutions. It is a vexed question, for instance, how far democracy entails naive majoritarianism, although it seems obvious from the historical record that all reasonably democratic governments—in fact, all counterfeits of democratic government—have institutionalized majoritarianism at some level (or pretended to). By the same token, no specific formula for legal or legislative arrangements is inherent, although Americans, perhaps parochially, tend automatically to associate democratic government with written constitutions, representative legislatures, the division of powers, judicial review, jury trials, fixed terms for elective office, and so forth. There is also an abiding sense that individuals, rather than tribes, cults, guilds, cartels, or communities, are the true building blocks of society and that institutional safeguards of individual rights must be put in place for a democratic polity to function. However, the democratic ideal per se remains latitudinarian in that it merely sets out some rather general limits for the structure and performance of governments without providing any clear blueprint for their organization. The one indispensable component is some mechanism for assuring, in theory, that individual voices can not only be heard, but can also exert some measure of influence on the decision-making process.

It is noteworthy that the idea of democracy still retains its force despite a historical record, now quite extensive, that demonstrates that majoritarian tyrannies can arise, sometimes with stunning swiftness, out of what seem to be democratic institutions or democratic social movements. This possibility is especially clear in the case of societies riven by ethnic, religious, or racial differences. Does this fact hint at a fatal flaw lurking in the very notion of democracy? The answer very much depends on whether one believes that authentic democracy can exist in the absence of a prior commitment on the part of an overwhelming majority to a consensus that the majoritarian will must be thwarted, from time to time, in the interest of certain higher values. It might also be argued that to count as the genuine article, a democratic polity must commit itself to a strong notion of individualism, for how else could it deter the recurrent impulse to degrade or disenfranchise people on the basis of their membership in disesteemed groups? Experience clearly warns us to distinguish between democracy and naive populism, and makes clear that the capacity for democratic citizenship is not simply an attribute of uninstructed human nature.

Nonetheless, the raw idea of democracy as such remains a commanding one. Nowadays, we don't accept that grave political crises can be authentically and stably resolved unless some presumptively democratic mechanism, usually an election or plebescite based on universal suffrage, is part of that resolution. For most political intellectuals, non-Western as well as Western, it is axiomatic that democracy of some kind must be woven into the ground plan of political life.[3] When political views clash, advocates of each assert (with whatever sincerity) that the superiority of their position lies in its greater consonance with democracy. Socialists argue that democracy can only be fulfilled by extending its reach into most economic matters. Free-market conservatives retort that only marketplace individualism is consistent with the autonomy appropriate to a true democracy. Chauvinists of various breeds, whose programs by definition renounce the universalism inherent in most democratic theory, nonetheless appeal to a presumed desire for democratic institutions within the favored group. It is still possible to find thinkers who are principled anti-democrats, but the very wariness, delicacy, and circumspection with which they are constrained to express their opinions demonstrates that the idea that democracy is indispensable is preeminent in contemporary social thought. Though hypocrisy doubtless abounds in human affairs as it has always done, the notion of democracy remains a mythic presence in virtually all political discourse that is meant to be taken seriously, from the most wistfully idealistic and universalist to the most pugnaciously nationalistic and exclusive. To be avowedly undemocratic is to court failure and irrelevancy in almost any political context.

Turning to science, we find an idea whose institutional embodiment is very different from the governmental and constitutional arrangements through which democracy is presumably made concrete. Even by the most generous count, which would accept physicians, engineers, computer programmers, science teachers, and the like as practitioners of "science," only a small minority of adults, worldwide and in every nation, belongs to the scientific community. Inclusion, moreover, does not follow from the simple desire to be included. Membership is selective, essentially by definition, and selection is keyed to certain competences. The filtering process is sometimes offhand and informal, but it is usually rigorous, and frequently severely rigorous. No society, however egalitarian, has ever eliminated the sense that science is an elitist calling, that it demands raw intelligence and special skills that far exceed what is to be expected of the average person. Although improvements in the effectiveness and comprehensiveness of science education may ultimately succeed in developing scientific talent in more people, drawn from a wider range of social and economic backgrounds, it seems doubtful that scientific competence will ever become widespread, let alone universal.

Moreover, science is not a plateau whose members stand at uniform height above the general competence of society, but rather a steep-sided peak rising from a base already quite elevated. Within the scientific culture, one finds, in general, that a disproportionate amount of achievement and progress comes from the work of a relatively small number of extraordinarily able individuals. The history of science is often popularly presented as a record of heroic individualism. While there is considerable distortion involved in this imagery, there is also a great deal of truth. This can make the very existence of science seem somewhat miraculous. One might well imagine that a species very much like the human race could have evolved, but without the capacity to produce singular individuals whose talents outrun those of ordinary people to the extent manifested in an Archimedes, Newton, Pasteur, or Einstein. Without such unaccountable talents, it seems doubtful that anything like modern science could have arisen. For while science is undoubtedly cumulative and able to assemble the relatively modest contributions of millions of individuals into a prepotent intellectual structure, it is also saltational and needs the periodic impetus of unique, intuitive visions to drive it onward. It is difficult to account for prodigious talent in purely evolutionary terms. Its existence might even be regarded as a miracle by those susceptible to such explanations.[4] Nonetheless, it seems to be an ongoing, even a reliable, miracle—an important reason for believing that the progress of science is open-ended and, barring a hideous global catastrophe, is likely to continue indefinitely.

Yet if science seems ineluctably elitist, both in the rigorousness of its recruitment processes and in its extraordinarily sharp gradient of achievement as a function of talent, it has been historically associated with a democratic, even an egalitarian mood in the surrounding culture. In part, this is an accident, or so it might seem, of history, geography, and geopolitics. The idea of democracy as the core political ideal of our culture grew side by side with the idea of a systematic, cumulative, expanding, and ever more accurate science. By the same token, the authoritarian presuppositions that had to be defeated for democracy to emerge as our primary political paradigm were closely linked, and sometimes identical, with the obscurantist articles of faith that science had to sweep aside in order to gain its place at the center of our contemporary knowledge system. In some cases, this is utterly obvious. The entrenchment of dogmatic religion was (and, to some extent, still is) an important prop of a social order based on hereditary caste and class. Simultaneously, it was wedded to an epistemology that automatically excluded both the modes of inquiry on which science depends and the conclusions about the physical and biological universe to which it inexorably led.[5] Consequently, religion had to be annulled and diluted as doctrine, or divested of its political power and shunted to a subsidiary social position, for either democratic politics or science to thrive.

Beyond this, there are further linkages, harder to correlate with specific historical mechanisms, between the respective ways of thinking about the world that authorize democratic politics, on the one hand, and the standard notion of scientific validity, on the other. The most obvious relation is that both viewpoints arose and became entrenched, more or less, in the same place—western Europe—at approximately the same time—somewhere between 1600 and 1800, give or take a year or two. They diffused in tandem to other cultures—first within Europe, then more widely—and took hold of the imaginations of various peoples at more or less the same time. Aside from their intertwined historical and geographical trajectories, however, there has been an ongoing symbiosis between the scientific viewpoint and the emerging democratic ethos. Some commentators, in singing the splendors of science, treat this as near-axiomatic, as though it followed unproblematically from the meanings of the terms.[6] I think this overstates the case.

In my view, the mutual reinforcement of the two systems of thought has been a subtle and sustained process whose historical particulars are not always easy to trace and which has been largely unmarked by signal, well-remembered events. Scientists, like poets, artists, and, for that matter, beekeepers, have always included among their number individuals with strongly held political ideas who are inclined to urge them upon the society at large. Yet the spectrum of ideas thus represented is no narrower than that to be found in the much larger community of nonscientists. If the political consensus within the scientific community of a given nation at a particular moment has been far more enlightened than the point of view that officially prevails, this has usually been in circumstances where scientists have had, at best, limited access to the foci of power and the organs of opinion.

Scientific literature, in the narrow sense, is usually silent on political questions. The code of scientific etiquette long in place simply does not countenance emphatic political advocacy or ideological *obiter dicta* within scientific texts, and scientists tend to resent their appearance even when they agree with their sentiments. How, then, has the developing ethos of science managed to nudge societies in the direction of liberal values, by which I mean opposition to absolutism and to arbitrary authority, negation of hereditary privilege, openness to a wide spectrum of opinion, and desacralization of mere tradition? The process is one of slow diffusion, during which the liberal ideals propagated rapidly lose most traces of their original association with scientists and the scientifically well informed. Nonetheless, one can postulate a plausible mechanism. To begin with, successful science, in almost every instance, owes little to what might be called anti-liberal values. It finds small virtue in blind appeal to authority, or in the restriction of speculative thought, or in the imposition of a hierarchy not founded on meritocratic values. Repeatedly, scientists have found themselves vexed by pressure to conform to unexamined assumptions, whether those of religious authority or those of ingrained tradition. Moreover, some of the most resplendent achievements in the progress of science have materially challenged the postulates of conservative belief systems, so that to place one's faith in the validity of a given scientific advance is to abandon one's faith in the reliability of tradition per se.

Even temperaments conservative by inclination are thus to some degree unmoored by this process from unthinking allegiance to articles of faith. Attitudes consonant with important axioms of liberal social thought thereby come to be naturalized among the scientifically literate, though of course this process is sometimes grudging or partial (and never uniform). It is not all that difficult, in any age, to find distinguished scientists in whom the bleakest anti-liberalism prevails.[7] All the same, it is a fair guess that during the past three centuries, the presence in society of ever-growing numbers of people conversant with science, its methods, and the frame of mind necessary to understand and to advance it has had the effect of seeding the educated population of Europe and North America (and ultimately, the rest of the world) with kernels of liberal opinion. These have generated attitudes toward authority, liberty of thought, and meritocratic principles that pointed toward the consensus that now underwrites democratic ideology.

There were, naturally, celebrated individuals whose advocacy of the nascent liberalism was clearly tied to their knowledgeable enthusiasm for science—Voltaire, the Encylopedists, Lessing, Franklin, and Jefferson, for example, come easily to mind.[8] In some cases, scientific prestige gave added weight to political advocacy. But more important, probably, were thousands of now-forgotten individuals of modest talent and small fame whose influence never went beyond a local circle. It was through them that a wide stratum of important opinion came to be permeated by certain conceptions of knowledge and certain congruent conceptions of social organization. The widespread European enthusiasm for eighteenth-century Freemasonry gives concrete organizational evidence of this phenomenon. Presumably, this was replicated on a smaller, more informal scale in many ways.

I am by no means asserting that the increasing power and prestige of science constituted the principal, let alone the sole, impulse behind the swelling liberal consensus. Clearly, economic, demographic, agricultural, and strictly political developments were at least of equal consequence. In fact, the most important role of science in the creation of modern political and cultural norms may simply have been to enable the development of vastly more productive industry and agronomy, thereby creating conditions of prosperity and security in which egalitarian ideas lost something of their previously utopian character. Nonetheless, to the extent that the ethos of liberalism, pluralism, and democracy entails a certain sense of how the truth, in all sorts of matters, is to be sought for and confirmed, how evidence is to be weighed, and, above all, what preconceptions about the material nature of the world are to enter into these judgments, natural science has clearly been the source and model. A crucial effect may have been largely negative— to dispel the special prestige of religious, hereditary, and state authority— but even on this basis, the debt of democratic normative values to scientific epistemological values remains immense. The science historian Margaret Jacob summarizes an important historical truth: "[T]he willingness of experimental scientists from Boyle onward to give relative autonomy to the church and to participate in voluntary associations (such as the Royal Society) inadvertently paved the way for the evolution of civil society, and it in turn cleared a path for more egalitarian forms of social behavior and political life both in Britain and the American colonies."[9] A more acute and dramatic example is furnished by recent events in China. Surveying the role of scientists in the Chinese pro-democracy movement, China scholar H. Lyman Miller finds:

> For scientists such as Fang [Lizhi] and Xu [Liangying], the antiauthoritarian norms of science translated easily into a classically liberal politics. The message these scientists carried into the larger political arena defended above all the sanctity and worth of individual autonomy and conscience above the claims of state and society. They advanced a pluralist politics rooted in appeals to reason. They called for all of the freedoms attendant to liberal politics—freedom of speech, assembly, the press, and so forth. Above all, they placed sovereignty squarely among China's citizenry, not in the state itself.[10]

In the context of India, a pro forma democracy still smothered by an oppressive blanket of traditional prejudices and retrograde social norms, Meera Nanda observes: "The oppressed Others do not need patronizing affirmations of their ways of knowing, as much as they need ways to *challenge* these ways of knowing. They do not need to be told that modern science is no less of a cultural narrative than their local knowledges, for they need the findings of modern science, *understood as transcultural truths*, in order to expose and challenge local knowledges."[11]

Given the repeated historical tendency of science and its value system to nurture a liberal-democratic ethos, the inverse question is of interest as well. To what extent are the norms of liberal democracy required for natural science to flourish? The answer here is clearly contingent on one's notion of flourishing. The most efficiently totalizing and ruthlessly authoritarian of political systems in history have arisen in this century, and their singularly bleak efficiency has been the product of the technocratic pseudorationality that science helps to empower. Moreover, these systems have sustained and even exalted their own scientific subcultures, eagerly granting the scientist privileges, prestige, and immunities sternly denied to the general run of citizenry. Just as eagerly, many scientists in such societies have grasped those tainted gifts. Disappointing as it may be to one's sense of fitness, often they have done so without diminishing the quality and value of their scientific work.

A tender political conscience is not a prerequisite for scientific distinction or outright genius. Albert Einstein, who by now may safely be called the greatest scientist of this century, appears in our secularized hagiography as a notable paragon of humane politics and sterling ethical conscience (at least in pubic affairs), but this is rather a matter of our good luck than a clearly necessary condition of supreme scientific insight. If special and general relativity had first issued from the mind of a Nazi fanatic or a Stalinist timeserver, they would have been just as valid, although the circumstances of their gestation would have been depressing to contemplate. Despite the fact that it has been unfashionable for some decades to say so, it is possible that gross character flaws preclude one from creating art of consummate greatness.[12] Science is more philistine. A powerful scientific imagination is no guarantor of moral integrity. Success in science attests to few virtues beyond persistence and the capacity to work very hard, at least in spurts. The value of scientific ideas is completely independent of the moral soundness of the people who invent them.

Nonetheless, the historical record seems to show that in the long run, science as an institution eventually finds itself locking horns with authoritarian regimes despite the best efforts of those systems to woo, encourage, and reward scientists. The clash of cultures is too essential to be smoothed over indefinitely. The more valuable science grows—the more requisite to the instrumentalities of power it becomes—the more it seems to the authorities that direct control is essential. As science accrues prestige within a culture, the subcultural values that allow it to function draw scrutiny from the larger population, and the desire to introduce those values into civil affairs grows correspondingly. Hence, if its position is at all insecure, the political elite finds itself increasingly wary and skeptical of even the limited autonomy ceded to scientists.

In recent years, China has provided a clear instance of this phenomenon. According to H. Lyman Miller:

Despite the highly stratified nature of scientific communities, science professes an ethos or ideology that is inherently anti-authoritarian. In co-opting science and its ideals in the service of reforms, the Deng leadership opened the way for scientists to judge the regime itself according to those ideals. At the same time, the regime's legitimacy and authority was tied explicitly to its claim to a scientific methodology and body of principles proven in practice; hence the party leadership could not but be extremely sensitive to changes in interpretation of science's findings about the natural world. Issues of scientific judgment seemingly remote from political concern had the potential to rapidly become controversies about the correctness of prevailing Marxist principles. Out of such controversies emerged clashes over who was competent to judge the validity of scientific arguments, over the relationship between science and Marxism, and, ultimately, over the relationship between knowledge and power.[13]

The autonomy that science needs in order to function well may indeed be granted by an authoritarian system for a limited time and a limited purpose. But it seems to be the historical tendency that even within such well-warded hothouses of knowledge production, a scientific subculture will grow up that not only reveres the internal social forms that enable it to perform its task, but seeks eventually to extrapolate these values and to have them prevail within the larger social world. Envious of the singular prestige of science (though they themselves have nurtured it) and wary of the subversive appeal of a liberatory ethic backed by that prestige, those who hold power find themselves impelled, however reluctantly, to check the growth of scientific autonomy and to resort to forceful means to remind scientists of the boundaries that are meant to contain them. Rarely can this be done tactfully enough to avoid demoralization of the scientific community or to escape the unintended consequence of creating principled and intellectually vigorous opponents of the given order.

To this point, all I have said seems celebratory to the point of complacency about the historical role of science. The view I have sketched is usually denoted "Whiggish" by political, cultural, and intellectual historians, in that it portrays history (at least the recent history of western European cultures) as a march of progress, leading monotonically to a more enlightened social order and to an increasingly accurate grasp of the principles underlying the natural world. In recent years, Whiggery has acquired a sour reputation. Those espousing it have been taxed with failure to recognize the enormous harm to non-Western populations which accrued from the unabashed expansiveness of the West, and with blindness to the rich spectrum of "local knowledges" that Western science, with its monomaniac insistence on the universality of its own brand of rationality, threatens to obliterate.

A fashionable view among academics these days, popularized by thinkers such as Michel Foucault, is that the apparent advance of fairness, justice, and human dignity has been a charade behind which lurk ever more subtle and ever more effective modes of repression and authoritarianism. Likewise, the advance of scientific knowledge is widely held to be a mere evolution of social consensus in a direction serviceable to the forces, manifest or covert, that structure the social order. The idea that either social values or epistemological methods represent a genuine advance over what reigned in prior centuries is decried as a naive illusion. The further idea that science, per se, represents a transcultural method of astonishing accuracy for coming within sight of the truths of the natural world, and that it therefore stands above previous or extant culture-bound systems of knowledge, is met with the rebuke that such universalism represents "cognitive colonialism."

I think that this current genus of academic nihilism is vain, captious, and ultimately unavailing. I have already written about this topic at some length.[14] I don't intend to speak of it systematically in what follows, and insofar as I shall speak of it, it will be more as a symptom than as an interesting body of thought. For as thought it is weak, shallow, grossly tendentious, and decidedly uninteresting. In terms of the historical issues, especially as regards science, I have been and remain an unabashed "Whig."[15]

True, the injuries done to numberless non-Western peoples by Western arms (or Western microbes) comprise a record of suffering and cruelty beyond hope of reparation. Nonetheless, all the ideals that seem to offer any hope for the future (including the very ideals under which the bloody history of Western imperialism is arraigned) are the product of intellectual forces that first took root and flourished in the West and bear the ineradicable stamp of that genesis. This is not a particularly gratifying fact to the sensibility that prefers to view historical actors in terms of undivided good or evil, but so far as I can see it is nevertheless a fact, though a deeply ironic one. The postures that have taken the fancy of would-be scourges of Western culture, from which vantage all things Western are seen as indelibly tainted by the blood-guilt of a thousand expansionist adventures, may provide some transient sense of high virtue to those intellectuals who assume them. All the same, I doubt that much will ever come of them in terms of relief of actual suffering or advancement of justice. Those parts of the human community that have suffered most from the depradations of European civilization and its scion cultures have, as their most hopeful recourse, the forthright adoption of the best that can be distilled from the European political ethic. Their most worthy revenge would be to impel Europe and its daughter cultures to live up to that very same set of values. This is, again, replete with irony and gives slight satisfaction to those for whom justice must always have a retributive element. But it is the only viewpoint, by my reckoning, that currently offers much political hope to the world.

Likewise, in view of the more than casual complicity of Western science in developing the instruments of Western hegemony and non-Western degradation, it might seem no more than simple justice to view Western science as a deeply flawed system of thought, ephemeral in its intellectual dominance over the world. This, indeed, has been the theme of the most vocal and pugnacious elements of the new academic field called "science studies."[16] In its unmuted form, it holds that all sorts of knowledge systems are as epistemically valid as modern science, that other "ways of knowing" deserve to be regarded as equally true, especially if they arise within societies or social groups that the Western world and its ruling classes have grossly mistreated. Indeed, from this point of view it sometimes seems that Western science is the *only* form of knowledge that is to be treated with skepticism!

This attitude, stripped of its quasi-humanistic political rationale, is shared by millions of other Westerners. The idea that outsider cultures have special access to mystical wisdom or preternatural powers is an old one in the Western world. It was rife in the culture of ancient Rome and renewed itself in early modern Europe, where interest in occult practices bred a fascination with the Hebrew Kabbala and the pseudo-Egyptian lore of Hermes Trismegistos. In the eighteenth century, it was manifest in Rosicrucianism, a credulous offshoot of Enlightenment Freemasonry. At the turn of the twentieth century, it appeared, often allied to spiritualism, as enthusiasm for the supernatural powers of swamis, fakirs, Tibetan lamas, and so forth. In today's culture, all these themes endure (with some new darlings—the sages of North American and Australian tribal cultures—now added) and are hardly absent from even the most educated strata.[17] The term "New Age" (though nothing about it is particularly new) embraces most of this eclectic concoction. Within the supposedly sober scholarship of science studies, it is not terribly hard to find wistful echoes of this subculture, often mounting to strong endorsement.

I find virtually nothing to admire in this enthusiasm for relativism or for the kind of polyvocal epistemology that regards science as just one narrative among many, and not even *primus inter pares* at that. I believe that science—Western science, if you must, but really, the worldwide science that descends from ideas and practices first developed in western Europe—is an enormously strong and stunningly accurate way of finding out how the natural world works. Its truths, though they may be incomplete in certain ways and occasionally in need of reformulation in an enlarged context, seem to me stable and enduring, well beyond what any system of social convention could ensure. The reason is simply that they are very closely molded to the contours of the real, objective world. Notes E. O. Wilson: "Science, its imperfections notwithstanding, is the sword in the stone that humanity finally pulled. The question it poses, of ultimately lawful materialism, is the most important that can be asked in philosophy and religion. Its procedures are not easy to master, even to conceptualize; that is why it took so long to get started, and then mostly in one place, which happened to be western Europe."[18]

From this, it follows that other "systems of knowledge," whether of Andean shamans, Tibetan necromancers, or Tennessee fundamentalist state legislators, are much weaker reeds for anyone with an honest interest in the actual constitution of the universe to lean upon. To be blunt, when there is evident contradiction between any of these bodies of belief, on the one hand, and reasonably well-confirmed science, on the other, one must make the default assumption, in the absence of a mountain of contravening evidence, that the scientific belief is right and the alternative system simply wrong in the most straightforward sense. Stated thus nakedly, this is a credo that makes even some highly educated people—scientists certainly among them!—edgy and unhappy, even when they have no particular quarrel with scientific rationality and no particular urge to endorse the claims of a conflicting belief system.

It is a view that seems somehow impolite in the face of an ethos that incessantly stresses the virtues of an open mind, and which enjoins its adherents to be at all times deeply respectful of other human cultures. Nonetheless, blunt as it is, I think it is a proposition that bears frequent and confident repetition, one that would, in a far more perfect world, be a widespread, if not universal, article of belief.[19] It grants what science unequivocally *deserves* after centuries of striving for—and achieving—increasingly accurate, increasingly comprehensive, and increasingly subtle knowledge of the natural world. It is what is owed to science for its accomplishments—and for its promise. Says Sir Karl Popper (and rightly so): "Thus we may say that Bacon's Utopia, like most Utopias, was an attempt to bring heaven down to earth. And so far as it promises an increase of power and of wealth through self-help and self-liberation through new knowledge, it is perhaps the one Utopia that has (so far) kept its promise. Indeed it has kept it to an almost unbelievable extent."[20] If this be Whiggery, let us make the most of it!

I raise these points and make these strong, though fully justifiable, claims because in what follows, I intend to examine the nature of what to me is an unsettling, and indeed, a deeply depressing phenomenon. Despite all that I have said about the close correspondence between science on the one hand and liberal-democratic political ideology on the other, it seems to me that contemporary societies—more precisely, those that have deeply naturalized the liberal-democratic ethos—find themselves still greatly ill at ease with science, puzzled by it, intimidated by it, and unsure how it fits into the scheme of politics, policy, and civil society. If, as I believe to be true, the historical alliance of these two great systems of thought was more than merely opportunistic, the congruences between them have been and remain less than perfect. To put it another way, the triumph of liberal institutions (and, despite marked skepticism that all is for the best in our current political frame, I certainly think it was a triumph) has by no means set the stage for the easy and ungrudging assimilation of the scientific worldview into the outlook of society as a whole. As I noted above, the body of scientifically competent people, though nowadays huge by historical standards, is nonetheless a tiny fraction of the population. Moreover, its culture is emphatically meritocratic. Even worse, from the point of view of deeply embedded popular feeling, it is elitist outright.

Regarded coolly, elitism, in the literal sense, ought not to affront democratic sentiments.[21] It means, simply, ceding certain difficult functions to those best fitted by training or by natural gifts to carry them out. Playing the *Hammerklavier* Sonata to a paying public, like playing above-the-rim basketball for an NBA team, is the province and privilege of an elite. So, for that matter, is playing with hadrons at Fermilab. In some areas—sports is certainly one of them—popular culture accepts this brand of elitism gracefully without any cavil (aside, perhaps, from grumbling at astronomical salaries). Perhaps, in the case of sports, most people (or at least most men) have tried their hand at it—playing basketball in the playground, competing with high-school squads or on college intramural teams, participating in health-club pickup scrimmages, and the like. They have, in most cases, obtained some satisfaction from the process, realized the joy of honing their own skills, come to terms with their own limitations, and developed an informed appreciation for the outsized talent of the phenomenally gifted. Alas, the analogous experience with respect to science does not seem to be commonplace among nonscientists.

Though there are exceptions, the usual run of high-school science curricula is more burdensome than gratifying. The same is true for most college-level science survey courses intended for nonspecialists. There is a certain *de haut en bas* feeling to most of these encounters. Typically, students (and, at the high-school level, not infrequently teachers) are baffled outsiders looking in, trying to master modes of discourse and habits of thought that do not grow out of direct personal experience. It is a commonplace of psychology that there is a sharp difference between "knowing that" and "knowing how." It is the sort of thing that most of us have run into in learning how to drive a car or ride a bicycle, use a computer, or speak a foreign language. At the beginning, with the help of a friendly instructor, we have some kind of idea of *what* it is that we are supposed to be doing. However, even if we are letter perfect at this level, we find ourselves clumsy, nervous, and unsure in the actual performance; we don't really know *how* to do it. Only through experience does the theoretical knowledge become know-how. Indeed, often this is marked, physiologically, by the transfer of control over our performance from the cortex to the cerebellum, a more "unconscious" or "automatic" part of the brain.

I submit, conjecturally, that something similar is involved in "learning" science, that in part it is a matter of certain mental functions becoming habitual and "unthinking," even though "higher thought" is obviously involved.[22] It is a matter of absorbing a corpus of insights as a reflexive component of one's mental apparatus, so that it may be invoked and put into use without conscious reflection. A laundry list of facts, and even of the theories that supposedly bind them together, even when recalled perfectly, is not the same thing as comprehension at the level of mature insight. Without the development of that peculiar brand of insight, science simply does not cohere as a way of knowledge. Instead, it remains a file cabinet of odd facts. Even those who study science seriously for a number of years, with the initial intention of making it their profession, sometimes fail to naturalize, within their own minds, the inner, unspoken logic of the discipline they aspire to. Needless to say, this can induce recriminations and lasting resentment.

What the culture has to offer laypeople in terms of informal instruction in science is often enlightening and entertaining, and frequently it induces friendly feelings toward scientists and their elaborate if mysterious repertoire of conjuror's tricks. Rarely, however, does this acquaintance go beyond the most undemanding superficialities. Lacking the enforced discipline of classroom work, it produces understanding even more evanescent than that coming out of once-over-lightly survey courses and such. Personally, I am enormously fond of the museums, displays, educational broadcasts, and so forth that are devoted to science. I think they can be wonderful recruiting devices for bright young people with a bent for scientific work. Yet they are severely limited in their ability to increase the scientific sophistication of the general public. Sophistication, to state it again, is not just a matter of direct technical knowledge. Professional scientists, in the main, are narrow specialists, and by inclination highly averse to offering "expert" opinions in areas at any distance from their own. Nonetheless, it seems to be the case that they can comprehend the logic and methodology of even unrelated scientific disciplines with a facility and accuracy that elude most laymen, even the most earnest. As geneticist Steve Jones puts it, "Most scientists are quite ignorant about most sciences, but all use a shared grammar that allows them to recognize their craft when they see it."[23] This is because the method of approach of one specialty, while not a template that can instantly be transferred unmodified to another, is an instance of a general pattern of thinking about the phenomenal world and of analyzing its regularities which prevails throughout science.

I do not claim that there is any readily formulated "scientific method" that stands apart from the actual practices of scientists and that can be inculcated *in abstracto* as a preliminary step in scientific training. I claim merely as a "Darwinian" fact that, inasmuch as scientific success in any specialized area depends upon making one's thinking as transparent to objective and unremitting natural reality as possible, engagement in scientific work produces similar habits of thought across the enormous spectrum of scientific interests, a process analogous to convergence in biological evolution. It is direct working experience, rather than catechism, that is responsible for this effect.

Some critics of science speak as though there were some formalized scientific worldview in which fledgling scientists are instructed and to which they must pledge fealty in order to be admitted into the select circle. This is wildly false. Typically, a scientific apprenticeship consists of little but a sequence of highly technical and very specific courses and laboratory experiences, growing rapidly in difficulty and degree of specialization over the span of a few short years. Almost nowhere does the student meet with a formal overview of the subject. History, methodology, and the place of the given specialty within the overarching framework of science as a whole are, for the most part, thoroughly neglected. The questions that obsess philosophers and historians of science are never encountered in the context of serious course work, and are rarely encountered at all. The remarkable thing is that there nonetheless develops, among those professionally committed to science, a certain comprehensive sense of how knowledge works and how the world works. It comprises a corpus of background assumptions varying very little across the vast panoply of specialties, subspecialties, and sub-subspecialties that comprise the scientific endeavor.

To say it again, the best explanation is evolutionary in spirit; certain ways of looking at the world allow one to assimilate specialized knowledge comfortably and, eventually, to make progress in scientific research. Attitudes of mind that diverge too greatly from this norm are self-defeating. Consequently, a lingua franca comes into being that binds together most of the community in a tacit consensus as to "what science is all about." But it also enables scientists from widely disparate areas to understand one another's way of working with a reasonable degree of insight, even in the absence of specific technical background knowledge. I am not suggesting that a particle physicist, say, can transfer, at will, into molecular embryology (though some have doubtless done so), nor the other way around, but I do contend that there will be a fair degree of mutual intelligibility between the physicist and the biologist even before anyone does any special cramming.

The other side of this coin is that people without an intense background in serious science—even highly intelligent, very well motivated people—find it hard to follow scientific discussion, that is to say, not only highly specialized, technical discussion, but also the sort of thing that is intended for a "general" scientific audience. Similarly, the various popular or semipopular science publications—*American Scientist, Scientific American, Science Spectrum, The Sciences*, and so forth—are basically sold to, and read by, professionals, or at least by serious students who hope to become professionals. This is despite the fact that such journals are written and edited for maximal clarity and carefully stripped of daunting technicalities. More popular magazines—

Discover and *Natural History*, for instance—are further diluted, to the point where very little "hard" science is to be found in them. They are enjoyable, and valuable, but keep readers quite distant from the cognitive atmosphere that prevails in most scientific disciplines. Even so, I suspect that readership is still disproportionately drawn from the professional or apprentice scientific community. A glance at the letters column of these publications clearly reveals that professionals dominate.

This readership may be taken to represent a more comprehensive fact: if laymen have a certain difficulty in comprehending even generalized scientific discourse, they have far greater difficulty in contributing to it. A discussion—about some aspect of science in relation to public policy, say—may well start off at a level that seems to invite comments and suggestions from nonscientists. But more often than not, technical details and methodological intricacies turn out to be inescapable, and the honest layman starts to feel that matters are getting beyond him, and that he is being exiled, bit by bit, to the role of dimly comprehending spectator.

In such situations, silence is the strategy that best comports with dignity. Well-meaning scientists, honoring their own democratic ideals as best they can, may try to slow the pace of the discourse, to distill the common sense at the heart of the scientific discussion into a form more readily absorbed by the inexpert but determined citizen. Sadly, this process can only go so far, especially in the case of active controversies where the issues are still obscure and refractory to even the most able professionals. Laypersons find themselves confronting a bewildering array of unfamiliar categories, the relations among them a mystery. What is dead certain, what is plausible, and what is wildly improbable among a flurry of competing claims is an issue where their intuition falters.

Again, I want to contrast the nonprofessional's helplessness with the situation of a scientist in an unrelated field who tries to follow the same discussion. The specimen scientist might not "know" much more than the nonscientist about some of the technical material involved. Yet scientists are attuned to the intellectual rhythms of scientific discourse in general and have some good intuitive sense of what sounds solid and what rings false. They can sense, with fair accuracy, which positions are grounded in systematic research and good methodolgy, as contrasted with those offering little more than speculation, or worse. Even more important, they probably will have mastered a trick that the nonscientist still finds elusive—that of forming opinions that are provisional and involve assigning probabilistic weights to the open possibilities, while acknowledging gaps or inconsistencies in the relevant data, and anticipating the entry of new information or new ideas. In short, scientists know how to form opinions that are set up in advance for revision or self-correction. Nonscientists have a hard time with this, at least in situations where science confronts them.

Beyond the frustrations of trying to deal with the difficulties of a subject that simply won't yield its secrets easily to the approximate and common-sensical approaches of everyday life, nonscientists often find science offputting at a psychologically deeper level, one haunted by phantoms and fears that have resonated within our culture for centuries. Although I am resolutely Whiggish in my sense that Western—and, indeed, non-Western—society has grown progressively more mature in its capacity for generating reliable knowledge, as well as in its ability to articulate (if not to effect) a sound political ethic, I would never claim that these strengths are uniformly well reflected in the attitudes of ordinary citizens. As one shrewd observer, moved to reflection by the group suicide of the Heaven's Gate cult, observed:

> It is a sobering fact that the most intensely scientific century in all human history has also been a time when tens of millions of people have perished at the hands of leaders who were unwilling to alter their presumptions according to objective evidence. Historians are still arguing about whether more people died because of Hitler's delusions than because of Stalin's or Mao's, but the death count amassed by just these three pillars of unblinking zeal is sufficient to rival, if not surpass, the multitudes whose lives were saved by science. . . . Though science is stronger today than when Galileo knelt before the Inquisition, it remains a minority habit of mind, and its future is very much in doubt.[24]

The universal acceptance of heliocentric astronomy is often cited as the classic instance of the triumph of sustained rationality over embedded tradition. One must reflect, however, that so far as "average" opinion is concerned, the idea that the Earth and the other planets move around the sun is accepted not because there is any sense of why there is more evidence for it than for its ancient alternative. It may be condescending to note that neither observational astronomy nor celestial mechanics is a deep-seated aspect of the intellectual equipment of most people, but nonetheless, it is true. Most individuals in industrial cultures accept the idea because it has been incessantly repeated since childhood and because there is no apparent emotional cost to accepting it. The fact that it displaces humanity from a supposed central position in the scheme of things has long since ceased to perturb the psychic equilibrium of the culture, probably because that kind of centrality was always too abstract to inspire widespread reverence.

On the other hand, the theory of biological evolution, especially as regards human origins, still provokes pain, rage, and defiance in many quarters. This is true despite the fact that, in any scientific sense, evolution is as thoroughly established as the picture of the solar system due to Copernicus, Galileo, Kepler, and Newton. Both are equally familiar to the public at large and are vouched for by the same scientific authority. Of course, certain influential religious groups, fundamentalist sects in particular, deplore the concept of evolution and make strenuous attempts to exclude it from basic science education or to supplement it with thinly disguised religious fables.[25] But religious groups—largely the same ones—also object, pro forma, to many aspects of our culture, such as premarital sex. The vast majority of people not only finds these mores acceptable—uncontroversial, in fact—but also regards those who denounce them as ridiculous prigs. For instance, virtually all of the popular sitcoms that infest the airways these days feature young, attractive people who enjoy active, and sometimes casual, sex lives, without benefit of clergy, and even without the dire premise of connubiality lurking in the background. Although religious traditionalists of various persuasions no doubt fume and mutter at this portrait of universal lubricity, it is clear that any organized attempt to decry it would be laughed to scorn. On the other hand, no producer of this kind of mass entertainment would ever endanger his bottom line by offering a sympathetic character— a high-school biology teacher, say—who is shown as routinely and sensibly advocating and defending standard, scientifically uncontroversial, evolutionary ideas. Any such depiction would clearly provoke anguished protests from the fundamentalist lobby, which would be answered, in turn, not by mockery, but rather by diffident muttering about evolution being "just a theory." It is clear that evolution remains "controversial" at the level of mass opinion, not merely because certain impassioned cultists denounce it, but also because, unlike heliocentric astronomy, it provokes anxieties about the status of humanity in the natural world. The idea that the human species arose by degrees from nonhuman predecessors remains deeply unnerving to many. Its religious opponents thrive, in this culture, precisely because they are able to play on this discomfort, even to the point of recruiting the passive assent, and sometimes even the active support, of many who are, in other respects, indifferent to the underlying theology.[26]

This is a particularly sharp and concrete instance of a diffuse and general unease, felt throughout the culture, that has arisen over time in response to the overall view of the universe—the world-picture—that science assumes for its practice and almost invariably reinforces by its findings. Rarely is this philosophical overview presented to the public when the achievements of science are put on display for general approbation, or even when they are being roundly denounced. It is not, nor has it ever been, the official position of "Science" or of any concrete scientific organization. There is no enforcement apparatus in place to assure that scientists become or remain faithful to it. Nonetheless, it is firmly in place as part of the discursive style and methodological strictures that are embedded in scientific practice. Moreover, it constitutes the baseline from which most scientists who choose to venture into a formal statement of their philosophical views embark.

I shall outline this viewpoint only briefly, forgoing any lengthy attempt to flesh it out or to defend it (for I believe it is defensible—and correct). Two key words that must arise in any accurate characterization of this viewpoint are "monism" and "reductionism." The former simply refers to the view that there is essentially only one kind of "reality," one kind of material existence, governed by its unique and invariable set of laws or, if you prefer, regularities. Thus, there is no split between "material" and, for instance, "spiritual" reality. If angels exist, on this view, then they must be comprised of the same kind of substance, behaving in accord with the same principles, as that from which aardvarks are made.

With regard to the ancient "mind-body" problem, a monist would maintain that, insofar as any sense is to be made of the notion of "mind," it must be understood as a physical function of a physical body. "Confidence in the unity of knowledge," E. O. Wilson reminds us, "rests on the hypothesis that all mental activity is material in nature and occurs in a manner consistent with the causal explanations of the natural sciences."[27]

Many recent critics of Western science have cast the seventeenth-century philospher René Descartes in the role of principal villain, on the understanding that Cartesian dualism divided the world into intelligent, comprehending, and masterful "subjects"—self-conscious, efficient "knowers," exemplified by (white, male) scientists and philosophers—and passive "objects," whether of "nonhuman nature" or of "culture," which are subservient to the categories imposed by the theorizing subject. This, allegedly, has been the philosophical sin responsible for our civilization's propensity for treating animals and forests as mere things, and for demeaning women and people of color. While not wanting to dismiss the historical logic of this characterization, which is useful for certain purposes, I wish to point out that it badly misreads the history of Western philosphy. Descartes' dualism represented a late, postmedieval attempt to rescue the world of thought from the monism toward which it was apparently heading, not least because of some of Descartes' own ideas. It sought precisely to establish the viability of a spiritual reality, manifest in the human soul, that could be said to exist apart from the quotidian material reality of inanimate objects, animals, and the physical bodies of human beings. It must be said unambiguously that within the mainstream of Western thought, especially scientific thought or thought heavily influenced by science, the dualist program failed, although its failure is not attached to any single great name. Descartes is, in this respect, a great dead end, undone by the relentless march of scientific monism.

Reductionism is a closely linked notion. It holds that even the most complex objects and processes exist, when all is said and done, as concatenations of simpler objects and processes, which, in turn, may be reduced to an even more basic level, and so forth, down to a presumably fundamental layer of reality. As E. O. Wilson puts it: "Thus, quantum theory underlines atomic physics, which is the foundation of reagent chemistry and its specialized offshoot biochemistry, which interlock with molecular biology . . . and thence, through successively higher levels of organization, cellular, organismic, and evolutionary biology."[28] This is a picture of the world that reflects the overwhelming consensus among scientists about how things work. It would be premature to declare the picture more or less complete. For one thing, we aren't sure yet whether the most fundamental entities now known to particle physicists—quarks and leptons—are truly "elementary"; some evidence now suggests that they are not. Nor does the acceptance of "reductionism in principle" enjoin "methodological" reductionism on scientists or anyone else, though this point is widely misunderstood. Zoologists, after all, study zebras, not the quarks and leptons of which zebras are presumably composed. The laws and regularities they observe are laws and regularities of zebra anatomy and behavior, not laws of physics. No sane person would suggest that it should be otherwise. Nonetheless, I venture that there are few zoologists who won't cheerfully concede that zebras are, in fact, constituted of quarks and leptons, and that their properties, including those of most interest to zoologists, are ultimately determined by what goes on at the quark-lepton level (or whatever level might turn out to be even more fundamental).[29]

These two principles of monism and reductionism might be united in a single term, "materialism," although one must take some care to use this locution with an awareness of how it has shifted in meaning over the past few centuries.[30] There is yet another idea often associated with the scientific viewpoint—determinism. This, roughly, is the notion that the future is completely determined by the state of affairs that now prevails, if, by "state of affairs" we encompass the state of the entire universe (whatever "the universe" happens to be). Likewise, the state of affairs now prevailing is an inevitable consequence of the state of affairs one minute, or one year, or one century, or one millennium ago. For more than two hundred years, the best mathematical models of fundamental physical law suggested that the universe is, in fact, deterministic. Since 1925 or so, this view has been under strong challenge because of the emergence of quantum mechanics, one of science's most astounding success stories. I think there is a strong argument that the supposed death of determinism is premature, though I will not vex readers with this discussion.[31] The rival "indeterminacy" view is, however, not a notion that easily fits in with naive or intuitive understandings of free will. In my view, *neither* philosophical position on these fundamental questions of physics offers much in the way of comfort to those who want to believe that the universe takes special account of the desires of humans or the fate of humankind.

The status of all these notions within the walls of a philosophy department is not my present concern. I do assert, however, that monism and reductionism are, de facto, part of the worldview of the scientifc community, and thus, in some sense, of science itself. This is manifest at several levels. First of all, in the ordinary, day-to-day workings of science, the language used, which is to say the categories and assumptions invoked explicitly or tacitly, incorporates the monistic and reductionistic view. To put it another way, the language and usage of scientific discourse simply has no room for nonmonistic or *explicitly* antireductionistic views.[32] Even beyond unthinking linguistic habits, however, monism and reductionism have a strong grip on scientific thought. Any theory that openly and specifically challenged them would be ipso facto subject to the most withering scrutiny. It would take mountainous quantities of direct evidence and enormous explanatory power to force most scientists to take any such notions seriously, even in the most tentative, hypothetical fashion. This is not to say that there have never been scientists with such inclinations—merely that they have been rare and their fate, to this point, unhappy.

Finally, there is the matter of scientists who venture into the mine fields of philosophical speculation. While this has not been a uniform group by any means, the overwhelming tendency among the most distinguished thinkers has been to accept monism and reductionism in some version.[33] Antimaterialism, in general or specific form, tends to make scientists uncomfortable. This, despite frequent allegations to the contrary, is not really a matter of dogmatism. Rather, it reflects long, amazingly successful experience in generating reliable knowledge and ever more comprehensive explanatory systems. Materialism, with its methodological and ontological restrictions, has worked splendidly. Attempts to evade or contradict it have invariably come to grief. Finally, there is the question of the basic worldview espoused by those scientists—only a small minority, to be sure—who delve into philosophy with attempts to draw larger lessons from what science has learned. Though there is no "standard" viewpoint, and scientists, no less than other thinkers, range over a wide spectrum of ideas and predispositions, the great majority of such scientist-philosophers are firmly, even emphatically, materialist in outlook.

The scientific consensus on these points, fascinating as it is, is not my chief concern, however. Rather, what I want to address is the sense that has leached into the culture that these dicta—monism and reductionism—are integral to the way in which science addresses the world. I don't mean to suggest that these terms, in themselves, would mean much to the average citizen, or even to one with a reasonably sound education. But the essential practical significance of these terms—that electrons are to be taken seriously, whereas angels are not, that any belief that qualifies as "scientific" must conform, in the entities and principles it invokes, to what science takes seriously—is quite clear. Even more clear is what, to most people, is a highly distressing implication. Science, bluntly, has no room for human values, purposes, ethics, or hopes. Science stands apart from what is most present and vital to almost everyone in ordinary, day-to-day life. Science does not—cannot—endorse what we most desperately want to believe, nor refute what we most want to reject. It cannot sustain moral judgments, let alone the spectrum of hopes with which people have always tried to defeat the ever-present certainty of personal extinction. That science is a matter of "is" rather than "ought" is, for many scientists, a point of great pride. However, for many folk of ordinary, or even extraordinary, intelligence, it is an ominous thing to realize, particularly in a civilization where science is more and more central to knowledge and understanding, and where it has a continuing record of vanquishing all rivals.

All this would matter relatively little, in the scheme of large civil questions, if science were merely the esoteric preserve of the unworldly and obsessively curious. In some cases, rather amazingly, it still has that character. But even the strongest defenders of the ideal of a pure science detached from practical concern—science as philosophy, contemplation, and disinterested understanding of the world—now grasp that science flourishes chiefly because it is regarded by its practitioners, as well as by those who provide it with material support, as an essential component of the general welfare. The Baconian prophecy has been vindicated. Science is an important foundation—in our age, probably the most important—of power, power to remake the world, to evade its misfortunes, to wring wealth and comfort from it, to cheat its inevitabilities as long as possible. Pure science, or, as it is sometimes called, "curiosity-driven science," endures, and often, to its credit, it drives a hard bargain with the surrounding society, procuring a place for itself as the price for helping to maintain a scientific enterprise that lavishly rewards more mundane interests. Still, everyone is conscious of the bargain. We all expect that the contours of science, well into the future, will be shaped to a great extent by what the ambient society needs and wants, and that the fate of science is closely linked to how well it manages to meet these expectations. The distinction between pure and applied science, between science and technology, is possible to make, and, for some conceptual purposes, it is important. But, to the extent that science engages with politics and with societal expectations, it is of secondary concern. Society conflates all these roles, an ignorant solecism annoying to many scientists. But ignorance, we know, is sometimes more prescient than wisdom. On countless occasions, pure science, done out of the most disinterested curiosity, has, to the great amazement of its creators, rapidly been conscripted for important, immediately practicable, and immensely useful technology.

There is an unsteady balance between what our society expects of science and what it fears from it, between the admiration nonscientists feel and the resentment that stings them, between the hope they find, at the most crucial moments of their lives, in a benign and powerful science, and the helplessness that afflicts them as they realize how little they comprehend the sources of that hope. Despite the edenic visions of the wilder fringes of the environmental movement—Deep Ecology and the like—science is so deeply intertwined with our social and economic structure that no hope exists of continuing our civilization in recognizable form without it. Though some suggestions have been made to the effect that science must become less absolute in its claims and more humble in its intellectual demeanor,[34] most such rhetoric simply ignores the fact that science cannot go forward without inspiring immoderate love and enthusiasm among its practitioners.

Recently, scientists themselves have become increasingly aware of the ambivalence—indeed, the hostility—with which they are widely regarded. Astronomer Alan Hale (co-discoverer of the Hale-Bopp comet) expresses the consequent distress:

> Based upon my own experiences, and those of you with whom I have discussed this issue, my personal feeling is that, unless there are some pretty drastic changes in the way that our society approaches science and treats those of us who have devoted our lives to making some of our own contributions, there is no way that I can, with a clear conscience, encourage present-day students to pursue a career in science.[35]

To accede to the relativist demand (for that is what it amounts to) that science discard its privileged status as an especially accurate way of learning about reality is not only to defer to questionable philosophy but, as well, to yield the core assumption that drives scientists to endure the considerable pain and travail of learning their craft and practicing it with rigorous honesty. If what is understood to be at stake is not the truth about the natural world, but merely the construction of a narrative intriguing to those whose taste runs in certain directions, most scientists, including those frankly engaged in very practical projects, will rapidly lose heart. If science gets downgraded to a species of clever, opportunistic, ad hoc tinkering, then it ceases, in some spiritual sense, even to exist; and so, the tinkering itself falters. Science, in order to function well, must think well of itself. Institutional self-confidence is a necessary condition of success, because nothing less will impel scientists to make the extravagant investments of time and effort that science demands. There are many personally diffident scientists, but few who are diffident *about* science. Consequently, the aura of Faustian hubris that clings to science can never quite be dispelled, regardless of how diplomatic or how genuinely modest individual scientists may be in addressing a wider public. (It might nonetheless be a good idea to provide the average scientist with some rudimentary training in modesty, or at least the ability to counterfeit it.)

In the recent past, hundreds of important public issues have emerged in which science and technology play a crucial role, and in which accurate comprehension of the underlying science is the key to sound policy. The mechanisms of public decision making must be accurately tuned to scientific knowledge. They must assimilate what science has to say, relate it to the far wider spectrum of concerns that politics has to take into account, and in the end produce decisions and plans that accord with responsible scientific judgment, among other things. This, however, is merely a maxim, not a very good description of the realities of the ongoing political process. Real-life political machinery is only fitfully competent at taking account of relevant scientific knowledge, and there is a strong case for thinking that, in this regard, things are getting worse rather than better. High barriers stand between laymen and the kind of scientific sophistication necessary to integrate knowledge into policy, and probably they grow higher from year to year. The irritants that estrange nonscientists are not losing their virulence. The ideal, never more than vaguely formulated, of a society that has come to reasonable terms with its most potent intellectual engine now seems more vexed and less realizable than ever. Misunderstandings abound. Politicians, bureaucrats, and the general public demand clear, unambiguous answers from science in situations where science simply cannot provide them. Almost as often, clear answers are rejected or ignored because they run into interests or prejudices that society has not yet learned to tame. In short, the notion of making science fit well into democracy, of using the powers of the former to enhance the values of the latter, is a hope without a practicable mechanism to put it into effect.

Politics, even at its democratic best, is a method for processing opinion, for triangulating the world in terms of what people happen to think about it, not for understanding the world through the lens of a pristine epistemology. Whatever we may ask of a democracy, it is clear that at some level it must be answerable to the opinions of its citizens, at least over the long run. However cynical one might feel about "public opinion" and its vulnerability to prejudice, myth, rumor, or demagoguery, a government that veers too far from what it urges, that persists in ignoring rather than persuading it, that contrives to dispossess it of its central political role, is violating one of the founding principles of a democratic polity.

Yet so far as science is concerned, public opinion seems, in many cases, obstinately impermeable to scientific good sense. The fit between the intellectual habits of most laymen and what is required for reliable scientific judgment is depressingly inadequate. Even if we restrict our attention to the population of idealistic, well-educated, and thoughtful men and women, those who make some effort to be more than casually aware of what science and technology portend, we find severe shortcomings. In fact, it is tempting to speculate that in this area, the maxim that "a little learning is a dangerous thing" has special force.

Nonetheless, democractic institutions, if they are not to wither into empty forms, must respond to the imperatives of opinion reasonably swiftly and without making invidious distinctions among opinion holders. This applies to public debate on issues in which science has a significant role as much as to any other area of contention. Of course, no society of any size embodies anything like perfect, plebiscitary democracy. In practice, democracy means democracy heavily mediated by representative legislatures and judicial oversight. But no reliable mechanism exists within the structures of any democratic government to enable it to filter out uninformed or poorly informed opinion when scientific issues are on the table. If legislatures or judges choose to defer to scientific competence, applying standards that would seem on the whole reasonable to the scientific community, this is serendipitous. Nothing restrains them from doing the opposite, and this misfortune has certainly been known to occur. Often the most heated issues, those that evoke the most feverish expressions of public opinion, are those in which conventional political calculation is most likely to defer to shallow knowledge or determined ignorance.

The problem of meshing science with the imperatives of democratic government has preoccupied a number of idealistic people in recent years, scientist, ex-scientist, and nonscientist alike.[36] Some of these have been thoughtful and scientifically well informed, others perhaps more susceptible to the temptations of ideology.[37] The general theme taken up by almost all of these thinkers has been the necessity to "democratize" science, to open up decisions to nonspecialists and those without scientific credentials, in the name of a more equitable society or one less susceptible to out-of-control technology:

> If the first order of business concerns democracy and freedom, are technical experts especially qualilfied to offer answers? Probably not. After all, if today technological decisions are made undemocratically and with inadequate attention to their structural political significance, it is certainly not because scientists, engineers, economists, and other experts grasp the situation, are clamoring to inform the world, but are unable to get a fair hearing. To the contrary, business, government, and the press all routinely solicit the opinions of technical experts, but the latter typically remain oblivious to technologies' politically relevant aspects.[38]

Foremost in the mind of most of these critics is the elitist nature of science and its propensity to develop close tactical relations with other institutions—the military or corporate capital—similarly elitist, at least in the sense of being narrowly controlled by privileged specialists. It would be frivolous to dismiss these concerns as simply the effusions of warmhearted but softheaded leftists. No less than the rest of humanity, scientists are susceptible to having their objectivity warped and their honesty corroded by the intoxications of power and money. Case histories of such corruption are not hard to come by.

Nonetheless, I believe that when all factors are given appropriate weight, it turns out that the crusaders for ostensibly greater democratic control of science and technology are looking at the wrong end of a deeply difficult problem. Simply put, they fail to address the vexing question of how sharply the ability to think accurately about scientific questions falls off as one moves away from the community of professional scientists. Some critics disguise this problem by invoking currently fashionable academic theories that purport to undercut the privileged position of scientific knowledge, redescribing it as the parochial viewpoint of one particular narrow community.[39] Other communities of "knowers" are said to be equal, if not superior, to scientists when it comes to judging local circumstances where life experience supposedly outweighs the analytical and generalizing powers of science. For the most part, this represents the kind of romantic wishful thinking that only the hothouse ambience of the university can sustain. Upon close examination, concrete examples tend to fade into special pleading.

The less wistful elements of the movement to democratize science, on the other hand, are skeptical of the rather mystical notion of special "local" knowledge. They concede, at least in principle, that competently done science enjoys a substantial edge over its epistemological rivals in generating understanding and in guiding plans for action. Therefore, they see the urgency of elevating the public comprehension of science. Their chief idea for accomplishing this is that, with the assistance and guidance of suitably idealistic and dedicated scientists, the average person can, indeed, absorb enough technical knowledge to justify the claim for a high level of citizen involvement in all sorts of questions concerning technology and science. Although far more plausible than the radical version, this hypothesis still fails to address the depth and width of the gulf that divides scientific competence from earnest dabbling. The urge to preach "democratization" as the panacea for all defects in the relation between science and society is so strong that often it bypasses the need for a careful examination of the obstacles that exist to the presumed universalization of scientific understanding. In a word, just as scientists can be corrupted by money or power, idealists can be corrupted by hope. The latter sin is perhaps the lighter, and one reproves it reluctantly. Nonetheless, it can be a damaging sin, a canker on the very hopes that inspire it.

As I hope to show subsequently, the advocates of a democratized science often neglect not only the difficulty of imparting an adequate grasp of science to their intended proselytes, but also the danger that an upsurge in enthusiasm for populist intervention in science policy might well veer in directions that are politically unappetizing as well as scientifically foolish. There is already ample evidence that this is so when the issue is evolutionary theory and the persistent opposition of religious conservatives. Similar forces may be in play in the case of "alternative medicine," although here the issue is muddled, since many of its backers think of themselves as progressives and believe they are advancing progressive interests by challenging medical orthodoxy. Even if there were no disquieting political echoes in the clamor for a supposedly democratized science, there is more than ample evidence that on the issues where it has been most strongly felt, science, and consequently the public interest, has been poorly served.[40]

All ideas are hostages to opinion. Ideas are, so far as we know, brought into existence only through human agency. Ideas contend with one another only insofar as their human champions carry on the fight. Tautologically, then, the fate of ideas is always at the mercy of the rule of opinion. In the case of scientific ideas, however, there is an important distinction to be made between mass opinion and scientific opinion. It is the habit of many contemporary thinkers to deny that such a schism exists, or that there is any reason to prefer what scientists have to say about their own specialty when it clashes with the notions of a supposedly grassroots movement. Yet it is a fact, and an unwelcome one, that opinion in the general community takes form through processes that have little in common with the mechanisms through which scientists reach consensus. The probability that on a given issue popular opinion may diverge sharply from what scientists generally take to be the truth of the matter is not negligible, and may be alarmingly high. Which opinion will decide the fate of an idea then becomes a question of more than philosophical importance.

In what follows I shall try to analyze the standing and the prospects of science in a society that is steeped in a democratic ethos, professes to admire science, and expects great things of scientists, but which, notwithstanding a massive educational system, comprehends science rather poorly. I do so with the sense that what I have to say will bring little gratification to most readers, since the news, as I see it, is bad. To put it plainly, despite three centuries of coevolution, despite frequent episodes of mutual encouragement and support, the culture of modern democracy and that of modern science are in many ways incongruent. Orthodox history of science regards certain developments as the most sweeping and fateful of triumphs: the Copernican revolution that culminated in Newton's synthesis, the Darwinian revelation that humanity is an adventitious consequence of the convolution of biology and history, the relentless explication of biological process, including those of the human organism, in terms of chemistry and physics.

The depressing, though often unspoken truth is that these are regarded as sovereign insights only within the relatively tiny community of the scientifically well educated. In the larger society, even Copernicus, Galileo, Kepler, and Newton are more accepted than understood. It is hardly necessary to remark that Darwin, as a historical figure and as the symbol of an idea, is widely reviled. The ongoing revolution in genetics and molecular biology,

while doubtless deserving of intelligent ethical scrutiny, has often been received with what amounts to superstitious terror. Even worse than ignorance
or resentment of what science has achieved is the incomprehension of how
these advances came about.

Science, though it has been the chief force shaping the material conditions of modern social life, fits only uncomfortably into the matrix of contemporary society. Politics, law, culture high and low, finance, business,
religion—virtually all the organs of our collective existence—are surrounded
and nourished by an amniotic sea of scientific ingenuity without which
many of their unreflective grounding assumptions would collapse into sheer
fantasy. Yet none of these subcultures has a deep understanding of, nor much
in the way of sympathy for, the scientific worldview. The interface between
science and public policy works fitfully and unpredictably at best. The advocates of a democratized science, for all their undoubted ideals and selfless
passion, have thoroughly, and in some cases willfully, ignored these deep
discordances in order to construct a simplified moral landscape. Their overriding imperative is to subvert the presumed alliance between the arrogance
of a pampered scientific establishment and the greed of a profit-driven economic elite. Frequently enough, this has fostered a willingness to elide the
judgmental standards and the systematic rigor of scientific inquiry, both in
specific emotionally charged controversies and as part of an overall mood
of truculence toward the presumably oppressive institutions of modern industrial society.

What is proposed, it often turns out, is not so much the democratization
of science as the supplanting of science by a mélange of viewpoints and
methods in which populist enthusiasm or even quasi-religious dogma will
be anointed with the cultural authority of the "scientific." For all my sympathy with the political instincts of the would-be democratizers, I can endorse neither their methods nor their immediate aims. If, as I shall claim,
the scientific view of the world is still in crucial ways alien to the surrounding culture, if it has never been truly naturalized into the body of assumptions with which most people, including the intelligent and well educated,
address the world, then calls for democratization—which, with slightly different rhetoric, come from the right as well as the left—are at least premature and possibly disastrous. To me, it seems that any movement to
restructure the relation between science and the wider society must wait upon
a systematic and—to the extent that any such thing is possible—dispassionate
analysis of how science is actually positioned in society. The following chapters constitute an attempt to do this, at least at the level of a preliminary
outline that brings out some of the most salient problems.

Before undertaking this project in greater detail, I owe the reader some account of my assumptions and perspectives. Since politics and political atti-

tudes are certainly central to the inquiry, I have to say something about my own political viewpoint; the reader will subsequently judge whether it has skewed or tainted my analysis. Fortunately, a credo first formulated by the sociologist Daniel Bell summarizes my own political *weltanschauung* rather neatly: like Bell, I am a socialist in economics, a liberal in politics, and a conservative in culture.[41] This leaves me with the smug assurance that I begin from a position that displeases almost everyone in some respect. In terms of epistemology, I am, as previously stated, a convinced advocate of the notion that science—the actual scientific practice that has developed over three hundred years and which is embodied in the work of millions of living scientists—is, by an immense margin, the best way we have of coming to grips with the reality of the world in which we find ourselves. Thus I shall assume, without apology, that in cases of conflict, a reasonably well established scientific conclusion trumps any challenger. I am sympathetic with the pugnacity—and the distress—expressed by the distinguished science writer Timothy Ferris:

> I think it's time for those of us with an empirical bent to adopt a "No more Mr. Nice Guy" stance when we encounter people promoting superstition and pseudoscience. I recently heard, on [the] San Francisco public radio station, an hour-long discussion of astrology during which a pair of astronomers stated that astronomy merely represents a "different world view" from astrology. Missing was any declaration that astrology is demonstrably hogwash.[42]

Whatever flaws this book turns out to have, a "Mr. Nice Guy stance" shall not be among them. I shall not frequently appeal to the theories or supposed findings of recent academic work in "science studies" or the like. Although there have undoubtedly been occasional insights from that quarter, most of its effusions have been deeply contaminated (as I have argued elsewhere) by faddishness and ideological obsessions.[43] Too many of its champions have been seduced by the narcotic of painless heterodoxy. As a steadfast monist and reductionist, I also reject supernaturalism in any of its variegated forms, including the most popular, religion. Atheism—in this society, the conviction that dare not speak its name—is the correct term for this position, and I accept the label without demurral, trusting that the pious reader will not be shocked into ignoring the burden of my argument. I am sure that this invites the charge that I am, in fact, professing the "religion of science," which, according to the cliché, is the fashionable religion of modernity.[44] Frankly, I don't think this holds water. Confidence in the accuracy of science is a very different thing from religious faith. Nor would I accept that "scientism," the elevation of science to the status of religion, is a common creed in our day. Scientists, for one thing, are as a rule too knowing and too skeptical to make good worshipers, even at a shrine that they themselves have created. They are far too aware of the foibles of scientists!

I agree with Richard Dawkins: "Science is not a religion and it doesn't just come down to faith. Although it has many of religion's virtues, it has none of its vices. Science is based upon verifiable evidence. Religious faith not only lacks evidence, its independence from evidence is its pride and joy, shouted from the rooftops."[45] Such nonscientists as might be tempted to become science worshipers are altogether too rare to form a significant congregation. They are swamped, numerically, by devotees of parascience, pseudoscience, and other forms of obsessive nonsense, all inducing very different emotional resonances than honest-to-goodness science.

Another strong possibility is that I shall be charged with concocting a smug apologia for science, one that ignores the damage to the environment wrought by technology, the weapons of massive overkill produced by scientific ingenuity, and the unfair—even viciously unfair—concentration of political power that results from elite control over scientific institutions and research agendas. I do not mean to slight the importance of any of these manifest evils; indeed, they are part of the story I am trying to tell. Nonetheless, insofar as I am unhappy with the populist program for neutralizing these threats, despite its currency among many intellectuals, I quite expect to be taxed with such sins, in any case.

The best I can do, the best I can hope to do, is to lay out, with as much honesty as I can muster, an analysis of a situation in which I take no joy and for which I can offer no straightforward remedies. I deeply wish that the history that has nurtured science in its growth to its present enormous intellectual power had brought forth a society capable of meshing gracefully with it and receiving its insights with full understanding and the wisdom to make benevolent and generous use of them. I wish that the conflicts and incongruities that beleaguer science, that subvert its great potential for good, and suborn it to do ill, could be abolished by a transparent political program capable of generating wide popular support and proof against unintended consequences.

Reality, alas, is highly immune to mere wishes. The uneasy relation between science and contemporary culture is deeply etched by history, custom, and prejudice, and will not be quickly resolved. The first step in that direction consists of recognizing how troubled our situation is and how much work will be required to put it right.

ONE

Culture

Jeremiads come cheap. It is all too easy to portray a culture as degraded, debased, and fallen away from its own finest standards. Insofar as I want to investigate the aspects of this culture that render so problematical its relationship with the scientific subculture it sponsors, exploits, and fears, I might be best advised to heed my own cautions and avoid too much rhetorical brimstone. Nonetheless, if I am to be true to my own strongest instincts, I must say something about the overall tonality of our contemporary life, something that will, indeed, constitute a jeremiad, recurring to the ancient trick of praising the past in order to denounce the present. The point is not to assert that there once was a time when science sat more comfortably within the enclosing society. Rather, it seems to me that in an earlier age the culture, though far more naive in many ways (scientific and otherwise), nonetheless possessed certain habits of thought and intellectual discipline that, had they endured undiminished, might well have formed the foundation of a more sophisticated and mature mode of dealing with the ideas and methods of science.

In saying this, I can only speak with some small confidence of American culture, historically and at present. Intuitively, I feel that the same sort of thing might well be said about the culture of western Europe, although I'm less able to supply telling examples. Readers who feel that my hypothesis is borne out, at least in part, by my argument can then judge how far it may be extended to a non-American context. If, as is commonly assumed, culture has become internationalized to a great extent, and if American culture is in some sense the prototype for the international version, at least some extrapolation will be warranted.

I want to consider how matters stood in the middle of the nineteenth century. My examination is necessarily brief and impressionistic. In putting forward my first exhibit, I'll indulge a pet obsession, Civil War arcana. Consider the following passage:

To say that none grew pale and held their breath at what we and they there saw, would not be true. Might not six thousand men be brave and without shade of fear, and yet, before a hostile eighteen thousand, armed, and not five minutes' march away, turn ashy white? None on that crest now need be told that *the enemy is advancing*. Every eye could see his legions, an overwhelming, relentless tide of an ocean of armed men sweeping upon us! Regiment after Regiment and Brigade after Brigade, move from the woods, and rapidly take their places in the lines forming the assault. Pickett's proud division, with some additional troops, holds their right; Pettigrew's their left. Their first line, at short interval, is followed by a second, and that a third succeeds; and the columns between support the lines. More than half a mile their front extends;—more than a thousand yards the dull grey masses deploy, man touching man, rank pressing rank, and line supporting line. Their red flags wave, their horsemen gallop up and down; the arms of eighteen thousand men, barrel and bayonet, gleam in the sun, a sloping forest of flashing steel. Right on they move as with one soul, in perfect order, without impediment of ditch, or wall, or stream, over ridge and slope, through orchard, and meadow, and cornfield—magnificent, grim, irresistible.

This is from an eyewitness memoir of the Battle of Gettysburg. It was composed by Frank Haskell, at the time a lieutenant in the Union army, who was posted with the defenders who met and withstood the great Confederate assault usually known as Pickett's Charge.[1] In my opinion, the term "Homeric" might well be used without embarrassment to describe Haskell's account, such is its grandeur and nobility.[2] In fact, it might be said to transcend even Homer, in that it is not a literary invention, primarily, but a factual picture, composed shortly after the event. That it is not as familiar to Americans as the Declaration of Independence or the preamble to the Constitution or Lincoln's great speeches (though those, alas, are hardly familiar enough) is a source of amazement—and chagrin—to me.

I want to focus not on minutiae of Civil War history, but rather on the tone and mindset of a culture in which such prose as Haskell's could seem natural, proper, and unsurprising. Haskell was a Wisconsin lawyer and dabbler in politics before the war, and, for his time, a very well educated man (though of modest socioeconomic background), but hardly an important, let alone famous, figure in his own right.[3] The passage just cited comes not from a formal essay, but rather from a letter written to the writer's family in the weeks after Gettysburg. This isn't necessarily as casual a piece of writing as that description might make it seem. It is thousands of words long, and very systematically composed. Such "letters home" were often published in local newspapers or read aloud at public meetings. Possibly, Haskell may have had even more formal publication in mind. Nevertheless, the mere existence of such writing, as well as its target readership, tells us something important about the culture in which it arose, and about our descendant culture.

I propose a small mental experiment. Try to examine what might happen if a prominent contemporary figure, a politician for instance, were to address today's public in such resonant periods, with similar diction and a comparable standard of eloquence to guide him. There is little doubt that, on the whole, such a performance would be received not only with surprise, but also with dismay—and a good deal of resentment. In fact, it's likely that the meaning of his statements, as such, would simply escape many people, especially if the delivery were purely oral. If a courtroom lawyer attempted such exotic rhetoric before a jury, his client would likely be in deep trouble. It's not that politicians or lawyers are inexpert in using language to sway opinion. As a rule, they are quite canny and inordinately sensitive to the effect of language. They function, however, in a soundbitten society. They rightly assume that any sentence containing more than a dozen words bewilders its auditors, and that any invocation of a vocabulary beyond Basic English runs the risk of alienating them.[4] Frank Haskell, though consciously striving after literary grace and an elevated style, would never have worried that readers or hearers of normal intelligence might have trouble understanding his letter or savoring its rhetorical effects. The kind of eloquence he reached for was one of the givens of public discussion of matters of importance. Politicians, far from eschewing it, sought to cultivate it. It was a necessary component of the thaumaturgy of electoral politics.

Rhetoric, of course, is a constant factor in social intercourse, and no society has ever devised a style that is insulated from abuse or cynical manipulation. The style that came naturally to Haskell, and which strikes many contemporary ears as fulsome, was, in fact, susceptible to hyperbole and lent itself easily to exaggeration. On many occasions, no doubt, it veered into outright rodomontade. The traditional American comic image of the fast-talking huckster, from Twain's Duke and Dauphin through the orotund personae favored by W. C. Fields, trades on the highfalutin absurdity into which such an opulent style can easily descend. It offers an easy entry to fustian and hot air. Nonetheless, it is not inevitably a dishonest or insincere style, no more than the flat, neutered, affectless style that infests our airwaves is a guarantee of forthrightness and clarity.

Inarguably, however, such a rhetorical approach demands more of readers and listeners than the language inhabiting CNN or *USA Today*. It requires greater attention to vocabulary, and allusion, and enforces, by its syntax, far closer attention to the logic and design of sentence and paragraph as they flow past the eye or ear. Even the bluntest, clearest, most direct prose created within the ambit of such a style—that, for instance, of Ulysses S. Grant, a man notorious for his plainspokenness[5]—demands of those to whom it is addressed a clarity of perception and an attentiveness that is foreign to our current expectations.

It is not only rhetoric that has become infantilized today; it is logic itself. Rhythms of speech are rhythms of thought. The recoil of our culture from any sentence that pesters its hearers with subordinate clauses, or discursive asides between commas, is also a flight from patterns of inference and logical connectedness that do more than ascribe a simple predicate to a simple subject.

The society to which the rhetoric of a Haskell or a Lincoln came naturally was highly imperfect by our current notions of democracy. Racial equality was the pipe dream of a small minority of whites, and widely scorned, even by abolitionists, as going against the grain of human nature. Sexual equality was at best an embryonic notion, not yet familiar enough to provoke widespread opposition. Nonetheless, at least as far as the white male population, North and South, was concerned, the practice of democracy was in some sense more pervasive and far-reaching than it has since become. For one thing, the megacorporation, with its enormous leverage over economic life and its immunity from political process, had not yet come into being, although the time of its ascendancy was not far distant. The great accretion of governmental agencies stuffed with countless bureaucrats had not begun. Local affairs dominated the political consciousness of most people, and these were often addressed with a town-meeting directness that has since become very rare, even in places where town meetings vestigially survive. The education enjoyed by Lieutenant Haskell—he was a Dartmouth graduate—was uncommon, and a few years of rudimentary and inconsistent schooling was much more the norm for most people. Yet class distinctions, on the whole, were far less pronounced than in our own time, where insular suburbs harbor gated communities, and the gap between rich and poor yawns wider every day. Plutocracy had not yet approached its current apotheosis.

The interesting thing is that this more rough-hewn age aspired to a level of discourse, in its public affairs, that mocks our own poor pretensions. It was expected that weighty matters had to be expounded and discussed in weighty language. The ability to navigate through such language was an index of one's capacity to deal with those matters at all. Newspapers were filled with verbatim accounts of trials, debates, and legislative deliberations. All these activities were carried on by men who prided themselves on eloquence in speech and writing. The surrounding society accepted these terms of discourse, admired the skills they demanded, and tried, at least, to nurture these skills among its members. Women, on those rare occasions when they fought their way into public debate, did so by mastering the conventions of public rhetoric. The same applies to the black men and women of the abolitionist movement.[6] Small towns throughout the country had their own clubs or athenaeums, where literature, poetry, and philosophy could be recited, read, and discussed. Much of this activity was doubtless naive, concerning itself with what we now take to be ephemera and unaware of what was evolving in the metropolitan world of European capitals, universities, and salons. Nonetheless, a society that has become addicted to the click of the remote as it switches from *Baywatch* to the *700 Club* is in no position to mock the artlessness of an earlier time. That artlessness was an aspect of a disciplined seriousness of purpose that, for all its provincial gawkiness, had its eye on the sublimities of Western culture, however badly it sometimes misjudged them. It was a society that an upstart Cockney like Charles Dickens could gleefully mock. But it was also the society that produced Emerson, Hawthorne, Melville, Thoreau, Whitman, Poe, and Dickinson. Aspiration was high, and accomplishment was beginning to measure up to it.

The British novelist Anthony Trollope, a more sympathetic observer than Dickens, compared Harvard University to Oxford and Cambridge, and noted that "the degree of excellence attained is no doubt lower than with us."[7] Yet he also went on to observe:

> But I conceive that the general level of university education is higher there than with us, that a young man is more sure of getting his education, and that a smaller percentage of men leaves Harvard College utterly uneducated than goes in that condition out of Oxford or Cambridge. The education at Harvard College is more diversified in its nature, and study is more absolutely the business of the place than it is at our universities.[8]

Even more impressive—because it deals with the education commonly available to all and not just the Harvardian elite—is Trollope's enthusiastic account of a New York public school:

> The female pupil at a free school in New York is neither a pauper nor a charity child. She is dressed with the utmost decency. She is perfectly cleanly. In speaking to her, you cannot in any degree guess whether her father has a dollar a day or three thousand dollars a year. Nor will you be enabled to guess by the manner in which her associates treat her. As regards her own manner to you, it is always the same as though her father was in all respects your equal.
>
> As to the amount of her knowledge, I fairly confess that it is terrific. When, in the first room which I visited, a slight slim creature was had up before me to explain the properties of the hypotenuse, I fairly confess that, as regards education, I backed down, and that I resolved to confine my criticisms to manner, dress and general behaviour.[9]

This is, of course, mere anecdote, and the cynical will see in it a description of yet another Potemkin village. The conjunction of excellence and democracy it describes is heartening, but must be contrasted with the widespread illiteracy of the time.[10] Nonetheless, it accords with the earnest aspirations of an era and cannot be dismissed as merely illusory. It represents the authentic highmindedness of a society. Even when that highmindedness is more than faintly comical, one deeply respects it.

Religion was, of course, thoroughly embedded in the shared assumptions of the culture. The standard version was Protestant, and fervently so, although the sectarian diversity inherited from the uproars of seventeenth-century

England prevented any single church establishment from having more than local ascendency. Roman Catholicism, though faintly barbaric to the eyes of Puritans, Anglicans, and Evangelicals, never endured more than brief persecution, nor did Judaism. The worst-afflicted faith was the thoroughly native Mormonism of Joseph Smith, which for a time was kept under close watch by the army. On the other hand, irreligion and outright skepticism, though never popular, were accorded far more respect than in contemporary, supposedly secularized, society. It's hard to conceive that in our day, a journalist could achieve even modest success by scoffing at religious credulity. Contrast the successful journalistic career of the openly infidel Ambrose Bierce a century ago. In nineteenth-century America, the echos of the Enlightenment rationalism of the Founders yet lingered, and the bland, one-size-fits-all piety that suffuses our own public life had not yet developed.

Science, during this period, was still an exotic import, with creative work in any field rare or nonexistent. Fascination with technological innovation had, however, put down deep and early roots. As all of our schoolchildren know even in these days of spotty historical education, the United States was, in the middle of the nineteenth century, one of the great, if not the greatest, center of practical invention in the world for purposes of agriculture, industry, communication, and transportation. Names like Whitney, Morse, Fulton, and McCormack bear witness to this fact. Religion itself partook of this spirit, as the history of the Shakers attests. The course of the Civil War itself was deeply influenced by the brilliant achievements of engineers like John Ericson and Herman Haupt. Thomas Edison, even more than Washington, Jefferson, or Lincoln, was soon to become the archetypical American.

However, "pure" science lagged badly in comparison to directly useful technological creativity. It was still an era where hands-on experience and shrewd intuition made up for a lack of deep theoretical understanding. The early examples of Franklin and Rumsford (the latter, alas, a Tory and an exile) as important figures in basic science generated no immediate successors. To the extent that "world class" science was represented at all on these shores in the antebellum period, it was through imported figures like Louis Agassiz, who flourished at Harvard, and James J. Sylvester, who definitely did not flourish at the University of Virginia.[11] Even the first of our great scientific institutions, the Smithsonian, was founded by a sympathetic foreigner.

Nonetheless, within a generation of the Civil War, a substantial portion of the wealth generated by the burgeoning industries and the ever more efficient agriculture of the reunited country went into building up research universities on the German model (Johns Hopkins, Chicago) and converting the genteel northeastern colleges into centers of intense scientific work. Important nonacademic centers like the American Museum of Natural History in New York City and the Woods Hole Marine Biological Laboratory in

Massachusetts also accelerated the move toward serious research. At the same time, state universities were starting to lose the provincial and nakedly utilitarian character that had marked them at first, forming the base for an expanding research establishment. As if on cue, scientists of international renown emerged to gild the reputations of these new undertakings. These include Albert Michelson in physics, and Jacques Loeb and Thomas Hunt Morgan in biology. (One must also mention, though without much honor to any formal institution, that strange internal exile, Charles Sanders Peirce.)

What was it within the culture that sustained this remarkably swift transformation of the United States from an inconsiderable scientific backwater to a world center of scientific creativity? It is obvious that the prolific wealth of the Gilded Age was an indispensable condition, but this cannot explain the psychological factors that allowed the society to evolve from one that harbored virtually no working scientists to one that routinely nurtured thousands of scientific careers. Clearly, education had something to do with it— if, by education, we understand not only what is formally inculcated by curricula and programs of study, but also the cultivation of appropriate habits of mind, favorable attitudes toward learning, and the willingness to strive obstinately in the face of difficulties, frustrations, and setbacks.

In other words, we must look to the *moral* dimensions of education, where we understand morality to mean something quite different from priggery or conventional piety. Hopelessly old-fashioned words like diligence, fortitude, and craftsmanship must be resurrected for such an analysis. I want especially to emphasize the last, since craftsmanship, appropriately transposed to the intellectual plane, is a crucial aspect of day-to-day scienctific work which is often neglected in accounts of what scientists do. The folklore image of science as one uncanny, intuitive bolt from the blue after another, as well as the more recent and even less accurate accounts, popular among postmodernists, of science as a kind of unthinking zombie programmed by social codes, obscure the artisanal aspects.[12] Scientific work largely consists of the patient shaping of experiments, reasoning, and exposition, from which the publicly announced result is eventually formed. If there is anything "socially constructed" about science, it is the ethic that fixes the commitment to work in this way in the minds of its practitioners.

Even beyond this, doing science requires a realization that substantial achievement rarely comes easy, that apparent serendipity is usually the fruit of endless hours of seemingly unavailing hard work. Also needed is a sense that reality is dense, but not impenetrable, and that coming to terms with it requires a correspondingly dense network of ideas. In many areas of life, even of the nominally "intellectual" life these days nurtured by our universities, a rhetorical gesture or verbal flourish in the general direction of an idea is taken as adequate evidence that one has thought about that idea. This is welcome generosity for those interested in the quick-and-dirty compilation of extensive publication lists. In scientific work, however, it leads swiftly to inanity.

In sum, what I'm claiming is that the rapid emergence of first-rate science in the United States was conditioned upon the widespread internalization of a kind of intellectual puritanism, the willingness to be unremittingly hard on oneself, often driven by the sense that compromise with this ideal is a sign of deep unworthiness. A culture that sustains this ethos is not necessarily gracious, and it may often seem obsessed, but the notion now current that it is perforce neurotic and oppressive is not only uncharitable but dangerously wrong as well. Moreover, it is not even necessary that such a code be universally in force at every level of society for science to flourish. It suffices that there be a critical mass of individuals with this temperament (as well as with the obviously necessary intellectual gifts) and that it can renew itself from one generation to the next. The United States constituted such a society at the end of the nineteenth century. Earnest high-mindedness, conjoined with the simple willingness to work hard and the maturity to submit to the discipline that mastery of a craft requires, infused a large enough cohort of young people (let's be frank: young men) to sustain a creative scientific community ready to step upon the world stage.

Since it is my declared intention to indulge in a jeremiad, it will come as no surprise that presently I shall assume the obligations of the form, burdening readers with a list of deep contemporary shortcomings, in contrast to which the virtues of the past will shine resplendent. Though this is a seductively pleasant task for the jeremiadically inclined writer (which is why the breed flourishes), let me postpone it for a moment in order to propose that many of the admirable aspects of nineteenth-century society—which, as I have argued, fecundated the beginnings of serious scientific work in this country—continue to thrive within the contemporary scientific community. The craftsmanship of today's scientists, irrespective of field, lacks nothing in comparison to those of their forebears a few generations ago. Whatever makes research work "good science"—accuracy, clarity, pertinence to broad and basic questions, and capacity to stimulate further and deeper work— has been as naturalized in this country as anywhere in the world.

Given that I believe that our culture has become more ambivalent toward science—less supportive, less comprehending, and even more hostile—over the course of the century, what accounts for the fact that the internal culture of the scientific community, its adherence to the ethical and behavioral norms that make true science possible, has become stronger, if anything, over the same period? I submit that the answer is largely demographic. The United States had the good fortune to receive several waves of immigration which, in one way or another, proved bountiful sources of scientific talent. Some of these were populations arriving with intact cultural norms that naturally directed their brightest young people into intensive engagement with scholarship and science—for example, Jews who arrived from the 1890s to the 1920s, and Chinese (along with others from the Far East) who have been steadily arriving for the past few decades. These groups have more than adequately supplemented old-stock Americans as recruits to scientific careers, even as the core of our national culture allowed its educational ideals to wither. Another enormous boost came from the influx of gifted European scientists fleeing Nazi persecutions during the 1930s and '40s. Finally, America benefits from a continuing global "brain drain," which brings trained scientists and promising graduate students to this country in large numbers. They are attracted by the twin lures of a commodious lifestyle and the chance to affiliate with the American scientific community—still the most powerful, as well as the most productive, in the world. Thus, whatever its other problems, American science does not lack able recruits. To the extent that there is a shortfall, it is one that could easily be remediated by a minimally more openhanded attitude toward the funding of science.

Nonetheless, we live in an era when science, as a whole, finds itself increasingly at odds with the ambient culture in ways both familiar and novel. Though much has recently been made of the shortfall in research funding which now incommodes so many able scientists, this is, I believe, merely the token of a deeper cultural unease. Congress could easily grow more munificent.[13] That would hardly be surprising, since any realistic calculation of the national interest seems to demand it. Unfortunately, the rift in the culture has deeper sources and will remain.

This is an odd sort of rift, and merely to characterize it as hostility between the scientific community and the great mass of nonscientists is deeply misleading. Scientists, most pollsters tell us, are widely respected and admired. Among all professions, they come very near the top of the list in terms of public esteem. Certainly, they fare better than politicians, lawyers, and businessmen. This admiration is based on the still-powerful vision of science as the fountainhead of technological and medical innovation, the ultimate source of all those devices that make life comfortable, safe, and—not least—diverting. Nonetheless, the admiration is shallow and distorted. It coexists with an appalling ignorance of what science is and how researchers go about expanding its scope. It comes from a culture whose schoolchildren, along with the general population, lag distressingly behind most of the rest of the industrialized world in basic scientific knowledge.

These surveys illuminate the point that, while most people vaguely and casually approve of science, their admiration is fogged by incomprehension. Their approval of science extends, all too often, to endorsement of the grossest cranks and quacks, who usually aren't embarrassed to describe themselves as "scientists." The public is happy to admire science as long as that admiration doesn't require understanding science deeply or developing the insight that enables one to see the difference between honest science and crude counterfeits. Indeed, the counterfeiters are quite versed in playing to the public's nominal "love" of science. Ironically, it is far easier for a scientist to lose public support by carrying on necessary and valuable work involving animal subjects than by advocating the notion that little men with big heads descend in UFOs to seize honest citizens for weird sexual experiments. What the polls describe in shorthand as "approval of science and scientists" turns out to be a wild mixture of attitudes and opinions, some of them quite inimical to the genuine ethos of science. As the anthropologist Christopher P. Toumey put it in a recent book, modern science is in a position where "regardless of where we assign the blame . . . its intellectual substance is alien to large parts of the general population, whereas the common symbols of this ethos are borrowed from a different vision of nature and science."[14] Moreover, the further one goes up the ladder of supposed intellectual sophistication, the more one finds that incomprehension of science has curdled into hostility. As Gerald Holton has observed, "Even among educators, scholars, and commentators of our culture, one now hears all too often scientific research described as being an unpleasant, soulless activity, merely 'logical,' 'linear,' 'hierarchical,' and devoid of all human passion."[15]

The cause of this unpleasant situation is multifactorial and, I suspect, very difficult to disentangle into distinct strands. Though it may be antediluvian to say so, among the important reasons is the dissolution of a standard of judgment that understood that difficulty and complexity are the price to be paid for authentic understanding. In the absence of successful engagement with these terrors, the right to judge, to approve or to censure is truncated, if not utterly forfeited. Toumey assigns the label "Old Testament Science" to the style of public perception which "respects without understanding," which deferentially takes science at its word merely because it *is* science. There is much to object to in this attitude, not least the intellectual passivity it tolerates and promotes. Furthermore, whatever virtues might recommend it, this way of engaging with science is terribly vulnerable to the advent of cynical and well-orchestrated pseudoscience. Nonetheless, it is preferable to its successor, a brand of populism—which might be called the doctrine of the "scientific priesthood of all believers"—that declares that every voice that wants to make itself heard on scientific issues is a voice worth listening to. The "Old Testament" attitude at least had the sense to recognize that there is in principle such a thing as genuine science, and that it is worthy of special respect.

Both Holton and Toumey are well aware of the slippage in popular esteem that has befallen science. Toumey ascribes this, in large measure, to an ancient conflict: the antagonism between scientific rationalism, on the one hand, and the instincts of the religious temperament, on the other. He theorizes that the increasing professionalization of science at the end of the nineteenth century had much to do with this:

Gone were the "autodidact amateur" scientists who had sustained the Protestant model. . . . The scientific research ethos thus had several distinctive values. It was *secular*, that is, uncoupled from religious beliefs regarding purpose or method. It was *rationalist*, employing a positivist attitude and such logical methods as organized skepticism. It was *naturalistic*, requiring that natural phenomena be explained in terms of natural laws or processes.[16]

This is an important point, in my view, and largely correct, although, I think somewhat too narrow in seeing the conflict only in terms of science versus religious traditionalism. It is a theme I shall pursue at length in a subsequent chapter. Here I want to stress another motif: the degree to which the *breakdown* of a traditional set of values, rather than its persistence, led to a decline in the cultural status of science.

It is illuminating to consider the status of religion itself, as we find it in contemporary America. Whether or not outward forms have changed, whether or not religious sects are of ancient lineage or as new as digital TV, there is almost universally a conspicuous break with the emotional and psychic resonances that have pervaded religious faith in the past. To put it succinctly, the sense of sin, doubt, and unworthiness that haunted believers— in the Christian tradition and others as well—has atrophied to the point of vanishing. The notion that faith arises only as the endpoint of a dire, self-doubting struggle has almost evaporated. The examples of such questing and such doubt, from Augustine to Kierkegaard (and which, if we throw in the scientists, include Pascal, Descartes, and Newton as well) seem hardly comprehensible in terms of contemporary "spirituality." What reigns is a much-cheapened notion of transcendence—transcendence as pure anodyne, where one is not compelled to wrestle with any angels, nor to struggle with the refractory material of the soul. Nearly all contemporary religions, from the most reactionary fundamentalist to the most theologically liberal and socially progressive, survive and prosper as mirrors of their congregants' untroubled sense of their own high spiritual worth. Their sanctuaries are lined with the thistledown of "self-realization" and "personal growth." Religion, American style, equals organized smugness, whether it's the smugness of a Pat Robertson fan anticipating the hellfire that awaits homosexuals or the smugness of an upscale Unitarian decrying the wrongheadedness of racism, sexism, and homophobia. The communicant is certain that while waiting for the gates of heaven to open, he may guiltlessly enjoy the good life among home entertainment centers, sport utility vehicles, and titanium golf clubs. Salvation (however defined) is not a problematic goal, but a given. This applies as much to exotic imports from the East and elsewhere as it does to the various strands of the Judeo-Christian tradition.[17] Religion as a market-driven constellation of choices, religion as a shopping mall, implies that proselytization can't succeed by the threat of eternal sorrow, but rather must promise a cheerfully buffed-up ego.

I don't insist on this truly jeremiadical point out of a tenderhearted regard for the fortunes of religon. As an atheist, I'm not much discomfited by the sight of religion stripped of its traditional dignity and moral earnestness. What is interesting is how the coarsened notion of spiritual attainment, the current opinion that salvation is to be had as easily and painlessly as a new pair of running shoes, reflects the declining recognition that one must *struggle* to achieve real insight or to burnish one's soul. Struggling to come to terms with difficult and intimidating ways of understanding the world is obsolete, as is the requirement that one must humbly serve an apprenticeship with those who have mastered these processes. By contrast, in the contemporary scheme of things transcendence is not merely supposed to be easily available, it is taken to be a birthright, demanding nothing of us.

The same kind of decline, perhaps springing from the same deep cultural factors, afflicts other realms of contemporary life that once stressed high-mindedness and the necessity for intense study and committed discipleship. The arts provide a telling example. Historically, they have been a primary repository for the ideals not only of craftsmanship, but of apprenticeship, of close and humble study of the past, of the notion that even the most revolutionary spirits must master tradition and come fully to terms with it before breaking with it. For many practitioners of what used to be called the visual arts, craftsmanship is a frightful and oppressive idea, to be derided in word as well as act. Increasingly, it has been accepted, and even insisted upon, that artistic merit reposes solely in the fact that the presumptive artist has seized upon some concept, some philosophical or (more commonly) political insight. Thereupon, he (or, emphatically, she, for much feminist art clearly falls into this category) need only indicate that concept by an approximate gesture, often merely verbal, for a "work of art" to come into existence. Transient witticisms, anemic aphorisms, mere unsupported claims of revolutionary virtue are the most important constituents of artistic accomplishment.[18]

The situation is hardly better in "serious" music. The postwar rigors of strict serialism, with its virtues of high intellectuality and its sins of brutal contempt for the sensuous element in music, are fading rapidly from their paradigmatic status, which was always precarious. An amiable, low-stress eclecticism has taken over. Conservatory graduates in composition often ape the simplicities of popular forms or leap onto the bandwagon of "world music," with its instant multicultural cachet. Or they may settle into a "conceptual" mode hardly distinguished, in its concrete realization, from conceptualism in the visual arts, completely divorcing their works from the traditional elements of musical art and rendering them merely whimsical at best. The guiding maxim in these artistic free-for-alls is that the simple act of declaring oneself an artist is a self-validating claim. An artist is whoever says he's an artist and a work of art is whatever he says it is (which may be no more than the declaration itself).[19]

What is involved in all this, I submit, is (to give it its crudest but most honest name) a deep cultural laziness. There has been a decline in esteem for, and familiarity with, discursive depth, density, and complexity. It is this that makes something as straightforward, by nineteenth-century standards, as Haskell's letter to the home folks seem to the ear of our contemporary everyman like an inscrutable message from an alien culture couched in a foreign language. It is a fault that amplifies itself because it breeds a concomitant sense of resentment against anything that resists the tug of its fallen standards. In parallel, the conviction arises that depth and difficulty are not only oppressive, but largely unnecessary. A rude populism now insists that any virtue that can't be attained by the uninstructed wisdom of the common folk is no virtue at all. This is reflected in the current academic enthusiasm for abolishing barriers between high culture and low, for regarding all "values" as "contingent," for deriding the "elitism" of those (increasingly vestigial) parts of the social organism that still cling to ancient notions of excellence and rigor. Politicians and media specialists inundate us in a flood of verbal offal. The traditional virtues of prose—words carefully chosen, sentences artfully shaped and gathered into coherent paragraphs, the whole matter organized into a consistent and fluent whole that makes rhetorical as well as logical sense—are not only rarely encountered, but mocked when they turn up.

As far as science is concerned, a popular mythology has grown up to justify the derogation of scientific authority. It is posited that orthodox science is, perforce, dogmatic and wrong, and that the kind of assertion that makes the orthodox shudder is not only gallant but also a manifestation of heroic genius in the Galilean mode. Suspicion falls not on the bizarre or hallucinatory claim, but on the one that has been painfully and systematically confirmed. The logic of *The X-Files* structures our discourse. If the experts are for it, it must be wrong. If the experts are against it, it must be the avenue to salvation. Thus, three hundred years after the witch craze died out, we find psychic healers in the operating theaters of great hospitals, "alternative therapies" not only subsidized by health care organizations, but freed, by government fiat, of the messy necessity of proving their safety and value in accord with traditional statistical standards.[20] Sadly, this phenomenon is not merely a demotic whim. It infests what are supposed to be the citadels of reasoned discourse—the schools and universities. Influential and well-established campus sects now exist that regard the special claims of science, as well as the traditional standards of art and literature, with flagrant disdain.[21] This is most dismayingly evident in the growth of a putative discipline called, ironically, "science studies," which has nurtured, in its brief life, the peculiar notion that knowing little about the substance of science ought not to be a barrier to making the most sweeping, and usually damning, judgments about it.[22] This is not terribly surprising, though it is very sad. It echoes a culture where expertise has been devalued, where the sweat and pain necessary to attain it has been disparaged, and where wisdom, in science as in all else, is ascribed to the claimant who can play best on the emotions and prejudices of his audience.

Not everyone who has acceded to the erosion of traditional values (esthetic or epistemological) has done so with an easy conscience. An ideological smokescreen has been laid down to obscure nagging questions. Among professional intellectuals, especially in an academic setting, the dismissal of traditional hierarchies of judgment, taste, and knowledge has been mediated by a loose network of doctrines and attitudes, often carrying nihilistic or at least heavily relativist overtones, for which "postmodernism" is perhaps the least deceptive term. Formulations, emphases, and vocabularies vary from one theorist to another, but the common assumption, more than justifying a unifying term for the phenomenon, is that standards of value and validity in all areas of human experience are invariably arbitrary, transitory, and conditional upon a tissue of usually unspoken assumptions. These, in turn, are said to represent the myths and anxieties of social power. Postmodernism, in this sense, has had a hand in forging the antipathy toward science that for now has such a salient role in intellectual fashion.

I don't intend to spend much time berating postmodernists in this volume (I have already done so elsewhere).[23] I do, however, want to bring up a few points that demand consideration, given the scope of postmodernist influence. First of all, there is the current claim, widely echoed in literature departments, the Modern Language Association, and so forth, that literary criticism, once a dilettantish, impressionistic, low-key enterprise, has been transformed by the advent of what postmodernists are pleased to call "theory" into a deeply serious discipline, fraught with rigor, intellectual density, and philosophical complexity. On this point, I am incorrigibly skeptical. The claim rests on the fallacy that verbal clutter and the interminable jangle of empty neologisms signify intellectual exactitude and authentic insight. My own experience in wading through this stuff is not extensive, but I have scrutinized enough examples to verify that this is a world where raw nonsense is more often rewarded than punished, provided it be presented in sufficiently jargonistic form. Praise, prestige, and perquisites have been lavished on the creators of work that, when examined coldly, dissolves into a slurry of errors and confusion. This is a severe accusation, but it grows out of analysis, not dogmatism.[24] As it turns out, what has been widely touted as scintillating intellectual fireworks consists largely of damp, pathetic squibs. This is evidence not of resurrected virtuosity in thought and argument but of its dismal opposite. It is a telling illustration of my general thesis concerning the abandonment of intellectual craftsmanship.

Beneath what is usually described as the frisky postmodern celebration of diversity, multiplicity, and the effacing of boundaries in all matters cultural and artistic, there lurks a besetting sadness. In great measure, this grows out of the failure of the wide-eyed political hopes of the intellectuals of the 1960s and '70s. These hopes have now been transmuted into contrarian post-modern whimsy, but the sour aroma of their curdling lingers. Even beyond this, there is a crepuscular tint to the postmodernist enterprise. It is a rather tawdry debater's trick to claim that your opponent really agrees with you, underneath it all, if only he had the guts to admit it—but in this case, I'll risk that onus. I think that, no less than cultural conservatives, postmodernist intellectuals are sensible of the cul-de-sac in which high culture has wandered (or been driven). The train-wreck status of contemporary art and music afflicts them too, as does the inanition of public rhetoric, the disintegration of unifying public myth, and the unending dullness and imprecision of the language that surrounds us, from beer commercials to the speeches of university presidents. The solution adopted by the postmodern ethos is simply to celebrate what cannot be evaded, to extol the fact that things fly apart, to flaunt the abdication of intellectual and artistic hierarchy.

The psychic rewards of this kind of thumb-sucking are obvious, but since many of the people who engage in it retain at least a smattering of their native intelligence, resentment and bitterness seep around the edges of the brittle postmodern pose. This is one of the factors behind the pervasive hostililty toward science which inflects postmodernist thinking, and the shameful willingness of intellectuals caught up in this subculture to enthuse over popular forms of superstition and anti-rationalism. If high culture has frayed and disintegrated, if it has compounded itself inextricably with common dross, why should not science share the same fate? Why not sneer and rail at scientists who won't concede that this is the case?[25]

Academic culture rarely dominates the wider culture of a nation. Sometimes it influences it, sometimes it echoes it, sometimes it stands in stubborn opposition. For the moment, as regards attitudes of trendy humanists (and a surprising number of ostensible social scientists) toward the natural sciences, the academy seems to be engaged in putting a high gloss on popular disaffection and incomprehension. From another angle, one might say that an important fraction of intellectual life has been penetrated by the ever-growing popular irritation at the requirements of careful, consistent, patient thought. It has absorbed the cultural tendency to enthrone wishfulness in place of logic, as well as the coarse demotic myth that one opinion is as good as another in all matters, expertise and cultivated skill be damned! These days, academics have discovered that significant brownie points can be had by writing tomes on quantum mechanics and chaos theory (for instance), despite having less grasp of those matters than a freshman physics major.[26] In doing so, they seem to be acting out in transmuted form, and with polysyllabic fatuity, the far more widespread popular enthusiasm for pseudoscience, astrology, parapsychology, UFOs, channelers, faith healers, and angelic prophecies. They seem increasingly sympathetic to even the most vulgar anti-science, provided it can be wrapped in the tinsel of "progressive" politics,[27] though even that requirement is sometimes waived.[28]

It may seem futile to arraign the drift of an entire culture, so much of which is unconscious and adventitious. Explanations for such a phenomenon are devilishly hard to pin down, and doubtless would be dauntingly intricate even if they were accessible. What purpose can it serve to accuse a culture, as such, of having become collectively lazy, or of abandoning the collective consciousness that once held it, with at least occasional success, to the ideal of intellectual rigor? How could one hope to find a formula for resurrecting genuine high-mindedness? After all, high-mindedness can be a dreadful bore, a cowardly evasion of real responsibility, and a smokescreen for outright evil. Yet without some tonality of high-mindedness in the ambient society, serious intellectual work, including science, finds its supporting cultural matrix melting away. Without the recognition that some matters are genuinely difficult, that claims of competence are not self-validating, that recognition and even honor is due those whose competence is demonstrably real, there is a danger that science may, over the years, become significantly demoralized. If, as a scientist, one has to compete endlessly with false claimants and charlatans of various kinds for public credibility, if the public lacks the concepts and the language for attending to such debates with informed judgment, if the culture increasingly comes to despise the very notion that there must be certain discursive standards in place for such a debate to be meaningful, then scientists will increasingly come to see themselves as freaks and aliens very precariously situated in society. It is idle to pretend that such a danger doesn't stare us in the face. But it is defeatist to declare in advance that nothing can be done about it.

Coda: I have been excoriating the degeneracy of modern culture and the delinquencies of contemporary habits of thought in a tone, admittedly, that might suggest to some, both those who agree and those who are appalled, a deep-dyed, reactionary nostalgia. I can say, with honest conviction and with the hope that I shall be believed, that nostalgia holds few temptations for me. Even aside from matters of health, comfort, convenience, and material prosperity, I don't believe that the culture of the last century holds any overall superiority to ours, least of all on the moral plane. Without romanticizing, one must recognize that this very culture, at this very time and for all its evident ills, is unprecedented in the degree of freedom it allows to vast numbers of people. All sorts of dead hands have been cast off during the past century. The strictures that inhibited not only behavior but thought itself in so many areas of life are largely gone. Though wealth is still inordinately dominant, considerations of ancestry, rank, class origin, and so forth have largely faded from the social order. Rigid notions—indeed, any notions—of what is befitting in consequence of one's sex, or race, or parentage have melted away. Religion flourishes; but it does so in kaleidoscopic variety, so that sectarian strictures on behavior and ideas have lost most of their leverage (although there certainly are areas—the abortion question, the continuing agitation of creationists—where desperate rear-guard actions continue). Who today lives in terror of thunder from the local pulpit? Money is still sovereign, no doubt, but given a modicum of wealth—well below the level of plutocracy—one can do whatever one likes in terms of ideas, dress, behavior, or sexuality, far beyond what would have been conceivable, let alone permissible, in any previous culture. The term "social ostracism" has pretty much lost its meaning. Snobbery still exists, of course, but now it exists in a thousand incompatible forms, and whose snobbery, if any, one chooses to be intimidated by is largely a matter of free choice.

In no way do I deplore this disintegration of social hierarchy and the constraints it places on individual freedom. In this respect, I am an extreme latitudinarian, if not an anarchist. I think the chief aim of politics should be the elimination of such barriers as still exist—mostly in the form of economic injustice and grotesque maldistribution of wealth—to extending this kind of freedom as widely as possible. Nonetheless, most of this essay has concerned itself with the historical fact, almost certainly no accident, that the withering of social hierarchy has been accompanied by the derogation of hierarchy in the realm of ideas, opinions, and thought. It doesn't seem to me inevitable that a latitudinarian culture must necessarily be a muddled intellectual free-for-all where feel-good (or sometimes, feel-bad) rhetoric consistently trumps logic. Nonetheless, the forces that have eroded the social hierarchy seem to have worked closely in tandem with those that have had an analogous and deplorable effect on the life of the mind. One serious consequence is that natural science, the highest triumph of the human intellect to date, finds itself, much to the surprise of many scientists, increasingly estranged from the fabric of social life. This cannot be healthy for science, and it is equally unhealthy, taking the long view, for society. Recognition of this fact is inherently dispiriting, but without such recognition it will be impossible even to begin the process of learning what to do about it.

TWO

Mathematics

Mathematics is the skeleton of contemporary culture. As with most such structures, its fate is to repose in obscurity at the core of things while citizens ignore it even as they throng the edifice it sustains. Anyone reading this is almost certain to have within reach many artifacts whose mere existence depends upon an understanding of the world that cannot be accurately expressed or encoded except in an essentially mathematical language. I take this point to be so obvious that example and argument are superfluous. I also assume that it is a point that will be forgotten by most people as soon as it is conceded.

In this sense, we are probably the most alienated culture that ever existed. We depend for our sustenance and survival on an interlinked network of ideas whose very language is hopelessly obscure to most of us. In no previous society were the commonplaces of life and work grounded in a way of conceiving the world so remote from the experience and understanding of the general population. The mathematical way of imagining how things are inhabits everything that we use or produce. It structures not only commonplace material objects but, as well, the contours of our social, political, and economic arrangements. Yet it is invisible except to a narrow class of specialists. Novelist Hans Konig frets over this realization in these terms:

> People at the start of this century were also living with an outburst of new technology. . . . But there was a big difference. Then, everyone with a modicum of education could see how these new things worked, if perhaps only in a vague way. . . . Semiconductors and microchips are a different kettle of fish. Their workings come from a world of advanced mathematics and physics where there are no James Watts with tea kettles to illustrate what is going on. The mystery of mathematics is that although it predicts and explains happenings, these same happenings have nothing to do with the logic of daily life.[1]

Our culture retains, though not without continual struggle, a sense of craftsmanship, a sympathy for the wisdom of hand and eye, an appreciation for the sensuous pleasure that may arise from bringing a well-made object into existence. We still manage to delight in calligraphers, carvers of decoys, quilt makers, and flower arrangers—though they stand under the unremitting threat of reduction to quaintness. On the other hand, most of the objects that come to us through a scientific understanding of the world are received much more coldly. We may dote on VCRs, but only for what diversion may be available to us through them. I very much doubt that anyone—aside, perhaps, from a few engineers active in its creation—has ever held a VCR in his two hands in rapt contemplation of all the elegance of thought that lies embedded in it. This is a shame. Even those who despise VCRs for the inanity they impose on day-to-day life might discover universes of beauty within the conceptual strands that unite to make possible the existence of such a thing.

Karl Marx, in trying to understand the economics of emerging industrial society, imagined industrial machines as repositories of stored labor power. It is an intriguing image, despite the fact that Marx's attempt to make analytical use of it went badly wrong, not least because Marx himself was a mathematical cripple. Yet it suggests another conceit, less specious perhaps, yet more paradoxical. Not only our machines of mass production, but the articles they produce—even the most quotidian, the most vulgar and graceless—are repositories of some of our deepest insights, insights which, when expressed as pristinely as we know how, take the form of mathematics.[2] This is a disconcerting thought on some level, since the cascade of cheap, tawdry, useless clutter that pours over all our heads daily is strong evidence, not symbolic but direct, of the degraded state of culture and the glut of useless things that we can no longer even dispose of without creating a greater mess.

Nonetheless, plagued as we are by the superfluous, if we look hard at any piece of routine junk and ask ourselves in all honesty how it came to be, we find that its genesis encodes truly beautiful ideas. It embodies patterns of thought that took tens of thousands of years to become naturalized in a corner of our culture. Grab a beer can. Think about what it embodies of our notions of geometry and efficiency. Every aspect of it had to be thought through, at some point in the last two or three millennia, with at least some cunning and often with the deepest inspiration. In many cultures, mathematical beauty is embedded in made objects as a result of a long, largely unconscious, groping toward optimality. Form has been sculpted by a Darwinian process in which the analytic abilities of human beings were only fitfully engaged. Only in our culture has optimality been built into the very logic of the productive process as the consequence of conscious, active thought. Even the most contemptible piece of mass-produced junk, therefore, bears the paradoxical traces of deep elegance, even sublimity. Mathematical beauty has been designed into it.

Mathematics holds a place in the popular imagination as something simultaneously fearsome and ridiculous. In the memories of most adults, only its echoes remain. These speak of agonizing boredom and the frantic attempt to cram one's aching mind with rote procedures, in the hope that they might bear up under the stress of an exam. For most people, almost all of this dissolves into a thinning haze of half-remembered formulas as soon as the pressure is off. The ideas that lurk behind the computational rituals, sad to say, almost never put down even shallow roots in the minds of the typical student. As to any lingering sense of elegance or beauty—well, that's a forlorn hope, at best.

It's safe to say that millions of people in this country have taken a calculus course of some kind, and perhaps some elementary physics based on geometry and algebra. Let me propose a modest problem that uses a fraction of that presumed knowledge. Recall the beer can, or rather an idealization of it as a perfect cylinder. If I require it to have minimal surface area for a given volume, what should its proportions be? Thinking of it as a real beer can, would it then chill quickly or slowly (relative to other possible shapes)? Once out of the refrigerator, would it warm up quickly or slowly?[3] I'm not trying to taunt readers with something they might once have known superficially but have thoroughly forgotten by now. But I do find it curious and frankly hard to comprehend that someone can put in a few weeks of work absorbing the ideas involved—they are quite straightforward—without permanently integrating them into his mental equipment. Of course, these ideas are products of thousands of years of thought and, certainly in the case of calculus, of one of the most stupendous acts of individual creativity in the history of the species. But once neatly formulated and explained with even a modicum of clarity, they are quite accessible. Moreover, they aren't the sort of thing that has to be "memorized" as one memorizes the periodic table of the elements or the bones of the foot. They are simply accurate ways of thinking things through in a simplified, idealized context free of irrelevant distractions. My gut instinct tells me that to comprehend them once is to acquire them forever. Learning them ought to be the intellectual equivalent of learning to ride a bicycle—how could anyone possibly forget?

My instincts, as years of experience have instructed me, are, alas, quite inaccurate. The kind of problem that I've just illustrated is, in the context of an ongoing course, rather cut-and-dried, and most students will get it more or less right, without too much frantic brain cudgeling. Rarely, however, will they acquire the permanent knack of handling such problems—unless, of course, they go on to take a few more years of mathematics on the way to a career in science or engineering. As to reproducing the reasoning that lurks behind the methodology—well, that seems to call upon gifts that are rare in the population I'm familiar with.

Of course, all this is the grousing of a long-suffering math teacher who has inflicted this stuff on a few decades' worth of long-suffering (and far more intensely suffering) students. I don't personally claim to have great pedagogical gifts. I am, I daresay, too remote and too impatient to establish much of a rapport with the average student. Also, when I stand in front of a class, my suspicion that I'm preaching to the deaf in many cases is probably all too plainly obvious in my face and in my voice. Nonetheless, something in me stubbornly insists that, uncongenial as my personal aura may be, once I've gone through something on the blackboard three or four times, any mentally intact human being who has been paying attention *ought* to have it written in his mind forever. No such luck, alas!

I'm not putting this forward as a snobbish instance of mathematical virtuosity. The level of abstraction and complexity where contemporary research mathematicians function is, to the layman, inconceivably distant from the kind of freshman stuff I'm now discussing. The people who are genuinely gifted at cutting-edge mathematical research are awesome figures, even (or rather, especially) to someone like me who has done a modicum of original mathematics. I wouldn't expect anyone but professional mathematicians to appreciate what they do or to have any idea of how to emulate them. My perplexity begins at a much lower level. How is it that even the first easy steps on the long, long road to deep understanding confound so many people so thoroughly? I think that everyone who has ever taught a semiserious calculus course has a similar sense of how recalcitrant the human mind can be in the face of what ought to be modest difficulties. I need only mention the intellectual pile-up that begins once the notion of limit (the essential conception underlying calculus) is introduced in a rigorous way. What is so frustrating is that all that seems to be involved is the inability to parse, both grammatically and logically, a fairly simple sentence. Specifically, elementary phrases like "if and only if," "for all," "there exists," and "such that" seem, when used even in small-scale combinations, to induce a peculiar paralysis of understanding.[4]

There is dreadful snobbery implied in taking note of this phenomenon, since the implicit moral seems to be that most of the human race—that is, those without significant mathematical talent—are in some sense intellectually defective. There can be few more graceless ways of making enemies than to utter such pronouncements. Although I wouldn't want to deny that mathematicians sometimes talk this way among themselves, they usually do so only whimsically, and often under the provocation of a particularly trying session with recalcitrant undergraduates. In all frankness, however, to make such an assertion in earnest would not only be arrogant and unkind, it would also be wildly inaccurate. Speaking personally, some of the most intelligent and astute people I have ever known have been all thumbs, mathematically. They seem to have little difficulty mastering complex bodies of knowledge, and they can draw upon it to make intricate and subtle arguments. They are eloquent in speech and writing and unintimidated by convoluted prose. Thus, to posit "mathematical ability" as a necessary element of intellectual acuity would be arbitrary and misleading. Mathematical gifts are no more necessary, in general, for intellectual distinction than expert ability at chess. (I am, to be honest, a wretched chess player, even though I'm a mathematician. The two talents are correlated, statistically, but they don't inevitably go hand-in-glove.) In fact, many scientists of the greatest distinction had scant mathematical talent. Darwin, for instance, though the creator of the greatest conceptual generalization in the history of biology, made little use of serious mathematical analysis.

Much science, even of the highest importance, can be done without recourse to serious mathematical modeling, and even the inevitable tabulation and statistical analysis of results can be subcontracted to software these days. There is a level of analytical and analogical reasoning where the defining precisionism of mathematics is not especially useful. How it comes about that intellectual acuity of this kind can coexist with mathematical clumsiness is a mystery of human cognition that eludes me utterly. I suppose that a near-visceral distaste for "dry abstraction" must have something to do with it. For some people, as I've come to realize, the inevitable formalism of mathematical analysis is hopelessly off-putting.

One commentator has noted of Fredric Jameson, one of the most highly reputed literary-philosophical intellectuals of our day, that "mathematics [is] for Jameson the very type of an ideality that can operate only through a denial of concrete reality."[5] For Jameson, the mathematical mode of apprehending the world renders everything "colorless" and drained of spirit. Jameson has plenty of company in this view, including humanists who differ from him about virtually everything else. The science journalist John Horgan, for instance, whose book *The End of Science* caught the attention, if not the admiration, of many scientists, seems to regard the use of abstruse mathematics as the sign that a science has grown decadent and disconnected from physical reality. As one mathematician reviewing Horgan's book put it, "In Horgan's view of things it is mathematics that 'subverts conventional notions of truth.'"[6]

The willingness to understand something purely within the terms of its relational and logical structure, the mental discipline necessary to exclude without pity everything that is extraneous and to concentrate solely on the essential, is a vital element of mathematical thought that strikes many people as disorienting and pointless. "Abstract" mathematics is one of the great test cases for the typology of human minds. Either one delights in piling one abstraction atop another to create the defining thought structures of mathematics, or one is appalled. Either one sees the results of such creativity as organic, beautiful, and gratifying to one's sense of harmony, or one sees them as embodying all the terrors of mental desiccation. There seems to be no middle ground. Mathematicians are, in terms of personality, behavior, and the whole spectrum of human qualities, as varied as any other group of intelligent people, yet the world at large ascribes to them a pronounced and uniform withering of ordinary human feeling. There is no objective justification for doing so. The stereotype arises because the dread most people feel in the presence of abstract mathematics and its severe demands is projected, understandably, upon mathematicians, who are, in everyday life, no more coldblooded than anyone else.

There is, so far as I can see, no immediate prospect for dispelling the antipathy toward the mathematical style of thought that reposes in most souls. This is a very different question than the perennial problem of improving the teaching and learning of mathematics. Most of the discussion, at this level, concerns the nurturing of certain easy computational skills, and the ability to apply these in a wide range of real life situations. To do so should be one of the primary aims of basic education, since the lack of such skills is paralyzing at the level of day-to-day activity. In contrast to the kind of mathematical skill I have been speaking of above, basic numeracy is an inevitable concomitant of intelligence. I've never run into anyone whom I regarded as "smart" in a general sense who had any real difficulty with everyday computational mathematics. The sticking point is not "discomfort with numbers" or a distaste for the quantitative. That sort of thing can, I believe, be overcome with the more or less willing cooperation of the average student, through routine pedagogical diligence. The horror of formalism and abstraction, which has little to do with aversion to numerical thinking, kicks in only at a higher level. Frankly, I don't know whether there is a pedagogy that can systematically defeat this phobia. If there is, I haven't any idea of what it would entail, despite having been through several rounds of "reform" in calculus teaching and the like. Some of this pedagogical effort may well pay off in incremental improvements in the performance of average students in routine courses, but I seriously doubt things will go much beyond that. Above all, the goal of "conceptual understanding" remains elusive.

Where, precisely, does that deficiency show up? Let me offer as an example a little problem that got an unwonted amount of publicity a few years ago. It offers some insight into the sort of situation where even very bright people can get hopelessly confused through lack of clearheaded mathematical insight. Here's the problem—call it the "Three Doors Problem"[7]—which I phrase perhaps too fussily in order to avoid quibbles. You are to play, repeatedly, a game involving three doors. A ten-dollar bill is placed behind one of them, and nothing at all behind the others. To make the game essentially a probabilistic one, let's assume you will play it 100,000 times. Placement of the prize is "at random"—without pattern or discernible system, so that calculation of probabilities is the essence of a correct strategy. There is a "game show host" who, in each and every case, knows where the prize is placed. In each round, you make an initial choice of door. Invariably, the host behaves as follows: he opens one of the other two doors—a door without a prize behind it—and asks you whether you wish to switch your choice to the remaining door. The round ends when you either switch or stick to your initial choice; if the door you've chosen conceals the ten-dollar bill, you win it; if not, you win nothing. Just to make the picture complete, you should know that when your original choice is correct, the host opens one of the other two doors at random. If it is incorrect, he picks the other incorrect door to open. Your problem, then, is to devise the correct strategy.

This is a rather simple problem of the kind that is routinely taught in elementary probability courses (which nowhere near enough students take). It involves no difficult mathematical concepts and no trickery or subtlety, merely the ability to visualize the essence of the situation sensibly. Here is the analysis. Two-thirds of the time your initial choice was wrong; in those cases, when you switch to the remaining, unopened door, you win. One-third of the time, your initial choice was correct and thus, if you switch in this instance, you lose. Consequently, if your strategy is to switch each and every time, you will win two-thirds of the time (and, of course, lose one-third of the time), so, in the posited case of 100,000 rounds of the game, switching each and every time brings you an "expected" gain of two-thirds of a million dollars, that is, $666,666.67 (rounding off slightly). By the same token, the strategy of holding to the original choice each and every time brings a gain of one-third of a million dollars—$333,333.33. Obviously, switching every time is twice as good as holding firm every time. Moreover, switching every time is better than any "mixed" strategy of switching X percent of the time and standing pat Y percent (where Y is obviously $100 - X$).[8]

This kind of exercise is standard fare for undergraduates ambitious enough to venture into the shallows of probability theory. The reason I bring it up is that it figured in a public controversy. A syndicated columnist named Marilyn Vos Savant, who bills herself as having "The World's Highest IQ," analyzed this example for her readers much as I have done, only to discover that many of them balked at accepting her explanation.[9] They insisted that once the host opens a door with nothing behind it, the situation reverts to a fifty-fifty proposition, that is, the probability that the prize is behind the door originally chosen becomes equal to the probability that it is behind the other, still-unopened door. Consequently, it is a matter of indifference whether you switch or stick with your first choice. Readers who have followed my explanation attentively (or who saw it for themselves without my help) will have no difficulty in discerning the flaw in this logic. Others are urged to reread my explanation carefully. The essence is that in two-thirds of the cases, the ten-dollar bill lies behind one of the two doors you haven't chosen, and thus the action of the host reveals the identity of this door. Notwithstanding what I insist is the transparency of this little problem, the controversy between Vos Savant and her stubborn readers raged for months.

Some of the protestors advancing the "fifty-fifty" argument even claimed to be mathematics professors, a situation that I find chilling. Eventually, the story appeared in the *New York Times*, which prompted another round of squabbles.

Candor compels me to confess my belief that a substantial number of my readers will find the "fifty-fifty" theory convincing, despite my best efforts to explain things clearly, and despite their best efforts to think everything through. They will, if experience is any guide, be remarkably stubborn in their belief—their delusion, for that's what it is. I say this without any disrespect to their sincerity or their overall intelligence, because I have observed this phenomenon several times in the flesh, and with some very intelligent people—a medical researcher in one instance, a college student in an honors program for the highly gifted in another. Their persistence was not a matter of hours, but of days and, in the end, I think they remained doubtful. Intelligence in a generalized sense is not the crux of the matter. What lies at the root of the difficulty is a too-confiding reliance on what seems compelling (though unanalyzed) imagery: those two as yet unopened doors, indistinguishable to the eye and hence having "equal claim" to the ten-dollar prize.

Before I let go of this example, let me pursue the story a little further insofar as it involves public events. First of all, there is the subsequent history of Ms. Vos Savant. I begin with the unreserved stipulation that she did a perfectly eloquent and clearheaded job of explaining the logic of the Three Doors Problem, and that the confusion of her readers is no reflection on her accuracy or clarity. However, one of her subsequent flirtations with mathematics was by no means so successful and raises some interesting questions. Most readers will recall that a few years ago, the eminent Princeton University mathematician Andrew Wiles announced that he had proved the famous Fermat conjecture, which had remained unresolved for more than three hundred years.[10] This proposition asserts that there is no solution to the equation

$$x^n + y^n = z^n,$$

where x, y, and z are positive whole numbers, and n is a whole number greater than 2. Because the conjecture is so old and well known, and because its content is understandable to laymen with only a casual knowledge of mathematics (in contrast to most of the questions that interest contemporary mathematicians), Wiles's achievement received enormous publicity.[11] Vos Savant, who is, after all, a professional writer, jumped on the bandwagon with a quickie volume "explaining" the Wiles result for the masses.[12] Not to mince words, the book was a hopeless and largely irrelevant mess. The author, "World's Highest IQ" or otherwise, has only a rudimentary mathematical background which, despite some rapid cramming, left her hopelessly ill-equipped to comment coherently about Wiles's work and its context.[13] Ms. Vos Savant is obviously a highly intelligent woman and had she pursued a career in mathematics, she might very well have been quite successful at it. Her success in dealing with the (very elementary) Three Doors Problem suggests this possibility, though by no means makes it a certainty. However, what is striking and sad about this little episode was her desperate desire to be seen as a serious authority on a matter where her knowledge was essentially negligible. Should this incident be taken metaphorically as an embodiment of the anxiety of many bright people when they contemplate the gap that separates them from serious knowledge of contemporary mathematics?[14] It is hard not to equate Vos Savant's attitudinizing, in this instance, with that of the readers of her column who stubbornly refused to acknowledge her perfectly correct explication of the Three Doors Problem. In saying so, I flagrantly risk the charge of snobbery and show-offishness. But the point I'm raising has a lot to do with the relationship of the general community—even the most educated and articulate segment thereof—to the bewildering technicalities of serious mathematics, and thus, to some extent, to the intellectual sinews of our highly technological civilization.

Snobbery notwithstanding, let me pursue the tale a little bit further. A few months ago, the *American Scholar* (the journal of Phi Beta Kappa) published a review by the journalist and essayist Jim Holt of a book called *Inevitable Illusions*, which addresses the popular misconceptions and errors of reasoning that result from the lack of clear thinking in many aspects of everyday life.[15] The Three Doors Problem and the controversy it stirred up figured in Holt's piece, which provided a correct analysis. Rather bizarrely, however, *another* simple problem having to do with elemenary probability was hopelessly botched.[16] A subsequent letter to the magazine corrected Holt's error with brisk efficiency. [17] However, the letter writer himself (a physician who teaches medical statistics in a medical school) made the mistake of "correcting" Holt's perfectly correct account of the Three Door Problem on the basis of the "fifty-fifty" fallacy!

What are we to make of a record like this? Again, I emphasize that all the people I have mentioned (and implicitly taken to task) are not only intelligent, in some general sense, but actually part of the educated elite. Once again, at the risk of being tiresome, I point out that the problems that occasioned all this embarrassment are simple and straightforward and involve no elaborate mathematical technique. The fact that bright people have so much difficulty with them is frustrating as well as depressing—but what does it prove? One obvious and probably true claim is that education in this country, even at the elite level, is insufficient in mathematical content. Mathematics is viewed as a narrow specialty, useful for certain technical professions, and fascinating to the small priesthood of professional mathematicians, but not, in general, valuable to the cultivated person, either practically or for the prestige it might confer as an intellectual ornament.[18]

Nonetheless, it should be clear that in the cases I have cited, the individuals involved apparently made special efforts to come to terms with the problems and examples in question, and that their attempts somehow badly misfired. This suggests a grimmer possibility. Could it be that the ability to do even elementary mathematical reasoning is really a relatively rare talent, and that it is rare because it depends on innate gifts that are themselves rare? Many people, of course, get through reasonably advanced mathematics courses at the college level. This, however, is rather a hothouse atmosphere, where certain more or less algorithmic techniques are stressed over and over. Here, success hinges on the ability to replicate standard solutions in connection with a small class of standard problems that have been made quite familiar through continual drill. To a greater degree than most instructors are willing to admit, much of this is rote work.

We math teachers expend considerable time, breath, and care on trying to convey some sense of the underlying logic of the computational techniques we teach. But I think, frankly, that our efforts reach at most a small percentage of students. What gets through, on the whole, is a repertoire of routines for calculation that can be mastered step by step and, a cool head allowing, can be applied accurately without any genuine comprehension of why they are justified. Perhaps the failure is all ours, as pedagogues. It may be that the talent is really there, latent in many people, ready to be awakened by techniques of teaching and learning that stubborn conservatives (like me) have never tried to develop or have ignored in our arrogance. So little is known about cognitive psychology (and its underlying neurophysiology) at this level that any such hypothesis is hard to refute. Perhaps it would be the better part of educational judgment to take the most optimistic hypotheses on faith. I certainly would welcome any hopeful sign. But personally, I am not particularly hopeful. The initiatives for reform of mathematics education that I have seen strike me as modestly promising at best. The worst of them, especially those emphasizing "gender" or "ethnicity," are by-blows of the silly pseudopolitics that has taken the fancy of the university community for a season. These would, in my opinion, make matters worse rather than better. Moreover, I suspect (and this is a suspicion only) that if anything can be done, by way of pedagogical innovation, to make the kind of mathematical common sense that I am talking about more widespread, it will involve a massive effort with very young children.

What kind of "common sense" might we hope for? Let me emphasize that I am certainly not talking about the depth of insight and experience that are necessary for an accurate sense of what contemporary research mathematics is all about. Perhaps there are some mathematicians who really want to be as famous as sitcom stars, but there can't be very many of them. It's not the first order of business of educational reformers to make terms like "p-adic algebraic geometry" or "spin cobordism" accessible to the general public. As a culture, as a society with problems that have to be addressed with something like collective wisdom, what we really need is the ability to scrutinize certain situations through the lens of generalization, formalization, and systematic calculation, which sound mathematical instincts provide.

One of the chief virtues of the mathematical turn of mind is that it reduces complexities to their skeletal essentials (which may turn out not to be so complex, after all). Some striking instances are provided by John Allen Paulos's little book, *A Mathematician Reads the Newspaper*.[19] What is illustrated here is that a "mathematical" way of coming at things strips away the inconsequential and clarifies the significant. It can, moreover, expose the fallacies that often lie embedded in the idiosyncrasies of commonplace discourse, as well as connect seemingly unconnected situations and create generalizations that can be extrapolated well beyond the immediate instance. It's not what the mathematician "knows" that matters, but rather the self-confidence and steady consistency, born of mathematical aptitude, that come into play as relatively elementary ideas are brought to bear on specific situations. It's one thing for a layperson to have worked through a few probability problems of the coin toss (or Three Door) variety in high school or college. It's quite another to apply the relevant ideas in "real time" to actual situations as they come up, where a haze of obscuring irrelevancies can only be dispelled by deep confidence in the abstracting schemata that formal mathematics creates.

I would be quite content—ecstatic, in fact—if only a handful of mathematical ideas became embedded in the tissue of the general culture, or even the demographically more limited culture of the "well educated." It would be wonderful if a much larger proportion of the population grasped geometry and trigonometry well enough to understand, if not to perform, calculations of length, area, and volume. A little elementary calculus would be good as well, not for detailed knowledge of its intricate computational tricks, but simply in order to have a sense of what "change" is all about, of the notion of what a rate of change is, and how it comes about that rates of change themselves have rates of change. How many people, even in the business community, can accurately parse a sentence like "The rate of inflation fell sharply in November"? Similarly, how many people know how to query an assertion like "Teen-age girls are suffering the greatest increase in AIDS infection of any group studied"? What does this mean? Does it signify that the number of infected teenage girls is increasing, in absolute terms, faster than that of openly gay men? Does it mean that the *percentage* of the stipulated population consisting of HIV-positive persons is growing faster than the comparable figure for other groups? Or does it mean that the rate of growth compared to the *already infected* population is large? All these meanings are inferable from the statement as it stands. A good part of the mathematical "knack" is to be alert to the necessity of challenging such propositions in the hope of supplanting them by less ambiguous (one hopes unambiguous) formulations. A surprising number of people who consider themselves well informed about AIDS (for instance) go around reciting statements like this one without the least sense of how incomplete and unclear it is. After all, if there were (purely hypothetical) five infected teen-age girls last year, and ten this year, that would represent a 100 percent rate of increase (figured as a percentage of the already-infected population) but *not* an enormous public health problem.

Above all, I think it very desirable for most people to have a sound grasp of the elements of probability theory and statistics. A reliable comprehension of all kinds of matters depends on this. Without it, it becomes virtually impossible to achieve an accurate view of the claims and counterclaims that often arise in connection with medical issues, environmental problems, governmental social policy, and so forth.[20] This is precisely what is so discouraging about the curious history of the Three Doors Problem. Stubborn misconceptions about this little example forecast equally ineradicable misconceptions about many practical—even vital—matters. The fact that deep flaws are often built into the methodology of experiments and studies does not always inhibit their publication, especially when the research is driven by self-interest or passionate conviction. All too often, claims of various kinds surge into public awareness much faster than scientific scrutiny can catch up with them. But even when that scrutiny is forthcoming, its conclusions cannot be absorbed or evaluated by a population to whom the lingua franca of the critics—statistics, probability, experimental design—is indecipherable.[21]

Beyond the realm of immediate practicality, beyond the necessity for public judgment to be grounded in appropriately sophisticated concepts, unfamiliarity with the mathematical mode of thought creates a more subtle form of intellectual infirmity. For most intellectuals—and this includes quite a few practicing scientists—mathematics is consigned to the realm of things that have no place in one's deep contemplation of the world. A sort of anti-Platonism therefore suffuses contemporary thought. A substantial corpus of archetypes and models, templates, perhaps, of some of the patterns that underlie reality—as much of it, at any rate, as will ever be accessible to us in our finitude—remains at best remote and vague to most of the people who are otherwise brave enough to venture into metaphysical speculation. What is the essence of the world, the true picture crudely mirrored by our coarse perceptions and the clumsiness of what we are pleased to call our common sense? The question is, of course, unanswerable. We still dwell, as we always have, in Plato's cave, faces to the dim figures on the wall. We shall never manage to crawl out. The severe proscriptions of Kant still constrain us, condemning us to forlorn uncertainty as to how much of the noumenal is revealed in the phenomenal. Yet, insofar as anything but futility awaits us in our ceaseless questioning, it is because of the evolutionary fluke that permits us recourse to a mathematical universe where rigor, abstraction, and pure idealization reign.

The tradition of speculative philosophy has died, by and large, as an aspect of high culture. Yet it lives, transformed and to many unrecognizable, in the work of physicists, cosmologists, and mathematicians, many of them quite unaware, in turn, of the degree to which they wear the mantle of Renaissance magus. During the long, and ultimately disheartening, debate over the funding of the superconducting supercollider, physicists were accused of abandoning the quest for practical and exploitable knowledge in order to pursue "philosophical" phantoms—the "theory of everything" of which the boldest and most hopeful dream. The physics community hastened to reassure bureaucrats and politicians, repeating the familiar litany—almost certainly true—that basic research leads ultimately to a deepened understanding of how things work, which, in turn, inevitably pays huge dividends in terms of technology. But in so doing, they yielded the spiritual side of the battle to the mongers of the vague amalgam of religious traditionalism and New Age puffery that constitutes what Americans are pleased to think of as "spirituality." Perhaps there was no choice. The "philosophical" inquiry pursued by the theorists and experimentalists backing the supercollider is effectively meaningless to most people. It is largely opaque even to most members of academic philosophy departments. Yet the supercollider truly *was* a philosophical project, the sort of thing that tempts one to abandon the relatively recent coinage "science" for the grandeur of "natural philosophy." The most staggering thing about it was not its (relatively modest) cost, nor its sixty-mile circumference, but rather the intricate beauty of the theories it was designed to confirm or to rectify into theories still more beautiful. All of this, however, was invisible to the vast majority, educated or not. The intellectual organ—mathematical imagination—necessary to make it visible lies undeveloped in them, if, indeed, it can be developed at all.

There is nothing wrong with a culture devoting substantial effort, and even comparably substantial wealth, to "philosophy," if that term is properly understood. That, I venture to suggest, is one of the chief justifications for "civilization" in the first place. Unfortunately, in our day the most penetrating and promising philosophy we can do is scientific work—whether in foundational physics or in the development of models of human cognition—which can neither be expressed nor conceived except in the thought-system of mathematics. To most people, this is the most tightly sealed of closed books.[22]

In the commonplace metaphors of our culture, mathematics stands above all for what is chilling, impersonal, inhuman. It is identified with the dictatorship of numbers, and the inexorability of systems blind to virtue and to beauty. Many recent commentators have lamented that the world is no longer "enchanted," that it has been stripped of mystery and of the numinous delight of intuition beyond understanding. Mathematics, so regarded, is mere quantity, and quantity is tyranny. What a horrid insult to our own visionary capacity! To those who in one way or another have developed the mathematical faculty, the world appears indeed "enchanted," not against the grain of mathematics, but through it. Mathematics is not "inhumanity" but an important part of what makes it worthwhile being human in the first place. It is the ability to discern in all the noise of experience, the pulse of something beyond. It is, in Einstein's phrase, "musicality in the realm of thought."

Into the indefinite future, our general culture appears doomed to remain largely dead to the mathematical way of thinking, not merely "innumerate" in the crude sense—that may be forestalled—but with a stunted and skewed imagination nonetheless. The costs to our civilization are hard to reckon, and indeed may be so deeply hidden that it seems like little more than the melancholy fantasy of a parochial mathematical sorehead even to suggest that they exist. Nonetheless, something tells me that ultimately those costs will work their way to the surface in ways that even the most hardheaded philistine won't be able, quite, to ignore.

THREE

Teleology

Stubborn as human beings tend to be about most things, they outdo themselves when it comes to imputing purpose to a universe which their most precise and exacting thinking reveals as altogether without purpose or design. Usually, the conviction that the universe *is* purposeful is embedded in and supported by a tissue of myth, legend, and supposition that constitutes a religion. If, as I believe, trying to ransom a prejudice by appeal to religious dogma is a profound error, it is at least an honorable error in making all its assumptions explicit. In debating teleological matters with those who hold a frankly religious viewpoint, one has the refreshing sense that the core of irreducible disagreement is reached rather quickly, and that the contradiction cannot be resolved or waved away. Except in the case of outright fanaticism (on one side or another), this leaves open the civilized possibility of agreeing to disagree.

Things become somewhat stickier when the conflict is with teleologically inclined people for whom religion or "spiritual reality" is purportedly irrelevant, who are willing, pro forma, to acknowledge the monistic view that the world is, in its essential working, undifferentiated. While disclaiming belief in a metaphysical level of reality beyond that which science can scrutinize, they nonetheless insist on intruding notions of purpose and of ultimate end into the discussion. Even more frustrating is to become entangled with people who persistently weave teleological assumptions into the fabric of their discourse without ever realizing that they are doing so. Thus, without any explicit recognition of the fact, their argument subserves a view of the world that not only assigns purpose to things in general, but equally unconsciously slides into the ontological postulate that some entity—usually "nature"—imposes this purpose. The frustration here is that you are arguing with people who insist that they share basic assumptions with you when, more or less unconsciously, they do not. This is an all too common quandary simply because the propensity to cast our stories of how things are in teleological terms is a very deeply ingrained propensity of the human mind.

We are, indeed, creatures urged on by habits of mind whose authority is not only cultural, but evolutionary.[1] We make sense of the world, and bind together the accounts in which that sense is codified, by narrative strategies, the concoction of tales that give shape to our ideas, link them to other notions, and make room for them in our memories. The construction of narrative is a human universal; no society is without its tales or without some ritual for solemnizing and giving special emphasis to the best and most important of them. There is a ceremonial aspect to recitation and recounting of stories, which often involves music and the heightened rhetoric of verse. In a literate culture, this is often apparent in the attention given to the physical aspects of the book—its esthetics as a concrete object, over and above whatever virtue its verbal contents might embody. No editor, at any level of publishing, wants to put out an ugly or anonymous-looking product! This is as true for the most abstruse of learned journals as for popular fiction where eye-catching colors are a selling point in a crowded mass market.

In my own field, mathematics, which is arguably the most abstruse of any and the most rigorously detached from lingering traces of the merely material world, published work nonetheless strives for the appearance of physical elegance, with typefaces, layout, and so forth carefully chosen. Even to someone for whom the ideas presented are inaccessible (as the symbols used to convey those ideas are likewise meaningless glyphs), a mathematics paper presents exemplary calligraphic elegance. To one who knows the subject, on the other hand, the very design is in some sense an echo of the embodied ideas. Indeed, on the psychological plane, the literature of mathematics shares the qualities of other literary forms, including fiction and drama. A mathematical proof of any length and conceptual depth has moments of suspense, confusion, tension, surprise, and, ultimately, resolution. It, too, is a narrative played out in psychological time, as much as any saga or tale of picaresque adventure. This is not to advertise the charm of mathematics, whose narrative energy will not, alas, spark much enthusiasm from most people. Rather, it is merely to emphasize the recurrent importance of narrative form in encoding what we have to know or choose to know about the world.

Inevitably, the narrative embodiment of knowledge conveys, along with mere information, the strictures of a surrounding cognitive frame that subtly insists that there is a purposive structure not only to the form of the information encoded, but also to the aspect of realilty the information reveals. Knowledge comes with a sense of "becauseness" built in. A is so because B is so. X exists in order to enable Y. The structure of the devices by which knowledge is expressed unconsciously conveys the sense of an underlying and undeclared purpose. The rhetoric of description intrudes purpose into the aspect of the world being described. Scientists, when on their philosophical guard, will insist that whenever "because" enters their vocabulary, it is merely a token of logical relationships in the theories they are developing, not a reference to a sequence of "becauses" grounded, ultimately, in the teleological foundation of things. If we say that planets orbit the sun *be-cause* of the way that gravitational attraction dictates their trajectories, we are, formally, saying nothing more than that a certain mathematical rela-tionship obtains between the posited nature of gravity, with its inverse square law (and so forth), and the geometry of planetary motion. Psychologically, the matter sits very differently in our minds. If orbits ensue because of grav-ity, we feel, it is because gravity is the more fundamental truth about the universe, closer, in some sense, to the "whyness" of things. It is possible to argue that gravity is entailed by the geometry of orbits—in fact, the first edi-tion of Newton's *Principia* seemed to argue that way, in some respects. In point of strict logic, the formulations are equally valid. Yet psychologically, they are vastly different. The customary way of putting it implicitly directs us to look for a more fundamental aspect of the way things are. That sense of one fact being more "fundamental" than another points back, through a chain of such relationships, to something inferred to be *ultimately* funda-mental. In turn, we cannot help visualizing that "basis of things" as giving direction—which is hard not to read as "purpose"—to all that unfolds from it. In other words, it is quite difficult for us to disentangle logical sequence—or narrative sequence—from a sense that things are *meant* to be a certain way. Who or what is exhibiting the intention may remain obscure, but the existence, however shadowy, of such an entity is always implicit.

The nearer our discourse approaches the everyday circumstances of hu-man life, the more we are tempted to infuse narrative with meaningfulness beyond its mere informational content. Accounts of human physiology, at every level from the molecular to the cellular to entire organs, speak, almost inevitably, of the "purpose" of the structures and processes they describe. Not to do so, although hypothetically possible, would constitute a species of obscurantism. Yet inevitably, such language suggests that a purpose is made manifest in the very existence of human beings themselves. Likewise, sociological or anthropological accounts of how families, clans, organizations, customs, attitudes, and myths all function to maintain the cohesion of so-cieties and, indeed, to insure their survival, suggest that human purposes are embedded in the nature of things, that they have a special status within the tissue of reality. Of course, careful thinkers will make distinctions be-tween mere idiom and philosophical doctrine, between locution and explicit assertion. They are aware of the difference between loose, comfortable meta-phors and more rigorous, if less colorful, language. It doesn't take much philosophical sophistication to scrutinize texts for incidental appeal to the pathetic fallacy, and to do the private mental redaction necessary to nullify the effect. Nonetheless, even elementary philosophical sophistication is a rare enough quality in this or any culture, and to keep it always in place as the ever-watchful filter of everything we read or hear is a tiresome chore indeed, if it is even possible. The relentlessness of our common habits of expression continuously bears down on our way of thinking, if only because it is never really absent from our way of speaking.

To speak of "purpose," or "intent," or "reason," as embedded in the fundamental nature of things, is so much a habit in this and every society that it is clearly futile to try to expel it. It is a category error, the imputation of humanlike qualities to entities, real or even merely hypothetical, that are inherently incapable of embodying them. Everyday discourse is, however, replete with category errors, and the recourse to teleology, to the postulate that there must be a humanly comprehensible purpose to the blind processes of nature, is hardly the most deleterious. The horror of a purposeless universe seems very deeply embedded in human psychology. Even at our most pessimistic, we prefer a universe that is malevolent to one that simply *does not care*. Merely speaking of an "indifferent" universe, we posit an entity that just *might* be other than indifferent. To say, "As flies to wanton boys are we to the gods; they kill us for their sport," is at least to impute conscious viciousness, a degraded form of purpose, but purpose nonetheless. The alternative—a universe incapable of taking notice of human action or fate, because incapable of taking notice of anything—is a horror almost beyond naming. Melville, in *Moby-Dick*, evokes it as "The Whiteness of the Whale": "Is it that by its indefiniteness that it shadows forth the heartless voids and immensities of the universe, and thus stabs us from behind with the thought of annihilation, when beholding the white depths of the Milky Way?"

No project for the radical reconstruction of society, no matter how extreme, has ever aimed toward a society comprised entirely of philosophers sworn to submit every aspect of their existence to the oversight of rigorous analysis. That goal is quixotic beyond the possibility of mockery—and I don't intend to pursue it! My drastically more modest goal is to explore the extent to which the predisposition to think teleologically, to envision purpose in everything, affects the way in which we, as a society, receive and filter the knowledge that science procures for us. Scientists are as afflicted with this habit of mind (if it is an affliction) as anyone else. They may, for instance, be religious, in a conventional or unconventional sense; the very essence of religion is to assume that a purpose lurks behind the way things are. They may be defiantly irreligious, and yet philosophically committed to a view of things in which the existence of human beings, with their all-too-human purposes, is something other than a matter of mere chance. Nominally atheistic Marxism provides perhaps the prime instance of such a philosophy, and has attracted at least the passing attention of many distinguished scientists. Even Charles Darwin provides us with an instructive example. He was as resolutely anti-teleological as any thinker who has ever lived, and, it almost goes without saying, his discoveries about organic evolution put in place the last of the great intellectual building blocks that enable those inclined to do so to deny that the presence of humanity in the universe has any cosmic significance. Yet even Darwin fell from time to time into the habit of speaking of evolution as though it led to the development of ever higher forms of life—better not just in the sense of dealing more successfully with the environment in reproductive terms, but better in some absolute ethical sense. The late Carl Sagan was resolutely proud of his religious unbelief and of his acceptance of a worldview that scorns to impute any human meaning to the universe. All the same, his great success as a public spokesman for science may have been as effective as the pronouncements of theologians and mystics in inducing large numbers of people to embrace the conceit that their private concerns echo those of a purposeful cosmos. Sagan's insistence, to the point of self-caricature, that we humans are made of "star stuff" resonated in the public mind as a quasi-religious guarantee of transcendence. Perhaps that is what Sagan wished for, in an unacknowledged corner of his mind.[2]

How does teleology enter into popular conceptualization of scientific questions? Usually, by lending an aura of moral worthiness or unworthiness to propositions that ought to be decided solely by hard evidence. Often enough, this comes about by construing "nature" not as the neutral given of scientific inquiry, but as an entity within which volition and benevolence reside. Let us briefly consider the peculiar authority of the term "natural" in the discourse of our culture.[3] It is not a neutral term. Though we may use the term "natural disaster" to speak of hurricanes, floods, and earthquakes, as an unadorned adjective, the term "natural" almost always connotes something worthy and good. This is evidence of the way many of us think. We continually invoke assumptions that are not made explicit, but which nonetheless sway us to an enormous degree. Use of terms like "nature" and "natural" often declares implicit belief in a moral hierarchy. In particular, it disparages the world that human beings have historically created—the intellectual as well as the material world—in contrast with a conjectural primal order of things. Rigorous thinking disallows this distinction; there is no "natural" way, as it were, to distinguish the "natural" from the "unnatural." To be whimsical, and perhaps paradoxical, the distinction between nature and culture is a cultural construct.[4]

Nonetheless, this is a distinction on which this culture insists. For many people, there is an enormous moral disparity between the "natural" and the "artificial." The natural is what was meant to be, the artificial or doubtful or evanescent ontological legitimacy. Usually, this view entails pronounced distaste and distrust toward science. This is more generally true of philosophies that try to impose a moralistic *telos* on our understanding of the universe, whether or not they deify "Nature" as such. There are, of course, some concrete reasons for these misgivings. Science is the ideological face of a technological and technocratic culture that has swamped the face of the earth and disrupted the ancient rhythms of hundreds of other cultures while despoiling their environmental patrimony. Indeed, Western guilt in this respect is an ancient theme in the culture and philosophy of the Western world, the very heartland of the scientific worldview. One finds it in Montaigne, Diderot, Herder, Wordsworth, Melville, Nietzsche, and Yeats, among our great writers. Hugely magnified by the horrors of the First World War, specific revulsion against scientific rationality dominated the thinking of intellectuals ranging from Martin Heidegger to Robert Graves. It was central to the gloomy prognostications of Oswald Spengler. As science historian Gerald Holton notes:

> Spengler holds that the thought style of scientific analysis, namely "reason and cognition," fails in areas where one actually needs the "habits of intuitive perception" of the sort he identifies with the Apollonian soul and the philosophy of Goethe. But asserting that an unbridgeable contrast exists between a pure "rationality" of abstract science and the intuitive life as lived, Spengler commits the same error as all such critics before him and after, of whom few seem even to have come closer to science than through their school textbooks.[5]

Spengler's unease is a deep, general queasiness that continually recurs, without any fixed place in the political spectrum, in twentieth-century thought. It has animated both progressive and reactionary social movements, both narrow nationalism and fervent universalism. The central theme, however, is that science has alienated us from the kind of creature we were meant to be. It has inundated us with knowledge we were not meant to have, while dissipating the wisdom, instinctive and natural, that was meant to guide us. The repeated word "meant" in the foregoing sentences is crucial. It is an extremely commonplace locution, and, for that reason, a very revealing one. It exposes clearly the deeply teleological cast of mind that is our inescapable psychological heritage. Confronted with the scientific worldview, as inevitably must happen in a culture deeply entwined with and utterly dependent upon science, this temperament sees an enemy. Anxious as it is to ascribe human meaning to all phenomena at every level, indeed to the cosmos itself, it is grossly disconcerted by the refusal of science, as science has evolved over the past three hundred years, to take part in the game. Jacques Barzun, the eminent humanist scholar, contends, "A general recognition that science is at the other extreme from art and cosmology; that at least a double view of nature is possible; that much of life is to be dealt with anthropomorphically, because *anthropos* is the customer to be served—this might begin to clear the ground of errors and stumbling blocks."[6] The inclusion of "cosmology" as antithetical to "science" seems bizarre here, although it is partially excusable in that the passage was written just before cosmology became a central interest of contemporary physicists and astronomers.[7] Nonetheless, it signals a belief that "cosmology" ought to mean a narrative in which the status of humankind is central, and where the unfolding of human history is the main thread of the story. But it is all too clear that cosmology, as most scientists understand it, has an entirely different emphasis. Physicist and Nobel laureate Steven Weinberg puts it thus: "I would guess that, though we shall find beauty in the final laws of nature, we will find no special status for life or intelligence. A fortiori, we will find no standards of value or morality. And we will find no hint of any God who cares about such things."[8]

This is the prevalent view among physicists and cosmologists. Yet there are a number of noteworthy members of that community who find it very hard to swallow, and who share Barzun's desire to make human fate the central theme of cosmology. Frank Tipler, a physicist of undoubted competence, has become famous (or, depending on one's viewpoint, notorious) for his advocacy of the "Strong Anthropic Principle"—the idea that the observed values of the basic physical constants, taken together, form such an improbable ensemble that the only way to account for them is to assume that the existence of intelligent life—human life—demands that they be as they are.[9] In other words, the universe exists to subserve human purpose! Tipler has carried this even further, with an ingenious argument that the design of the cosmos guarantees a kind of personal immortality, even, ultimately, a paradise where resurrected souls will reside.[10] Notwithstanding the clear emotional appeal of such theses, the scientific community seems loath to accept them—for the time being, at least. The ideas of Tipler and the small minority of scientists who agree with him seem most likely to be taken up by laymen for whom the proferred scientific rationale merely burnishes a creed they are committed to in any case.[11] Perhaps it will also do something to reconcile them to science as such. But in my view, such good effects are likely to be minor and transient. One needn't be an intellectual in the bleak Spenglerian mode to be made intensely uncomfortable by the style of thought that science, by precept and example, tends to impose on a huge range of questions. Science can do little to confirm the teleological stance or to encourage those who wish to reinvigorate it. Notwithstanding the efforts of Tipler and some other scientists to discern a place for human hopes, values, and morality in the grand scheme of things, the great weight of historical example, as well as the scientific ideal of strict logical economy, makes it unlikely that such philosophically inspired programs will display much staying power. Science is consequently doomed to its place in popular perception as the foe of a "meaningful" universe, simply because it can make serious concessions to the teleological viewpoint only at the cost of subverting its own deepest intellectual strengths.

The rift between the scientific worldview and the commonplace longing for some assurance that human existence is not merely a pointless accident insures that, well into the future, a substantial part of the population will contrive ideological and psychological defenses against science. Large num-

bers of people, perhaps amounting to a majority in even the most scientifically oriented societies, will remain alienated from science, though that alienation will often be covert and inarticulate, embodied in private reservations
rather than public manifestos. Perhaps, given the small number of people
who actually comprehend science reasonably well and who appreciate the
logic of its ungenerous attitude toward teleology, it would be more accurate
to say that science is fated to remain an ideologically alien presence in our
culture as a whole. In the corpus of assumptions and prejudices that
undergird our social existence, science is something of a foreign body. Consequently it draws antibodies to itself. This is not merely a case of isolated
individuals condemned to varying degrees of unease in the face of science's
implacable hardheartedness. There are significant social consequences attached. It is not just a question of atomized and solipsistic angst, because
an entire culture deeply rooted in history is condemned to pass through this
particular dark night of the soul. Such unhappiness is inevitably expressed
in social and political contention.

In contemporary America, this is most clearly reflected in the continuing
strength of creationist doctrine and its recurrent attempts to impose itself
on the educational system under one rationale or another. The conventional
view is that this struggle, tiresome and enervating as it is for biologists and
evolutionary theorists, is a confrontation between a biblical literalism that
has inexplicably outlasted its era and the enlightened modernity that is our
ostensible birthright. This is correct, insofar as the fight breaks into the headlines through school board meetings, court cases, and the antics of state legislators. Such squabbles are, however, merely the most public face of a deeper,
less articulate confrontation. I surmise that most of the people who support
the self-declared fundamentalists on this issue are not themselves literalist
when it comes to the Genesis myth or any other biblical episode.[12] Most
would be willing to view scriptural accounts as at least in part mythical or
allegorical, embodying edifying spiritual lessons rather than literal historical truths. To a degree, these folks accept that the Earth is much more than
6,000 years old, and that humanity appeared only at the end of a lengthy
sequence of geological eras, each millions of years long and marked either
by biological sterility or by a spectrum of life forms long since extinct. They
are even willing to accept the vaster chronology provided by contemporary
cosmologists, which includes the reality of the Big Bang. What perturbs them
about "Darwinism," then, is not its mere contradiction of biblical just-so stories, but its insistence on the completely random and adventitious nature
of evolution, an aspect that Darwin (but not Alfred Russel Wallace) continually emphasized. The not-so-hidden message behind conventional evolutionary theory is that the emergence of humankind, like the emergence of life
itself, is not a tale in which the purpose of existence may be read, but rather
a story that conveys no ultimate moral significance whatsoever. As Edward

J. Larson observes in his celebrated book on the Scopes "Monkey Trial," many of Darwin's nineteenth-century critics objected neither to his subversion of biblical myth nor even to his thesis of human descent from animals.[13] What they could not accept, however, was the Darwinian assertion of the purpose-lessness and adventitiousness of evolution, including the processes that led to *Homo sapiens*:

> There is a moral or metaphysical part of nature as well as a physical. A man who denies this is deep in the mire of folly. 'Tis the crown & glory of organic science that it *does* thro' final cause link material to moral. . . . You [Darwin] have ignored this link; &, if I do not mistake your meaning, you have done your best in one or two pregnant cases to break it—& sink the human race into a lower grade of degradation than any into which it has fallen since its written records tell us of its history.[14]

This unease continues to plague some of today's intellectuals. A typical *cri de coeur* comes from essayist Marilynne Robinson, whose religion seems to be undogmatic ethical Christianity and whose politics happen to be decidedly left wing:

> Since Malthus, to go back no farther, the impulse has been vigorously present to desacralize humankind by making it appropriately the prey of unlimited struggle. This desacralization—fully as absolute with respect to predator as to prey—has required the disengagement of conscience, among other things. It has required the grand-scale disparagement of the traits that distinguish us from the animals—and the Darwinists take the darkest possible view of the animals.[15]

Even liberal humanists without particular fondness for religion can be reduced to shudders by such a view.[16] Cultural critic David Denby, assuredly one of that breed, catches the mood in his chronicle of a journey to the Galápagos:

> I had resisted going to the Galápagos because I wanted to keep nature in its place. . . . It was the condition of our existence and I hated its dominion over me. That human beings had descended from the apes was no longer difficult to accept. But the notion that human existence is a mere accident—that the glittering jewel, consciousness, is just another adaptive mechanism—was a vile blow to one's self-esteem.[17]

Most scientists and responsible science educators insist that evolutionary theory ought to be taught in conformity to the rigorous understanding of science itself, that is to say, without teleological overtones. They nonetheless proclaim that since the presentation is ethically neutral and value free, it does nothing to sabotage the moral and religious instruction that parents might wish to impart. This point of view is undoubtedly sincere, but it is not particularly sensitive to the psychological resonances that permeate our culture. Stephen J. Gould has argued, as have many secular liberals, that "science" and "religion" address separate spheres of human experience, and that faith and piety have nothing to fear from the recognition of scientific fact (evolution, for instance) because the sacred realm they treasure is in some sense independent of the material world that concerns scientists.[18] In effect, Gould is inviting the faithful to resort to a dualistic view—one in which he himself does not believe—to escape the psychological pain of confronting scientific monism head to head. There is something opportunistic in Gould's formulation; he seems to be urging the weak-minded to find refuge in an ontology he himself dismisses, merely in order to induce them to go away and stop pestering the strong-minded, who have no need of such sops.

But for most sincerely religious people, even those skeptical of miracle mongering, it is not an easy matter to separate the spiritual from the material realm, as Gould urges.[19] To them, a religion that insulates the everyday world, making it immune to the direct working of the divine will, would be a feeble and cowardly religion indeed. They see the spiritual order as a superior one, not merely because it is uncorrupted by human sinfulness, but in the stronger sense that it is at all times capable of intervening directly in the material order so as to bring about justice, mercy, and the triumph of faith. Most people who pray at the bedside of a sick child ask for more than insight and resignation; they pray for a cure. Gould asks too much of them when he urges recognition of science as the lord of material fact. His formulation might appeal to a certain strain of theological liberalism, but this constitutes a negligible minority compared to the vast numbers who believe literally that God's majesty commands hurricanes and earthquakes.[20]

The idea that ethical systems may be constructed as mere human artifacts, in recognition that the very existence of our species is merely a matter of blind chance, is a rarefied notion. It is an intellectual exercise that may appeal to a sophisticated elite deeply attuned to Enlightenment values, but it is not easily swallowed by most people. Even those without sectarian doctrinal commitments quail at the idea that morality is simply an arbitrary product of human whim rather than a code of universal and objective laws. Despite a litany of denial from liberal religionists and conciliatory biologists alike, evolutionary theory, presented soberly and honestly, subverts the foundational view of morality. This metaphysical corrosiveness haunts the exposition of evolutionary ideas, even if it is never explicitly pointed out. Distressed parents who harp on this point are, perhaps, more perceptive than the scientists who seek to calm their fears.[21] On one side is the terror of Dostoevsky—"If there is no God, then all things are possible!"–on the other, the nihilistic arrogance of the Marquis de Sade.

"Anti-foundationalism" is a watchword of fashionable postmodernism, and in that guise, there is not much need to take it into account. But in a more important sense of the term, the evolutionary conclusions of biologists are anti-foundational on the moral plane.[22] This, it seems to me, is an ineluctable fact and, academic terminology aside, it is one that is recognized by our society as a whole. It is this rather than the compelling power of biblical literalism that creates the political space in which opposition to evolution survives and even flourishes. This is what anchors support for the movement to teach "creation science" in the classroom. People needn't be particularly enthusiastic about Christian fundamentalism in order to accept it to a limited degree as a stopgap alternative to a philosophical view that undercuts their deepest moral instincts. The image of morality as an adventitious by-product of a long chain of essentially random events is frightening and repellent. The notion of a "blind watchmaker" may be appealing to admirers (like me) of Richard Dawkins, but it is a specter provoking existential terror and moral vertigo in many decent, intelligent citizens who are not otherwise theocratically inclined.[23]

The most sophisticated representatives of the creationist movement are well aware of this pervasive unease. Anti-evolution propagandists like Philip E. Johnson and Michael J. Behe craft their messages to take advantage of it.[24] At the same time, they take care to disarm the suspicions that usually cling to biblical-literalist opponents of "Darwinism." My guess is that people like Johnson and Behe personally incline toward fundamentalism much more than their overt proclamations indicate, but irrespective of whether this is correct, their public strategy is to disclaim the specific authority of Genesis or any religious text as a source of positive information. They accept the time scale of the geologists and cosmologists, and the authenticity of the fossil record. In some cases, they concede, without much fuss, that humanity emerged as the endpoint of a long, long chain of ancestral forms, disconcerting as that may be to strict traditionalists who prefer to think that angels, rather than apes, are our closest relative.

But they insistently hammer away at the point that the history of life in general, and of humanity in particular, makes no sense unless a directing intelligence is assumed to have guided it. This superficially less strident version of creationism is usually called "Intelligent Design Theory."[25] Its strengths are both intellectual, in that it avoids commitment to the manifest absurdities of scriptural fairy tales, and social, in that it steers clear of a fundamentalist subculture that many Americans still associate with holy-rollers, hillbillies, Okies, and "white trash." The proclaimed target of the Intelligent Design theorists is "naturalism," that is to say, the idea that all phenomena within our ken, including the experiences we think of as most compellingly "human," result through the working-out of natural processes unmarked by the intercession of any supernal intelligence. "Naturalism" is clearly another name for monism and reductionism. It is the working philosophy of scientific practice and the personal philosophy of most scientists. Its enemies rarely bother to paint it explicitly as "atheism." This is because they sense that among their prospective sympathizers there are many who are not deeply committed to the idea of a traditional personal god—they may be pantheists or polytheists or Deistic in only the vaguest sense. But, as philosopher Robert T. Pennock puts it, "The creationism controversy is not just about trying to avoid being descended from apes, it is about trying to avoid an existential crisis."[26]

Intelligent Design proponents needn't personalize the battle or stigmatize their opponents as infidels, because the audience they wish to reach is fearful of an idea, not of individuals or factions. What these theorists offer is an escape hatch (decorated with plausible, if spurious, intellectual filigree) from the terrors of the worldview that underlies "naturalistic" biology. Whatever their long-term goal, their intermediate aim seems to be to make the world of educated discourse a little bit safer for teleological discourse. I do not wish them well. But, from a tactical point of view, they have a powerful psychological ally. They may yet prevail, even if 95 percent of the world's biologists and science teachers dig in their heels to keep the formal classroom free of creationism in any form.

Taking a hard line against creationism and its variants is no guarantee that teleological presumptions can't slip into scientific discourse through other doors. The habit of apprehending all we experience as a reflection of the purposes, obvious or obscure, of some willful entity is too deeply embedded as a habit of thought, and recurs, uninvited, in all the narratives we tell ourselves about what is going on in the world. The more emotionally laden a situation, the less we allow to the claims of intransigent skepticism. It is especially difficult to remain rigorously objective when we contemplate questions of health and disease. These strike so near our deepest fears that few of us can maintain an ironclad rationalism when we come face to face with grave illness in ourselves or those we love. The most severe materialist will, at such times, find himself slipping unawares into superstitious ritual. This might be dismissed as a kind of private mental tic, helpful insofar as it takes the edge off intolerable anxiety, harmless otherwise. There is, at any rate, little point in trying to exorcise it.

Yet, even at a level where we think of ourselves as deliberative and systematically rational, concern with personal well-being opens us to a view of the world in which we frame things in essentially teleological terms. Illness is a departure from how we were "meant" to be, a violation of the "natural" state of affairs. In fact, it is here that the uncritical use of a term like "natural" carries us farthest from reasoned analysis. We tend to think of disease processes as alien intrusions, anomalies in a scheme of things in which our natural condition is health. Here we are very close to personifying nature, thinking of it as a conscious force attempting to set things back on their proper—benign—course. This is not an entirely fatuous conceit, since the capacity of the human body for homeostasis, for restoring itself to equilibrium, is certainly real enough, as a matter of objective fact. If this is not "natural," then there is no point in ever using that word in earnest.

From a sober and thoroughly objective point of view, one can certainly talk of a state of functional optimality for organisms, including human ones. Therefore we might conclude that the aim of medical care is to bring a person's physical situation as near to the optimal as possible. But our tendency, in these matters, is to extend the notion of a natural state of health far beyond the tautological sense discussed above. Inwardly, and sometimes quite consciously, we have the habit of assuming that there is a category, which we might as well call "naturalness," of behavior, emotion, and material environment, which is, ipso facto, maximally consistent with our well-being. By contrast, alternative possibilities are very imperfectly consistent with our well-being, or even inimical to it, in that they are "unnatural." The intellectual shift, often quite unconscious, is to turn from defining as "natural" that which (as we might reasonably come to believe from empirical evidence) does the best job of maintaining and restoring health, to the assumption that some preconceived notion of the natural automatically accords with wellness. This may seem a subtle distinction, but in practice it can have profound consequences on a social as well as individual level.[27]

It is far easier, intellectually and emotionally, to devise (or accept) broad and picturesque notions of what is to be taken as "natural" in matters of health than to explore deeply the facts as empirical research—science—discloses them. This disparity is deepened by the fact that to really fine-tune the notion of optimality, we must take into account the substantial lack of uniformity in the human population. People are simply too diverse—in genetic heritage, developmental history, and so forth—for there to be a single, unqualified "healthy" lifestyle. Yet, when we externalize and reify our notions of the natural, we tend to forget such considerations. We create a social ideal in which there are elements of common sense, certainly, but which also embodies values and images that lack empirical validation but carry strong emotional force. We imagine ourselves as having the capacity to live long, largely untroubled lives—if only we can establish harmony with "natural" ideals. Sometimes these precepts correspond to what objective medical knowledge sanctions as sound practice. But often this is not the case. Why should it be? We, like all species, are an opportunistic and contingent product of evolution, not the embodiment of a Platonic blueprint. In our genes, we drag along a host of compromises and imperfections.

Our culture is still rife with its own brand of Platonism, however, and the ideal of a radiant model of human perfection obsesses us. This has a lot to do with the vogue for the natural—natural (or "organic") food, natural healing, natural alternatives to conventional medicine. There is supposed to be a benign force out there that knows us better than we—as rational, scientific beings—know ourselves. We are meant to find harmony with that force, but the contrivances of the too-rational mind get in the way. Sustaining one-self in good health, or being cured of a disorder, is not merely an instru-mental matter, to be guided by the specifics of scientific knowledge. It is also a ritual matter, in which that beneficent but veiled entelechy, nature, is being courted, conciliated, and even propitiated. This accounts for the fact that so much of what our culture conceives as conducive to health involves self-deprivation and even pain. Often enough, this will be accurate—I am not here trying to construct a rationale for unrestricted hedonism, self-indulgence, and laziness. But our cultural prejudices—perhaps going even deeper than culture—slyly insist on the curative virtues of sacrifice, self-denial, and pain in and of themselves.

Health-food diets are obvious, if not particularly invidious, examples of this. There is a presumed special virtue—a magical virtue—in turning aside from what we find pleasurable. Nature is supposed to call out to deeper in-stincts than mere pleasure. Vegetarians, while they are often motivated by humane feelings toward animals, find further justification for their abstemiousness in the idea that this is what nature *intends* for us. An en-hanced state of health is the supposed reward. There's no reason to deny that, as a practical matter, a vegetarian regime is probably better (for most people) than the fat-rich, red-meat-every-day diet that most of us fall into. However, if we insist on defining "natural" by taking our evolutionary his-tory as foragers and scavengers as a guide, we might conclude that it would be best to limit ourselves to vegetarian fodder (perhaps enlivened with a bit of fish or shellfish) for weeks at a stretch, but also to break with that rou-tine at intervals by gorging on rare steak for a few days.

Such minority predilections as vegetarianism are largely harmless (though there is danger in some versions) and, if they are a theoretical affront to an ideal rationalism, then they are upsetting only to the Quixotes who believe that a consistent and universal rationalism is a practicable project. But there are evolutionary lessons that really shouldn't be ignored. Evolution has shaped us as creatures capable of living and functioning well under harsh conditions long enough for reproductive success—thirty years, perhaps a bit more. Under the far softer conditions that culture and technology provide for us, unstressed by weather, hunger, disease, and predators, we can keep going a great deal longer than that. Yet, as we come ruefully to know, things start to break down when we hit our forties, irrespective of how kind life has been to us. We get ourselves patched together and perked up in various ways. Modern medicine, on the whole, is the best device for achieving this. But if anything is "natural," it is the entropy of life, the degradation of tis-sues, organs, and overall joie de vivre.

The great danger of the ideology of the "natural" is its failure to come to terms with this melancholy inevitability. The idea that we are "naturally" destined for some kind of perfection, for an immunity to time's ravages which evolution had no hand in shaping, becomes a dangerous idea, in a social sense, when it overrides the empirical and the demonstrably practical. This misleading intuition provides much of the social and political clout for "alternative" therapies. The devil of it is that such things—aroma therapy, homeopathy, healing crystals, to name a few—although relatively easy to refute in the strict sense of science, are beyond refutation at the cultural level. They seduce through the myth that we are meant to be stronger, sturdier, and more perfect than we can ever be. When the question of health is placed in a realm where teleological wistfulness comes to predominate, it is put beyond the supervision of the scientific frame of mind. That this is in the deepest sense unhealthy, for individuals and for society, is all too obvious.

It would be fatuous to claim that scientists, by dint of their specialized training or practical experience, are immune to the teleological predisposition. They can guard against it, as the best and most alert will, but to hope for absolute consistency is probably idle. Philosophers and, in a more malicious spirit, votaries of the new cult of "science studies,"[28] have pointed out that science necessarily makes leaps of faith, both epistemological and ontological. But it is important to note that science makes *moral* leaps of faith as well, or at least many scientists do. The prevalent postulate is that science is, of itself, a moral good—that is to say, greater knowledge always brings about improvements in the human prospect. Unfortunately, counterexamples spring to mind just as soon as this proposition is uttered. It is all too clear that scientific knowledge, in extending the destructive capacity of military hardware, has enabled the unprecedented slaughter that this century's wars have inflicted on a stunned humanity. Moreover, even apart from weaponry, the power of technology to degrade and even endanger life is plain to see. "Technology," notes the great physicist Freeman Dyson, "has not only done harm to poor people in our own American society; one can draw up an equally damning indictment against scientists for our contributions to the widening split between rich and poor on an international scale, to the worldwide spread of a technology that pauperizes nations and enriches elites."[29] This may be counterbalanced by the fact that for literally billions of people, the recent access to technology has improved the health, comfort, and security of existence immeasurably. But the final judgment is still in doubt.

Scientists invoke a standard defense: the problem here is not knowledge in its own right, but rather the overall configuration of power and authority in the societies within which science is embedded. The misuse of scientific knowledge is just an aspect of unwise or inhumane government, and thus there is no specific cure for it, beyond the much more general project of making our political institutions wiser and more enlightened across the spectrum. Least of all, on this view, is there a justification for special strictures on the practice of scientific research. It's fair to say, then, that the "leap of faith" I have been considering is often scaled down: "Given a fair and humane social and political system," we say, "all scientific knowledge is conducive to greater human welfare." This idea, I think, commands very general assent.

Is it *so*, however—even with the stated reservations? It seems uncontroversial if one has in mind a model where the only way scientific knowledge impinges on social and political well-being is through technological innovation. Presumably, a wise and humane political system will tend to generate wise and humane technologies. From this it follows that more knowledge and more accurate knowledge will produce still wiser and more humane technologies. But this, I think, views the effects of knowledge in general and science in particular altogether too narrowly. It's safe to say, three centuries after the scientific revolution, that knowledge can have psychological effects on a culture and can strongly shape its moral sense and its ethics. The idea that such effects are all to the good, at least in the long run, is one with which I'm in instinctive sympathy. Yet it is at best highly contestable.

The notion on which this essay is predicated is that the teleological view of the universe is an illusion. Bluntly, human existence is an accident, with no ultimate purpose. I believe this, as do most of my close friends. We're reasonably comfortable with this conclusion. But it would be fatuous to pretend that it does not inflict pain and psychic dislocation on millions, possibly billions, of people.[30] Science is clearly the primary sponsor and authority for this view of things. If one suspects that the ultimate effect of such a doctrine is widespread demoralization, that it will drive society and politics in unhealthy directions, then one must indict "pure" scientific knowledge as the main culprit. From this perspective, the science-affirming leap of faith seems much less tenable. Moreover, this point involves more than fading history. Day by day, scientific work adds to the effect as it blurs the distinction between life and nonlife, between life and death, between humanity and other species. All sorts of judgments that our culture used to be able to refer to absolute, dichotomous categories must now be sought amid a continuum of ambiguities. Dislocation, bewilderment, and irresolution are the inevitable consequences.

Despite a public image that denies this, scientists themselves can be susceptible to the discontents implicit in the kind of science that challenges comfortable moral axioms, and their discomfort can deflect and distort their scientific thinking. To take one obvious and ongoing quarrel, the longstanding question about the role of heredity in determining the competencies of individuals—and groups—is nowhere near being settled. For a time, in the first few decades of the twentieth-century, a strict, albeit none-too-accurate hereditarianism was in the saddle, with strong links to eugenicist and nativist movements. In the 1920s and 1930s, there was an about-face, partly due to the scientific solecisms of the eugenists, but also because of the assimilation of the scientific community into a climate of liberal egali-tarianism. It aligned itself with universalist values, values that emphasized the essential equality of individuals and of races. This dogma had its heavi-est impact in the social sciences, of course,[31] but the natural sciences, espe-cially human biology, were affected as well. Most scientists—there were exceptions—became very wary of lending their authority to the idea that genetic endowment equates rigidly to destiny. Even more, they came to dis-dain the notion that there could be significant differences in intelligence and other mental talents among the races or between the sexes.[32] This is hardly surprising, given that so many scientists were of Jewish, and, later on, Asian background. For better or worse, however, these reigning assump-tions were not, for the most part, backed up by a substantial body of tough-minded research. They were, in effect, teleological assumptions, implicit assertions of a belief in the moral basis of the world. They declared, basi-cally, that the natural order cannot be inconsistent with human notions of fairness. Consequently, there cannot be any serious disparity—certainly not among racial groups—in the distribution of valued talents.

Nonetheless, the underlying question of genetic determinism, and particu-larly of group differences, refuses to die despite an overwhelming desire among liberal—and many conservative—intellectuals that it should disap-pear. A recent flare-up was triggered by the appearance of Richard Herrnstein and Charles Murray's *The Bell Curve*. Even if this brouhaha subsides, there are certain to be recurrences. Personally, I devoutly hope that the hypoth-eses propounded by Herrnstein and Murray (not to mention William Schockley and J. Philippe Rushton) turn out to be dead wrong, and that clear evidence to that effect comes to light swiftly. The alternative is fearfully grim! But wishes are not controlling. My present point, however, is that many people, including scientists, talk, in this instance, as though wishes *were* con-trolling. The idea that the universe must somehow be set up to accommo-date one's deep-seated notions of fairness has overwhelmed the good scientific judgment of a large number of generally competent people. They scramble frantically to find any argument, no matter how makeshift, to ward off a morally unacceptable conclusion. What is unacceptable, moreover, is not merely the depressing proposition advanced by Herrnstein and Murray and their allies. It is even beyond the pale to suggest that there may be good reasons for considering the question an open one. Sad to say, the question *is* very much an open one.[33] The result is that eminent, kindhearted evolu-tionary scientists struggle desperately to deny this. Ironically, this compels them to behave, in some respects, like their creationist tormentors.[34]

The point of the foregoing is not to browbeat scientists for abandoning objectivity when their social and political commitments are on the line. I want, rather, to stress how pervasive teleological assumptions are, how strong they remain even in the face of deep commitment to the epistemological norms of science. Philosopher Michael Ruse, though a biologist by training and a deep admirer of biology and science generally, has nonetheless argued that evolutionary biologists have been deeply influenced by notions of "progress," even as they have made every conscious effort to exclude teleological dogmas from their work.[35] And this tendency can be remarked in other scientists as well, though it is probably less pervasive than Ruse claims. One may well daydream of a society composed of scientifically well-informed, philosophical grownups. Alas, this seems to be a fragile fantasy, hopelessly unlikely to be realized under any social or cultural circumstances that I can envision. Wishes, prejudices, and hopes are too ineluctably a part of our mental makeup for us to be fully successful in trying to exclude them from our vision of the fundamental ground plan of the universe. Individuals may be able to do so, from time to time, but it takes enormous will and concentration, and perhaps a good deal of sheer perversity as well. The propensity to find in the fundamental facts about nature some reflection of our own deepest moral desires will always problematize our relationship to scientific knowledge. Despite "social constructionist" theories of knowledge to the contrary, these desires probably cannot dictate the content of that knowledge over the long run—science is too autonomous and too methodologically resolute for that—but they certainly can and do govern how that knowledge is received, or rejected, by the larger culture.

FOUR

Credulity

Credulity, as I shall argue, is not entirely a passive weakness imposed on ignorance by the imperfections of education and the deceptions of the conniving. Rather, it is in many ways an active and deliberate choice, an act of manifest rebellion. Of course, the credulity I have in mind chiefly relates to the belief vested in antiscientific or pseudoscientific claims, ranging from the practice of hiring dowsers to the widespread inclination to accept tales of alien abduction at face value. This susceptibility bears a special onus, and, within scientific circles, comes in for particular scorn, because it represents not only easygoing softheadedness, but a perverse refusal to look at mountains of contravening evidence. Nonetheless, much as I might like to join many fellow-scientists in looking down our noses at a collection of presumed simpletons, there is, I suspect, something far too reductive in the view that this is merely a case of fools being misled by knaves. The willingness of the credulous to refuse the wisdom (and the presumptive consolations) of the scientific view may often be a far more subtle affair in which the roles of fool and knave (if there is knavishness at all) can be strangely confounded.[1]

How, after all, can one explain a story entitled "Using Advice of Psychic, Police Hunt for a Body" that appeared, not in a supermarket tabloid, but in the solemn pages of the *New York Times*?[2] The piece contained no hint of a raised eyebrow, no slightest suggestion that such eagerness to embrace a medieval worldview might raise legitimate questions about the competence and intelligence of the investigators. The "psychic" was indeed interviewed. But she was confronted only with the mildest questions, whose answers were then accepted at face value. She escaped the pugnacity that reporters usually direct at far less incredible claims from politicians, bureaucrats, and corporate executives. There was no hint of humor, no suggestion that anyone's leg was being pulled. Usually, it is safe to assume that reporters for the flagship of American journalism are intelligent and reasonably well educated people, and schooled to a certain skepticism. If the reporter was a fool—or, more likely, playing the fool—then it was foolishness of a special sort. At the time it seemed a bizarre caper for the august *Times*.

As it happens, a follow-up story leaves the impression that the initial version was a bit of drollery in which the reporter pretended to be more credulous than he probably was.[3] The later piece was entirely different in tone. It directly confronted the claims of the supposed psychic, and noted that in none of the hundreds of cases on which she has "assisted" is there unambiguous evidence of a correct solution. It quoted a few well-known skeptics and gave a tongue-in-cheek checklist of the "evidence" for the preternatural powers of the seer.[4] This at least made it clear how far ambiguity and self-deception have to be stretched in order to accommodate such claims. For whatever reasons—one can only guess at what went on behind the scenes—the *Times* and its reporter chose to redeem themselves belatedly from their original folly.

Nonetheless, the episode remains a perplexing and disconcerting one. The initial piece seemed to be squarely aimed at the susceptibilities of some *New York Times* readers. It echoed, and presumably hoped to exploit, a widespread discontent with the scientific outlook that is now well established within the literate population, even including subscribers to our newspaper of record. This restiveness is not merely the reflexive response of ignorance or of inexperience of the world. Lurking behind it is self-conscious rebellion against a suite of attitudes and philosophical premises that have characterized "enlightened" opinion since—well, the Enlightenment.[5] As philosopher Colin McGinn puts it:

> [T]he rationalist conception of reason clashes with a popular and misguided ideal of freedom. Logic, after all, constrains our thinking. We must obey its mandates. Yet people don't want to be constrained; they want to feel they can choose their beliefs, like beans in a supermarket. They want to be able to follow their impulses and not be reined in by impersonal demands. To suggest that there is only a single correct way to reason feels like a violation of the inalienable right to do whatever one wants to do.[6]

I personally am a fan of a debunking—a fan, rather than a practitioner. I like to watch from a comfortable distance as the fraudulent and the intensely self-deluded are unmasked, through careful preparation and tight methodology. Professional conjurers like the Amazing Randi and Penn and Teller are modestly famous for such work, but it can also be done well by the less theatrically adept. Even academics have had some success at the game! I follow these exploits avidly as a reader of *Skeptical Inquirer* (I own every issue ever published) and its slick new competitor, *Skeptic*. Whether the process of undeceiving the gullible is flamboyant or merely anticlimactic, it has a certain ritual value for me. It reaffirms the primacy of intellect, rationalism, and sequential logic in mastering, at least by modest increments, the uncertainties of a callous world. Yet I find myself shrinking from actual face-to-face conflicts with the credulous. As much as any contemporary academic urbanite, from time to time I run into believers in astrology, healing touch therapy, Freudian analysis, and similar delusions. My initial impulse to scream out objections usually dies before anything comes of it, drowned as much by cynicism and laziness as by tact. Indeed, I would be loath to claim that tact is a major factor in my psychological makeup, given my predilection for picking fights with fellow professors.[7] Nonetheless, confronted with a garden-variety fan of, for instance, the Psychic Hotline, I tend to change the subject, silently reflecting that people have been soothing their anxieties with such tripe for centuries. Rescuing a lone soul from delusion (assuming—fat chance!—I might succeed in doing so) isn't going to make much difference to the world, whereas the attempt will probably just augment the world's overabundance of resentment. I even managed to keep my mouth shut some years ago when a contractor I'd hired to dig a surface well on my country property hit me up for another twenty dollars to pay his dowser![8]

My view, therefore, is that humanity is unlikely to be rescued from even its coarsest superstitions any time soon. Some of these do no particular harm, certainly not on a systematic basis. Others are vicious and, given enabling circumstances, can work horrors. What most interests me for present purposes is the persistence of what might be called intellectualized systems of credulity, fantasies (for that is what they ultimately are) incarnated in highly structured, seemingly rationally oriented bodies of theory about the world. I am fascinated because to understand such systems reasonably well, and certainly to create them, requires a good deal of what is commonly and correctly regarded as intelligence. Thus they contrast sharply with merely ad hoc, hastily improvised piles of half-baked ideas, such as those apparently concocted by most of the jurors in the O. J. Simpson murder trial.[9] The conceit that William Shakespeare was merely a front for the true author of the plays and poems that bear his name is one example of an elaborately worked-out crank theory. It has seduced, among others, Mark Twain and Sigmund Freud—neither of them stupid, by any reckoning, and both inclined to close and cynical examination of human behavior.[10] When analogous systems contradict not merely received opinon but well-warranted scientific conclusions, it usually isn't just a matter of simple defiance. In fact, the proponents of unorthodoxy often attempt to invoke science, or what they understand to be science, on their own behalf. Intertwined with the imperfect or fraudulent science itself, one usually finds echoes of the myths with which our culture has surrounded science.

The most frequently heard is the Galileo theme, the idea of the lone, noble genius whose insight transcends that of his hidebound age and who consequently faces scorn and persecution for his advocacy of unconventional truth. Rarely has any cult hesitated to answer derision by resorting to this myth in some form. In recent years, this tactic has frequently been reinforced by appeal to a vulgarized version of Thomas Kuhn's famous concept of "paradigm shift."[11] The ready-made scenario, in such cases, is that the system in question has such "revolutionary" implications that they are "incommensurable" with the reigning paradigm. Hence, the generation of "routine" scientists who oppose the new idea must die off before it can triumph. We often find that such belief systems incorporate not an explicit commitment to nullify or destroy the scientific way of understanding the world, but rather a scheme to emulate or even to merge with it. It would be overbroad to acquit believers in such systems from a lingering urge to declare, at some level, *credo quia absurdum*! This is an eternal and recurrent temptation, and few idiosyncratic systems of belief are entirely free of it. But often this is little more than an undercurrent, just one theme in a confused brew of misgivings about contemporary science and its hierarchies. What is really essential in the believer's makeup is defiance rather than any special delight in the absurd.

It is important to distinguish commitment to highly elaborated, fully fleshed-out belief systems from casual endorsement of fascinating, if grossly improbable, rumors. I propose to call the latter "petty credulity" in order to stress the contrast with the "grand credulity" embodied in the former. Petty credulity can be excused, as well as condemned, for its unthinking, rather lazy nature. A curious and bizarre story comes someone's way and he chooses to believe it and pass it on (possibly with bizarreness-enhancing additions)—because it is amusing or (mock) terrifying, because it breaks up the stale routine of his usual round of thoughts, because it provides grist for the conversational mill, because, like a state lottery, it suggests the possibility that something really interesting might happen to him. But people who do this usually don't invest much effort in defending such beliefs. Even less do they do much of the work necessary to scrutinize and criticize these claims. They believe, not because it is absurd, but because it is fun, or because it is stimulating, or because it is shocking—sometimes all three.

Grand credulity is another matter entirely. There, all the outward machinery of ratiocination and even of skepticism is apparently in place. Science, though scientists and conventional skeptics might have it otherwise, is not a counterforce opposing the erection of such systems, but rather an inspiration. For the adherent, it is very desirable that such a system be elaborate, encompassing all sorts of purported supporting detail. Equally desirable are extensive terminology and far-reaching theoretical divinations. Such complication is a psychological device by which the believer affirms his sagacity, his depth of understanding, his courage, his originality, and his faith in future vindication.

The distinction between grand and petty credulity is not, moreover, simply a matter of belief content. One can casually believe, perhaps in a half-amused way, that an alien spacecraft crashed in Roswell, New Mexico,[12] in 1947, whereupon it became the subject of a frantic government cover-up. On the other hand, one can affiliate with a group of UFO believers which obsesses over this particular rumor and adheres to a highly elaborated version of the bare, original story. This version will be fleshed out with accounts of who and what the aliens are. There will be detailed narratives of how the supposed cover-up was carried out and how it was exposed, complete with heroes, villains, "documents," photographs, relics, and so forth. There will also be a "physics" of interstellar (or "interdimensional") travel, laid out in terminology borrowed from real physics and larded with references to the thinking of real physicists. From my point of view (and I hope that of the reader), the second kind of UFO belief is, on account of its baroque complexity and systematic paranoia, far more deranged than the casual, streetwise, UFOlogy of someone who thinks about such things only rarely. At least in the latter case, the beliefs in question are not deeply embedded in the believer's emotional life and are thus relatively easy to discard.[13]

If the "grand" version of UFO belief is fiercer—and crazier—it is not easy to specify where the craziness lies. Characteristically, the entire substance of the believer's self-esteem is implicated, which accounts for the passion with which the system is advocated and defended. Yet passionate defense of one's ideas is hardly diagnostic of mental pathology. It is the defining disease, if disease it be, of intellectuals as a class. In his study of the psychology of superstition, Stuart A. Vyse notes, "Although paranormal ideas are not rational in the sense of conforming to current scientific knowledge, they may, when deeply incorporated into one's world view or sense of identity, provide a sense of well-being."[14] The whole edifice of heavy-duty UFOlogy is a simulacrum of systematic rationality, and its version of rationality is quite impressive to many, including nonbelievers. Indeed, the recent media to-do over the purported "Roswell incident" (on its fiftieth anniversary) triggered yet another wave of systematic credulousness. Here, the supposedly cynical, disillusioned press corps was the chief actor—or culprit.[15] Gullibility was dressed up in the costume of hard-bitten skepticism—specifically, the axiom that the government can never be trusted to be above board in any of its dealings with the public. When the Air Force released a lengthy document that not only categorically denied the rumors of crashed spacecraft and alien cadavers, but also traced elements of the legend to their sources, the effort was not only ineffectual, but counterproductive.[16] In general, the media's attitude, conveyed by tone of voice or ironic wording, was one of narrow-eyed distrust—of the government. Earnest Roswell conspiracy theorists, by contrast, were treated with respect. Naturally, part of this was shadow play, cynicism within cynicism, an instance of journalism driven by careerism and greed to pander to the offhand credulity of the masses. The supposed cynicism of the press toward government is, let us remember, a highly inconsistent and selective mechanism. How often does an investigative reporter stalk the trail of an influential congressman to finger the people with whom he talks, socializes, dines, plays golf, takes vacations—

and consults on special-interest legislation? *That* kind of scrutiny of governmental behavior is not so likely to win the favor of owners of networks and newspaper chains. Nonetheless, easy as it is to dismiss the journalists who have played up the Roswell nonsense as carnival barkers in well-tailored suits, there is probably a layer of fitful sincerity there. To a degree at least, reporters are inspired by the sincerity, persistence—and, it must be admitted, the perceptible intelligence—of the UFO faithful.

One could not conceivably call such UFO cultists lazy, either, though the intellectual energies that they exert are of a constricted sort. Their methodology is not haphazard, though a lot of special pleading, as well as blindness to discommodious facts, is embedded in it. Confronting such a system and its believers, one feels that the entire apparatus is, above all, a device for banishing or burying personal feelings of impotence. It offers the chance to demonstrate mastery, acumen, and (as the believer may convince himself) steady nerves. It makes him part of a visionary elite, a fellowship of intellectual and spiritual voyagers.[17] It creates an emotional enclave where the wear-and-tear of daily life doesn't intrude. Inside these walls, one's concerns and beliefs, contentious and contestable as they may be, rise above the narrowness of petty, trifling quarrels. Even more, they forge a role for the believer—accepted as well by those he can convince to take him at his own estimate—that is deliberately analogous to and competitive with the role conventional wisdom assigns to the scientist. Like the scientist, the believer becomes an explorer of new frontiers, a synthesizer of bold new insights, and even a defender of intellectual integrity in the face of ignorant scoffers!

Such structures of belief may appear deeply flawed, indeed fatuous, to the unpersuaded eye, but they are hardly "anti-intellectual" in the ordinary sense of the term. Indeed, they tend to flaunt a self-conscious intellectuality. Some of them may be quite crude in their defiance of probablility, but this is not an inevitable characteristic. Theories like these need not be keyed to some grossly counterfactual account of the world and its history. Their champions are not necessarily café intellectuals wistfully and hopelessly aspiring to the Eden of academic life. Indeed, many of the specifically *academic* cults that have flourished over the past fifteen or twenty years, including some still sacrosanct on campus, are examples of grand credulity. Here, for instance, is a specimen that happened to cross my e-mail screen just as I sat down to begin this essay:

DELEUZEGUATTARI & MATTER
October 18–19, 1997
Philosophy Dept.
University of Warwick
Since the form depends on an autonomous code, it can only be constituted in an associated milieu that interlaces active, perceptive, and energetic characteristics in a complex fashion, in conformity with the code's requirements;

and the form can develop only through intermediary milieus that regulate the speeds and rates of its substances; and it can experience itself only in a milieu of exteriority that measures the comparative advantages of the associated milieus and the differential relations of the intermediary milieus.

When content and expression are divided along the lines of the molecular and the molar, substances move from state to state, from the preceding state to the following state, or from layer to layer, from an already constituted layer to a layer in the process of forming, while forms install themselves at the limit between the last layer or last state and the exterior milieu. Thus the stratum develops into epistrata and parastrata; this is accomplished through a set of inductions from layer to layer and state to state, or at the limit. A crystal displays this process in its pure state, since its form expands in all directions but always as a function of the surface layer of the substance, which can be emptied of most of its interior without interfering with the growth.

An invitation to submit papers addressing the specificity of the DeleuzoGuattarian contribution to the reformulation of a philosophy of matter. We are looking for papers to address this issue from a number of different perspectives with the aim of producing a research programme that sharply demarcates the DeleuzoGuattarian project and the enquiries that it initiates from other emergent critical apparatuses.

A philosophy of matter is produced out of pragmatic engagement with DeleuzeGuattari's many-headed critique of hylomorphism, atomism, and what in general can be called "identity theory," theories such as mechanics and energetics, where matter is flatlined under the prerogative of the physical. These lines of attack, and the concern to circumvent representation (the challenge thrown down to "signifier enthusiasts"), combine with the attempt to create an immanent grounding of the flows of energetic materiality, a concern less with "a matter submitted to laws than a materiality possessing a *nomos.*"

It will not be a question of making matter the matter of philosophy, but of constituting philosophy as immanent to matter in every case. To flesh philosophy out with zones of materiality is, we think, to take an impetus from DeleuzeGuattarri, to make of philosophy a pragmatics. It is a question of free energy, what can be done with the materials synthesized by DeleuzeGuattari?

Zones of Engagement include:

—code

—hylomorphism

—intensity

—machinic propositions

—number: ordinality/cardinality: production

—nomos/praxis

—semiotics

—singularities[18]

In citing the above, I am being peremptory and a good deal less than polite, as the academy understands politeness. I don't pretend to be an expert on Deleuze and his close collaborator Guattari (the welding together of their names in the text above is a conceit, not a typographical error). I have read just enough of their work to conclude that it contains little of philosophical value. To my philistine sensibilities, it seems to be just another heap of postmodernist mummery.[19] It's no surprise, therefore, that Deleuze's epigones and would-be commentators are so immersed in the mock euphony of their private language that meaning pretty much goes by the board. There seems to be no reason to resort to this kind of verbal chaff other than to forestall the intrusion of common sense and similar embarrassments.

Notwithstanding the august provenance of the announced event—Warwick is a very respectable English university and the conveners have respectable academic jobs (and salaries)—it is a piece of cult work. It is credulity on a grand scale indeed, if we take intricacy of rationalization, rather than popularity, as the measure of grandeur. Though it seems to breathe a rarefied atmosphere, it shares some interesting qualities with the kinds of credulous systems that play to the mass market. First of all, communicants of the sect share the delusion that they have special access to a layer of deep truth denied to the uninformed, to scoffers, and, of course, to adherents of rival sects. Note the remarks about "the specificity of the DeleuzoGuattarian contribution"—a polite way of saying that they're disdainful of rival claimants. Note, even more significantly, the writers' peculiar ambivalence with respect to science: on the one hand, they seem to be addressing a topic—matter and its manifestations—of particular interest to physical scientists. There is a reference, via crystallography, to scientific insights upon which the proposed theories are to be built. Scientific terminology echoes in the jargon. Yet the dense neologistic prose and the hierophantic tone suggest that ordinary scientists are too dull and earthbound to enter into these mysteries. This is a primal example (to use philosopher Susan Haack's terminology) of "sham" inquiry: "attempts, not to get to the truth of some question, but to make a case for the truth of some proposition one's commitment to which is already in evidence."[20] It might even be that "fake" inquiry—"attempts not to get to the truth of some question, but to make a case for the truth of some proposition to which one's only commitment is a conviction that advocating it will advance oneself"[21]—is on display as well.

To compare this minute academic cult to the vast UFO subculture may traduce scholarly self-esteem, but it is, I think, a sober and revealing analogy. The seeming light-years that separate the two cases are measured primarily in terminology and academic snobbery, which disguise the close psychological, indeed sociological, affinities. Frankly, I'd rather get stranded on a stuck elevator with a pack of UFO nuts than with a flock of deluded Deleuzeans. The tall tales of the alien-sniffers are far more entertaining and agreeable. True, the former might harbor a few quick-buck artists in their midst, whereas the latter will only host their scholarly counterparts who are content to be paid off in the doubtful coin of academic preferment. But I think it would be considerably easier to hold on to my wallet while among UFO buffs than to my sanity in the nattering presence of postmodernists. In neither case, alas, would a scientist find hopeful signs of intellectual maturity. And in both he would detect an envy, sometimes undeclared, sometimes phrased to seem its opposite, toward the accomplishments, as well as the intellectual prestige, of actual science.

The pseudointellectuality of academic cults may be more polished than that of popular superstitions, but it is driven by analogous mechanisms, and it founders on similar self-deceit. The quoted Deleuzean example, in its pomposity, is especially fatuous and marvelously droll, but it has many parallels within the academic precinct. Some are just as full of hot air, but many are less gaudy, content to employ plainer language if not clearer thought. Often, they are marked by fervent political motives, as in the case of Afrocentric theory and many versions of academic feminism.[22] Even beyond political ambition, however, they are driven by the deep urge to be seen as pioneers of new, earth-shaking truths. It goes without saying that these would-be epiphanies have to gratify a priori ideological conceits, but equally they gratify the desire, near-universal among intellectuals, to be at the forefront of a revolution in ideas.

Haunting all such movements is the plaintive question, "Why aren't *our* breakthroughs, our gallant, iconoclastic efforts to re-envision the world, given the same recognition accorded science—and scientists?" It is not necessarily the content of science that provokes this petulance, though sometimes this happens. Rather, it is the cultural authority which science commands, and the special prestige with which our society has garlanded it.[23]

Of course, most intellectuals outside the sciences find themselves asking a similar question at some point in their careers. Few are fully pleased with the pride of place that science has won in comparison with all other modes of inquiry. Most of them, however, easily resist the temptation to erect or affiliate with counterfeit systems of the kind that sabotage or nullify critical judgment. William Blake's *cri de coeur*, "I must have my own system or be slave to another man's!" undeniably appeals to many scholars, but only ephemerally. Nevertheless, the tendency of pseudologous systems of thought—or quasi-thought—to take root and flourish in the academic world has certainly been growing over the last decade or so. Neither jeremiad nor satire has done much to impede it.[24] The result has been a torrent of credulity, encompassing everything from deconstruction to "feminist epistemology" to the settled conviction that all cultures and races are kindly and noble, except for the European.[25]

In some of my earlier writing, I have been rather harsh with Professor Andrew Ross, director of American Cultural Studies at New York University, particularly over his book *Strange Weather*.[26] In this work, he investigates various parascientific and pseudoscientific subcultures and broods on their sociological and political implications. My criticism of Ross is directed toward his arrant, uninformed contempt for science and the scientific way of viewing the world.[27] But his analysis of New Age movements and their kin is not completely without insight. In particular, Ross recognizes the ambivalence of pseudoscientific cults toward legitimate science. On the one hand, the scoffing and offhand dismissal they encounter from mainstream scientists provoke resentment, anger, and the propensity to regard scientific orthodoxy as a blind, intolerant enemy; on the other, they greatly esteem the prestige and the accomplishments of science, as well as the imaginativeness and intellectual courage of great scientists. Resentment toward science notwithstanding, cultists yearn for a place within the scientific canon for their own favored belief systems. Professor Ross's own brand of intellectual priggery leads him to applaud such sects insofar as they defy scientific orthodoxy and its standards of truth.[28] To the extent that they admire science and wish to emulate it, Ross disapproves. That aside, however, his basic point is sound. In his perverse way, Ross has hit upon the deep ambivalence toward science that pervades even the most defiant countersystems. Science is bitterly rejected yet doted on from afar. The credulous perceive themselves as manifesting the same kind of courage that made science possible in the first place.

Why do people fall into such systems of thinking? The question is a convoluted one. Whimsicality or unalloyed perversity have little to do with it. No less than academics, the people attracted to such conceits are obsessed with finding reasons to believe that they have mastered at least some aspect of how the world works. They like to think that they have, by their boldness and diligence, created an important legacy of ideas for the human race. But why aren't they content to participate in, or at least to appreciate and support, standard science? Why the urge to contradict it? The blunt fact, I fear, is that science, especially in its foundational aspects, is extremely difficult to apprehend, let alone to master, at the depth necessary to make original contributions. Even impeccably trained and highly talented scientists function, for the most part, at a level where what they add to the overall stock of knowledge is incremental, or even fragmentary. Science on the heroic level, science of the kind that truly justifies the name "breakthrough," happens once in a while, and, from my admiring and frankly envious perspective, those who achieve it are the most fortunate of beings.

But the bulk of the scientific community consists of journeymen and minor masters, fated to be remembered no better than thousands of their peers. For most of us, this is still a pretty agreeable bargain with life. To be the first person to grasp even a tiny truth about the world is deeply satisfying and highly consoling to one's self-respect. However, the great majority of people who fall under the influence of crank systems or outright anti-science haven't the talent to reach even this modest plateau. Furthermore, it is doubtful whether it would suit them temperamentally even if they were to attain it. The recurrent irony is that science inspires such people chiefly through its most outsized legends—Galileo, Newton, Pasteur, Darwin, Einstein. The intellectual energy provoked by such iconic examples may be considerable, but it is wasted energy, incapable of being productively focused. Cranks are not lazy; indeed, they are, by definition, persistent to the point of obsession. But their ambition is nonetheless deformed because there it harbors the fatal belief that there is a shortcut, an ingenious way of evading the stringent demands that authentic scientists routinely submit to.

The Deleuzean rhapsody quoted above puts one form of this failing on display. Here, the work properly belonging to careful thought and exacting analysis is assigned to mere jargon. An unchecked flow of flummery and borrowed obscurities is offered in place of logic and argument. The reader is supposed to be impressed by a kaleidoscope of exotic terms (indeed, the writers surely impress themselves) and is urged to take their jangle for philosophical profundity. Again, this is symptomatic. A specialized in-group lingo is quite characteristic of grand credulity. It creates the impression that deep knowledge abides within the cult, and that it is accessible only to minds that have been appropriately fine-tuned. It reinforces the sense of specialness that binds believers together while it bewilders outsiders, who then can be dismissed as thick-headed. But, in fact, it isn't even jargon, properly speaking, for jargon is essentially a substitute for ordinary language, whose defining property is that it makes the obvious sound impressive.

Here, what is being obscured is not obviousness but emptiness and circularity. What we might call the Realist fantasy is also at work, if we understand that term in its Scholastic sense. Through the invention of names, descriptions, categories, reality itself is created in the susceptible mind. One reason that UFOlogy seems so convincing to its adherents is the elaborate taxonomy that its leading figures (hucksters and true believers alike) have created by elaborately classifying various types of supposed "aliens," "spacecraft," and so forth. The logical shortcut—short circuit, rather—thus created permits mere list making to do the work of the hard evidence that never quite emerges.

Astrology exploits a similar fallacy. The esoteric vocabulary and extensive, quasi-mathematical machinery of charts, conjunctions, alignments, and so forth is taken as a guarantee that there must be genuine substance beneath it all. Indeed, astrology is merely the hoariest example of a system that holds sway over the gullible by flaunting a terminology that seems closely linked to that of authentic science. After all, the very signs that astrologers use to denote planets are used by astronomers for the very same purpose. In a similar vein, terms from physics, like "energy," "radiation," and "quantum," are thrown around by a thousand sects (and rackets). An extremely popular book of health quackery is called *Quantum Healing*.[29] In it, the reader is assured that the nostrums it offers are confirmed by the deep insights of contemporary physics. The term "quantum healing" means nothing, of course—it is merely a huckster's coinage. But the resonance of the word is what counts.

Trumped-up vocabulary isn't the only illusory web-work upon which credulous systems depend. Equally appealing to acolytes are complicated systems of reckoning or interpretation, whether of texts or of supposed events. Biblical exegesis is, in our culture, the prime example, though all other cultures have mantic practices similar in their mumbo jumbo if nothing else. Again and again, over a wide range of examples, the requirement for solid evidence that the system actually generates valid results is evaded or deferred by the very elaborateness of the interpretive process. The underlying psychology is that anything that requires such intense, drawn-out, convoluted effort must inevitably be fruitful in the end. At bottom, that is why the casting of horoscopes by "serious" astrologers is such a baroque production. The thinking is that anything so exacting and painstaking must embody truth at some level.

Again, stupidity is not a precondition for falling into this delusion. Indeed, stupidity confers immunity to a certain extent, since it precludes a close appreciation of all the intellectual prestidigitation that goes into the work of "interpretation." Recall that Isaac Newton himself obsessively searched for the esoteric truths that were encoded in Scripture. And, as I write, a book called *The Bible Code* is topping the best-seller list—as nonfiction, of course.[30] Its authors happen to be mathematicians and computer scientists who are, in the narrow sense, quite competent professional scientists.

Consider the durable appeal of Freudian psychoanalysis and its variants. These cults have captivated many far-from-stupid people in the course of this century. They still have an extensive fan club among intellectuals.[31] Densely worded exegesis is their lifeblood. Freudian interpretation is such an elaborate, ingenious, and, it must be said, poetic procedure that questions of circularity, self-reference, and lack of independent evidence are put aside and eventually utterly ignored. The intellectual rhythms of the Freudian method are so seductive that they convey the feeling that a logical case is being built through hard facts when, in fact, nothing of the kind is going on. The "hard facts" are nebulous artifacts of the assumptions they are supposed to confirm. The long hunt for validation proceeds in a circle; but it is so long that the pursuers all swear that they are proceeding on a straight-line course toward enlightenment. It hardly needs saying that a very similar phenomenon appears among true-believing Marxists. People committed to schemes like these, and to far less respectable ones, are strong in the conviction that what they are doing is really scientific work or something like it. The complexity and difficulty of their efforts is quite real, but everything is built around an evasion, the refusal to look at thinking and evidence that go outside the wheels-within-wheels of the favored system.

This is especially painful to watch when the crank work at hand is pseudophysics or pseudomathematics. Like many mathematicians, I have been on the receiving end of long, closely written screeds (some of them beautifully printed and bound), claiming to have unearthed new and deep mathematical truths. Invariably, I find that an enormous amount of time and energy has been poured into something from which even the most elementary insight is missing. The writer desperately tries to emulate, even to outdo, a kind of thinking he regards as singularly profound and elegant. Outwardly, the text is marked by what might seem like the apparatus of mathematical exposition—items marked "Definition," "Lemma," "Theorem," and so on. Yet it is all hollow show. At an elementary level, the writer is impervious to the obvious, the sort of thing a decent mathematics student picks up in his teens. Even worse, the habit of continual critical evaluation and self-scrutiny, without which serious mathematical thought is impossible—the mathematical conscience, as it were—is completely short-circuited. Obsessively and pathetically, the crank grinds away forever, erecting one piece of meaningless apparatus upon another. The demanding intricacy of his self-imposed task blinds him to its pointlessness.

Nonetheless—to intone my moral once again—the people who get trapped in this kind of delusion are not in general stupid. A prime example is the philosopher Thomas Hobbes, one of the most insightful men who ever lived when it came to the psychology of power and politics and the mechanisms through which societies cohere. He was cursed, however, with the frantic desire to be a great mathematician. He was further undone by the absurd delusion that he was one.[32] His peculiar insanity is echoed ad nauseum by the thousands upon thousands of people who want to tell you where Newton went wrong in devising calculus or where Einstein went wrong in creating the theory of relativity. Equally ridiculous, though harder merely to dismiss, is a new type of academic, trained in the humanities or the softer end of the social sciences, who goes about pontificating on chaos, fractal geometry, indeterminism, and such without having mastered even a fraction of the mathematics necessary to make sense of any of them. The fact that these folks are given serious academic brownie points for such work constitutes at least a minor scandal.[33]

It is tempting to rail at the vast hordes of the credulous. Why, one is constantly moved to ask them, can they not be content with learning the science they are able to master, and delighting in the beauty and elegance it reveals? Scientists constantly tell us that their work leads to mystery, rapt contemplation, and profound wonderment. As a rule, they are eager to share these feelings with curious laypersons. Why, then, the widespread enthusiasm for crank belief systems of all sorts whose one unifying property is their frank contradiction of what science knows? There are obvious answers, already alluded to previously. Science can provide no guarantee of meaning on the level of human aspirations and values. Moreover, to be an outsider, on the wrong side of the walls that separate professional science from everyone else, means having at best a secondhand relationship to the facts and theories one is enjoined to admire. In contrast, the commonest forms of credulity validate one's deepest beliefs while offering the illusion of privileged insider status.

Furthermore, it is wrong to assume that the delusory nature of antiscientific belief systems is plainly obvious to all who encounter them. Such bodies of dogma don't carry warning signs, certainly not self-ascribed ones. Many of them, as noted, pretend to scientific legitimacy and speak a language chock-full of scientific terminology. But, even more important, they recruit from a population to whom credence in standard, orthodox science itself can seem hardly distinguishable from outright credulity. From the perspective of laypersons not deeply acquainted with science, contemporary scientific views are far from obvious, natural facts. They frequently contradict the intuitions, as well as the sentiments, of ordinary folk. As a rule, such views repose on evidence and on detailed arguments the public never sees at full strength. Publicizing these details wouldn't help much, because they are opaque to anyone without talent and training. Consequently, the findings of science may appear no more probable than many antiscientific belief systems, while in a practical sense they are far more difficult to assimilate.

Subjectively, scientific doctrine is not demarcated from its rivals by clear marks of epistemological sanctity. It is social convention that does the job. In short, scientific knowledge, as it presents itself to the lay public, is not really knowledge at all. It is, rather, something imposed from without by the fiat of a special class of people from which that public is clearly excluded. It is not something that the ordinary man or woman has learned and absorbed through intimate acquaintance and close scrutiny, and consequently it lacks inner authority.

Of course, science does have the authority of a civilization that nominally accords it a uniquely prestigious status. It is routinely and invariably associated with the amazingly productive material technology that surrounds us all. No one, it is safe to conjecture, is so credulous as to believe that a CD player or cellular phone is produced by star-roving aliens or bands of angels.[34] Yet when ordinary people encounter science in "pure form," through accounts of the theories or discoveries that excite the enthusiasm of scientists, it appears quite removed from familiar, or even prospective, technologies. It is unadulterated "theory."

For someone of limited scientific background to accept it, then, requires not only deference to authority, but also a leap of faith of sorts. Without that, abstruse science seems little different from antiscientific or pseudoscientific just-so stories. This should not surprise us. Professional philosophers as well as philosophically inclined scientists are still uncertain and deeply divided among themselves by the question of what, precisely, justifies the special epistemic authority that science commands. (This is not to say that many of them are eager to deny that it *is* deserved.) One consequence is that many people lack the intellectual equivalent of an immune system. There is no mechanism in place which automatically warns them away from the temptations of antiscientific belief systems. Their inclination to prefer science over its rivals is halfhearted, if it exists at all. Moreover, to the extent that they do prefer scientific thinking to nonscientific, the preference is circumvented by the fact that rivals to science are works of *bricolage*.

Crank theories often incorporate bits and pieces of science alongside unscientific and antiscientific elements. Sometimes their advocates include people with palpable scientific credentials. To those susceptible to such a chimera, it simply doesn't stand out as "unscientific" in the crude sense. The scientist who declares that it *is* unscientific comes across as no more a scientist than the self-proclaimed "scientist" who recruits on its behalf. Let's not forget that a leading organization of UFO believers was founded by a renegade astronomer who was at one time a government-endorsed expert on the "flying saucer" phenomenon.[35]

Of course, it is quite possible to make key distinctions between science and pseudoscience, notwithstanding the sneers of trendy postmodernists. For people thoroughly grounded in science, such discrimination is in fact reasonably easy. In demographic terms, alas, this kind of background remains, for all our elaborate systems of higher education, remarkably rare.

In consequence, extreme credulity, embodied through formal organizations or elaborate, if informal, associations, is not about to disappear from our social universe. It will be with us well into the future, even if science education in our country were to improve far more widely and rapidly than anyone believes possible. Grand credulity, as well as petty credulity, is woven into the fabric of our culture. It is, in many ways, the godchild of American pluralism. It is nurtured by the tacit social compact that looks with benign eye on a vast proliferation of religions, sects, cults, and creeds of all sorts. There is within the American version of the democratic ethos a strong inclination to assume that, in all matters, one person's opinion is as good as another's, that all opinions have the right to establish themselves institutionally and to proselytize without restraint, and that for any one of them to try to put the others out of business would be a gross violation of our sacred social tenets. Even science, that specially favored child of America's romance with the spirit of progress and prosperity, must abide by the rules of this particular game—or so most people tend to think. Thus we are condemned to a kind of scientific pluralism, or to put it more pessimistically, a perennial war where a swarm of claimants constantly rages against one powerful and legitimate, but ever more weary guardian of the shrine. Science will have to fight like the devil for every inch it gains in this contest, or even to keep from losing ground. The fight against credulity will be waged in an arena where courts and laws will be of little assistance, and where the rules of the fight will be dictated, to a great extent, by the old demons of human perversity, vanity, envy, and wishful thinking. It's not a war that most scientists deliberately signed up for, but they are conscripts all the same.

FIVE

Technology

About twenty years ago (1976–1977, to be exact) I was a visiting professor at Stanford University, which is located hard by "Silicon Valley," the complex of Santa Clara County communities that comprise the most important center of semiconductor technology—chip making and so forth—in the United States. While there, I noticed, albeit without particular interest or enthusiasm, that there was a local "microcomputer" subculture centered on devices that could be assembled by hobbyists from kits or merely from components. Every once in a while, I would drive past a shop catering to this avocation, usually a small-scale affair in a low-rent neighborhood, very much like the kind of store that serves the needs of model railroaders or ham radio operators. I never actually saw a hand-built computer, or met anyone who owned one. To the extent I thought about it at all, the whole thing looked like a mildly exotic pastime—much, indeed, like ham radio or model railroading. It seemed likely to appeal to introverted, sedentary characters with fine small-motor skills and the willingness to put up with a good deal of eye strain. It struck me, moreover, as a decidedly parochial phenomenon, dependent on the easy availability of electronic odds and ends and on the interest of a few hundred people who had some connection with the semiconductor business. The idea that I was looking at the genesis of what was to become one of the world's major industries never crossed my mind for an instant. If someone had offered to let me invest $500 in the idea of "desktop computers," on the understanding that the payoff twenty years down the line would be commensurate with the success of the concept, I would certainly have held onto my money.[1]

I have at least the consolation of knowing that in this case I wasn't any more obtuse than the general run of citizen or investor. In fact, I doubt whether many scientists, irrespective of field, would have found it easy to believe that within a couple of decades, incredibly fast, sophisticated computers would be as much a staple of middle-class households as electric toasters (or that ordinary folk would, like me, run through six or seven successive, increasingly powerful models in as many years, all the while keeping the same old toaster). I had, of course, wondered whether the convenience of computerized data processing might be exploited beyond the narrow realm of scientific users and of large-scale corporations that could afford to buy or lease expensive equipment. I assumed that something along these lines would eventually develop, probably in the form of computer companies selling telephonic access to their machines to medium and small businesses for the purpose of bookkeeping and record storage.[2] But it would have seemed the frothiest of fantasies to imagine that the time would come when no businessman or woman, no professor, no *student*, would be caught dead without a four-pound gadget (such as the one this is being written on) that embodies tens of thousands of times more computational power than the room-filling monstrosities I'd gazed upon awestruck as a high-schooler.[3] I suspect that at that point, even the most visionary computer hobbyist could not have seen far enough past his soldering iron to imagine such a time. As we now know, the canniest, or at any rate the most powerful, professionals in the computer business, in particular, the satraps of mighty IBM, were poor prophets as well. They were fatally incapable of seeing the future in terms of anything but increasingly more powerful (and probably more expensive) mainframes.

The price paid by IBM for its lack of prescience is by now one of the great cautionary legends of business and industry. This firm, which had, for years, been a byword for the efficiency and intelligence with which it dominated its market, missed one opportunity after another: the opportunity to develop a fully proprietary microcomputer design (instead of its easily cloned "open architecture" PC), the opportunity to write its own copyrighted operating code, the opportunity to acquire the modest firm whose operating system software it chose to adopt, the opportunity to create a user-friendly visual interface.[4] The epilogue of course, is the world we now know, where the modest firm just mentioned—Microsoft—has grown to dwarf the one-time industry giant that used to regard it as a minor vendor.

The moral of this well-known tale, so far as I am concerned, is the unpredictability of the course of technology, even for those experts who rule the heart of the relevant industries. Technological change seems as capricious as long-term weather. The chief difference is that we have at least some systematic and general ideas about basic meteorological processes. By contrast, whatever it is that generates long-term technological trends seems impossible to grasp or even to guess at. The personal computer revolution that remade the world was invisible to the most powerful computer company in the world until it was too late to stop a slide into mediocrity which even the power of its famous brand name could not arrest. During the crucial interval, the core of the growing personal microcomputer customer base was far more technologically literate than any sales manager wants to see. It knew that one 8088 microprocessor (to name the first of a now-famous series) was as good as another, just as one sheet-steel box was as good as another. Thus,

IBM's fate was sealed, much to the sorrow and surprise of its overlords—and stockholders.[5] However, investors and business analysts, as a class, were no more prescient—nor were scientists.

What is perhaps even more amazing is that the personal computer and software industry had barely begun to reach maturity when another technological surge shook it to its core, one that began just under its nose, if not within its bosom. As if through spontaneous generation, the Internet and, quickly thereafter, the Web swiftly emerged as further indispensable appurtenances to business, the academic world, and middle-class life in general. This time, the entire telecommunications industry—a whole flock of megacorporations, in fact—stood dumbstruck as a vast system matured and became embedded in the world's culture. Even as I write, a titanic legal and commercial struggle is taking form, triggered by Microsoft's massive attempt to make up for its earlier indifference to the Internet.[6] Likewise, telephone company executives haven't even begun to figure out how to deal with a competitor that is owned by everyone and no one.

Once more, the most striking thing about the Internet phenomenon is its seeming unpredictability, the fact that no corporation or consortium, and certainly no government, contemplated, much less planned it. What had merely been a convenient communications network for academics linked to the Department of Defense nucleated the uncanny growth of a system that developed by what almost seem natural laws of expansion, acting without human will or direction. The irony is enhanced by the fact that the spanking-new hardware and software companies whose products catalyzed the whole phenomenon were themselves among the most thunderstruck observers.[7] As one bewildered player in the Internet madness puts it: "The whole thing felt like an out-of-body experience in some ways. But there was a time when it felt like there was no question that we were going to make hundreds of millions of dollars. And then all of a sudden we weren't. But there was no logic here. The whole Internet business is a set of ploys, and it could as easily have happened as not."[8]

What all of this suggests is that technology, in itself, is a virtually autonomous entity, an "emergent phenomenon," as some philosophers are wont to say, of numerous lower-level interactions at a relatively microscopic scale. Those interactions are, however, what we usually think of as self-conscious, rational decisions by thinking human beings. Individuals plan, act, make commitments (including the commitment to invest time and money in novel gadgetry) on the basis of efficiency, comfort, or profitability. Sometimes the effect is amplified by the fact that the individual in question controls a business or some other organization. But in the cases I've cited, these organizations have themselves been modest, at least in the early stages of the snowball effect. The resultant of all these decisions, made for purely local reasons, is to shape and nurture a huge technological apparatus that no one in particular ever designed or planned.[9] Inexorably, it has brought about a vast complex of social institutions as well. Technology theorist Jesse H. Ausubel puts it this way: "I have argued that societal development is fundamentally evolutionary and thus to a high degree without purpose though with strict rules of choice at every stage. However, society does have directions in which it is driven. It moves in these directions largely without intent, particularly of the kind that characterizes political discourse."[10]

The history of technology seems to bear out this observation strongly. In the case of the Internet, we find service providers (grown, over the course of a few years, from undercapitalized pioneers to brand-new megacorporations), marketing services, Net-adapted libraries and information providers, and thousands of newly sprung-up interest groups whose members know each other only through the Net.[11] Legislators who, a couple of years ago, had never heard of the Internet now wax eloquent in defense or derogation of its folkways. Unprepared courts are suddenly confronted with the urgent task of devising precedents suitable to a terra incognita that has suddenly become a central part of the contemporary landscape.

It is tempting to see a close cousin of Adam Smith's "invisible hand" at work here, acting, at least in the case of the personal computer and the Internet, with astonishing celerity and seeming purposefulness, despite the fact that the only institutions arguably powerful enough to have had a measurable effect on the process were, during the crucial period, indifferent, scornful, or outright hostile. No one at any level of authority or responsibility has planned all this. Yet it has come to pass.[12] By noteworthy contrast, related revolutions meticulously planned, magnanimously funded, urged on by the politically powerful, have stalled ignominiously and died ingloriously. Here I have in mind the effort of "Japan Incorporated," that once-irresistible alliance of Japanese corporations and the Japanese government, to develop the so-called fifth-generation computer. At the inception of the project, gloomy American pundits foresaw the end of American computer supremacy. The inevitable consequence, they warned, was the final eclipse of American technological hegemony and the reduction of this country to a virtual satellite of Tokyo. Nothing of the sort occurred, of course. Japan's attempt to leapfrog over the rest of the world landed it in an embarrassing pile of ordure. The net result of this costly experiment was some bits and pieces of mediocre software. The inventors, at the end, couldn't even give the stuff away.

There is no reason for Americans to feel particularly smug in the wake of Japan's "fifth generation" misadventure. The United States has expended a comparable amount of money and talent, and even more time, in trying to come up with a usable "controlled fusion" reactor for energy production. The project started in the 1950s with high expectations, but it soon came up against the realities of plasma physics. Most of the effort has gone into magnetic confinement devices, principally tokamaks (as the currently favored configuration is called, in honor of the Russian prototype). While enormous progress has been made in controlling the propensities of such systems to become unstable, it is still unclear whether a truly practicable energy production device will ever see the light of day.[13]

The moral, then, seems to be that a technology which the ambient culture really desires, in its inarticulate and even unconscious way, will grow on its own, bursting like kudzu through restraints of politics, government, and the concerted power of monopolies. On the other hand, all the marshalled forces of a supereconomy seem doomed to futility when they try to bring into being a technological Leviathan that really doesn't want to be born. This, at any rate, is the litany repeated by many contemporary analysts as they contemplate the stories I've just outlined. The current fashion is to celebrate the Promethean fruitfulness of human ingenuity and enterprise when it is left to organize itself spontaneously. The same voices decry any form of large-scale planning or direction, whether from a statist system or an oligopoly.

The appeal of this version of the technological myth, both as narrative and ideology, is clear. It is like the legends of the great inventors—Whitney, McCormack, Edison, Bell—that saturated the textbooks of my youth, but with the fascinating difference that now we are all somehow included in the reigning definition of "genius." Or rather, it is the genius of the culture, rather than some special individuals, that appears to be the object of veneration. We need do no more than partake of the *zeitgeist* to qualify as heroes of the ongoing technological revolution. This is a uniquely comforting myth, reconciling individualism with a kind of collectivism, anarchy with corporate prosperity. The wisdom of our ethos brings forth all these technological miracles, but it is incarnated in billions of individuals, each with unique quirks, styles, and visions. The collective genius is articulated through the operation of unconstrained individualism. The paradox is resolved in the panoply of choices which an effulgent technology spreads before us, and through which we may all custom-build our own lives.

Having laid out the elements of the myth, as well as the paradigmatic cases usually offered as evidence, let me now intrude a measure of cynicism. It seems to me that technological stagnation is just as much a fact of contemporary culture as rapid technological advance, and that it is just as little planned or foreseen. To my mind, the automobile provides a perfect example. The essentials of automobile design, including engine and power train, were in place before the turn of the century. The gasoline-fueled four-stroke piston engine remains, in its basic form, pretty much what it was when the first gas-engined car trundled down the road. It has been improved by increments over the years, of course. But its fundamental flaws remain—it is noisy, smelly, and needs constant maintenance. Road vehicles also still run on rubber pneumatic tires subject to flats and blowouts. Car exteriors are still made of painted sheet steel easily marred, dented, or ripped in the lightest of collisions. The last significant design modifications—the self-starter and the automatic transmission—date from the 1920s and 1940s, respectively. In fact, the automatic shift is only a minor convenience which many of us prefer to forgo. Cars rolling off the assembly line today are different only in the most superficial respects from those of the 1950s. New cars are designed to be somewhat more fuel efficient (though that is now becoming less important once more), and they have some additional devices on board, like air bags, air conditioners, and CD players, that have nothing to do with the fundamentals of automotive design. Everyone has occasionally run into antique cars from the turn of the century as they parade down the highway, lovingly restored to their original condition. It's a charming sight— but also a sobering one when we contemplate the fact that they do pretty much the same job as our late-model vehicles, and do it pretty much as efficiently. Contrast the situation with that of airplanes, whose modern versions (aside from ultralights) bear only the vaguest resemblance to their ancestors of the first decade of the century.[14]

Somehow, the culture still accepts the basic design of the automobile without notable resentment. People will pay tens of thousands of dollars for minor cosmetic differences and slight improvements in performance and durability. But there is small demand for anything like basic redesign, and what demand there is usually reflects the concerns of environmentalists rather than the discontent of people who like cars. Cut-throat international competition doesn't seem to be able to bring about basic innovation, either. All industrial nations now produce the same automotive product, apart from superficial differences, and none seem eager to break away from the standard concept. It's not as if no reasonable alternative were ever considered. During the 1960s, in fact, American manufacturers experimented with turbine-powered vehicles. For a few years, race-car versions of these dominated the Indy 500 and other major competitions. Street-vehicle prototypes were produced and placed in the hands of ordinary drivers for year-long trials. They seemed to work out well, on the whole. Yet suddenly the whole idea of a turbine-powered car vanished. Turbine engines were banned from racing, and consumer versions ceased to exist, even for experimental purposes. To me, the whole episode seems murky, but typical. Other design concepts, quite appealing from the point of view of cleanliness and simplicity, have been equally ignored for decades.[15] These days, we're starting to see a few electric vehicles on the market, thanks mostly to environmental regulations rather than the presumptive spontaneous genius of our technocracy. By all reports, the best of these are not only clean and quiet, but vastly outperform conventional autos on the road, accelerating as swiftly as race cars. Nonetheless, it is far from clear that these will ever enjoy a widespread market, even if the problem of recharging batteries is solved.[16]

Perhaps an argument could be made justifying the stagnation of basic automotive design purely in terms of engineering and economic criteria. Nonetheless, history would still afford striking examples of technology suppressed or unexploited despite its obvious superiority. It is well known that shortly after the Tokugawa shogunate established itself (circa 1600) as the government of a unified Japanese state, it prohibited muskets and pistols (although it retained a few cannon for coastal defense). This was not a matter of chauvinistic prejudice against "foreign" technology; at that point Japanese gunsmiths ranked with the best in the world. The reason was the threat posed by the very existence of firearms to the rigid caste system of Japanese culture, which the shogunate hoped to use as a prop to its own stability. Guns were, quite literally, "equalizers," in that putting a gun into the hands of an ill-trained peasant made him at least as effective a warrior as a superbly trained samurai wielding bow and sword. The effacement of ancient social distinctions by mere gadgetry was not to be tolerated, and Japan thus put aside an element that was key to the emergence of modernity and colonialism in Europe.

One might be tempted to discount this as a tale pertinent only to a culture extremely distant from our own in terms of basic values, and thus without relevance to the role of contemporary technology. However, there is a similar, and even more puzzling, example that lies rather closer to home—from our own Civil War, in fact. Neither side in that conflict was at all diffident about resorting to technological innovation in order to gain military advantage. Both North and South plunged heavily into speculative military gadgetry, although the Federal side, with its much greater industrial base, had an obvious advantage. All sorts of schemes were put forward on both sides, some of them harebrained, some of them murderously clever. The Union developed railroad building and telegraphy into elaborate arts and put the first turreted iron gunships into action. The Confederate navy, for its part, was the first to use a submarine to sink an enemy warship. Yet, to its great cost, the Union largely rejected a mature and reliable technology it might well have monopolized and which would have enabled it to shorten the war considerably. Breech-loading rifles and the even more potent magazine-loaded repeating rifles had been reasonably well perfected in the 1850s and could have been turned out in quantity by Northern factories, along with the appropriate ammunition.[17] Had these been made the standard weapons of the Union infantry, and had tactics been developed to maximize their superiority to muzzle-loading muskets, the battlefield firepower of the North would doubtless have won the war much sooner. Yet, aside from those issued to some cavalry units, very few such weapons saw service until very near the end of the war.

The story of this strange reluctance to seize an obvious advantage has many facets.[18] These include the misplaced parsimony of the War Department and the conservativism and caution of a novice officer corps trained in obsolete tactics. Not least important was the common idea that military virtue and "manhood" for both officers and enlisted men could only be demonstrated through stand-up, mass-formation warfare like that of the Napoleonic era. Whatever the ultimate explanation, this episode unquestionably provides an example of an apparently modern, technologically oriented state putting aside a technology not only useful but vital to its interests because social factors render that technology too culturally alien to be assimilated easily.[19]

My intent here is not to analyze the peculiar immutability of automobile design, nor the triumph of a reactionary warrior ethos under the shogunate, nor the pigheadedness of Civil War commanders. There are dozens of historians and economists who could provide plausible accounts. From my perspective, however, the whole matter remains mysterious. Why should innovation, which, in the case of personal computers and the Internet, erupted irrepressibly even in the face of hostile and powerful cartels, have been so infrequent and superficial in the automotive industry? Why couldn't a warrior culture (whether samurai or Army of the Potomac) have made the minimal cultural adjustments necessary to maximize its military efficiency? One inference is simply that technology is an anarchic and unpredictable beast. The currents and countercurrents guiding it are largely unreadable, and long-term predictions are futile. The forces that come into play seem to go far beyond scientific soundness or the parameters considered by classical economic theory. Culture and history interact with these factors in unpredictable ways, as does raw, inexplicable prejudice. There is an undercurrent of irrationality to the history of innovation that nullifies pretty much all of the large-scale theories I have ever encountered. Marxists and free-market libertarians seem equally confounded by it all. The lesson is that the rigorous logic that necessarily underlies the scientific work on which technology draws is not necessarily transferable to the processes through which society adopts or rejects that technology.

Many commentators deny this point. They talk as though there were some kind of unitary logic governing science, technology, and the social systems that exploit technology. Frankfurt-school social theorist Herbert Marcuse, an inspirational figure for the 1960s New Left and its ideological descendants, remarks, "Technological rationality reveals its political character as it becomes the great vehicle of better domination, creating a truly totalitarian universe in which society and nature, mind and body are kept in a state of permanent mobilization in defense of this universe."[20] In a similar vein from a later generation, Jürgen Habermas asserts:

> The technical recommendations for a rationalized choice of means under given ends cannot be derived from scientific theories merely at a later stage, and as if by chance. Instead, the latter provide from the outset, information for rules of technical domination similar to the domination of matter as it is developed

in the work process. . . . The regulated feedback of technical rules is measured against the tasks set down with the social labor process, and this means they have been made socially binding.[21]

The tenet that science, technology, economics, politics, and culture are all deeply implicated in one another, and that they comprise a totality constrained within a binding ideology, is a widespread article of faith.[22] It is, indeed, a mainstay of current critiques of the scientific worldview. Although in the contemporary academy, such notions are usually disseminated under the aegis of the "academic left," they have a long, zigzag history, and have at times lodged comfortably within the reactionary right. They have as much to do with Nietzsche and Heidegger as with Marx. Some critics have taken to using terms like "technoscience" to signal their commitment to the idea that science and technology not only move in lockstep, but are merely different names for the same thing. In such circles, the notion that it might be worthwhile to make distinctions between curiosity-driven "pure" science and result-oriented development and application is looked upon as hopelessly naive. Or it is seen as a disingenious attempt to bolster the sanitizing myth of scientific objectivity.[23] It appears that reams of "long-winded sermons from mystical Germans" have done their work all too well in fostering such dogma. The idea of a specific modern "form of life" is deeply entrenched among intellectuals, especially intellectuals with scant direct knowledge of science. It holds that science and technology are refractions of the same basic underlying attitude toward the world, and that technology is not only complicit in a spectrum of great evils, ranging from nuclear weapons to species extinction, but that the very mentality that permits these horrors is a corollary of the scientific frame of mind. Supposedly, scientific thinking is a moral virus that was planted back in the seventeenth century by such miscreants as Bacon and Descartes.[24] The problems that ostensibly plague modern life, from grave to minor to utterly imaginary, can be decoded as evidence of the toxic effect of science on society and human values.

I think that in this instance intellectuals have greatly outsmarted themselves. The "scientific rationality" necessary to do scientific work, or even to understand it accurately, is admittedly a very different thing from the "humane rationality," the moral common sense, that must govern at all levels from personal to global if a swarm of evils is to be avoided. I don't think that scientific rationality has all that much to contribute to the humane version, beyond its power (not to be despised!) to distinguish the facts of life from wishful thinking or outright fantasy. But neither do I think that scientific rationality is particularly inimical to compassion and decency. Nothing within the scientific ethos inspires or animates racism, sexism, or any other form of domination and injustice. There are no sins particularly associated with the habit of viewing reality scientifically, nor any virtues that flow specifically from the rejection of science. I don't claim to have any idea how to inculcate what to me would be an acceptable morality in individuals or society, but I'm reasonably sure that love of science will not compromise the process. Perhaps it will advance it, if only to the extent of modestly heightening respect for honesty and consistency.

This, however, is a side issue, at least for the moment. If we leave aside the lofty question of how well science comports with ethical decency, we still must ask how far science—the systematic acquisition, expansion, and occasional rectification of our knowledge of the natural world—is intertwined with technology, the use of science for intervening in the conditions of our material existence as individuals and as a society. Those who are extensively versed in this debate will perceive that in the very framing of the question I have taken aim at a specific answer. Clearly, I reject the proposition that the real object of interest is "technoscience." This putative monster is supposedly geared to the productive (and ideological) needs of late capitalism. It is alleged to hide its true nature under a veneer of "pure," impractical science, exploiting the resulting image of the selfless, otherworldly, disinterested scientist in order to legitimize all sorts of selfish, very worldly, and far-from-disinterested goings-on. "Scientific objectivity," it is held, is merely a cover story under which the technocratic claws of the creature wreak havoc on the scientifically disenfranchised masses. "The inextricability thesis, to the effect that no principled distinction can be drawn between scientific, technological and socio-political activities, is widely endorsed amongst sociologists and historians of the sciences."[25]

It can't be denied that the prestige of science is frequently conscripted to justify the plans or decisions of technocrats. This sort of bait-and-switch occurs depressingly often. The airwaves, for instance, are crowded with advertisements in which narrowly commercial technical work is solemnly draped in the sanctified cloak of "science." The same thing happens repeatedly, often more subtly, in other arenas, and certainly the layman can be forgiven for confounding science with very different kinds of activity. The situation is further complicated by the fact that a substantial quantity of good, even excellent, scientific work is done in the course of research that has a concrete technological purpose. Scientists themselves drift back and forth between "scientific" and "technological" careers without much altering the kind of work they actually perform. However, the trick in analyzing all this is to bear in mind that the question of what constitutes science is an epistemological one, while investigating the nature of technology is a job for economics and sociology.

The claim is often advanced that one may not draw a firm line between epistemology and the effects of social, economic, and political factors. The latter, it is said, shape and often distort the criteria for "truthfulness" of putative scientific claims, and may favor one body of theory over a rival on grounds that have more to do with the commercial prospects of an employer than with detached consideration of evidence. It would be naive to claim that in a world full of deceit and calculation nothing of the sort ever happens. Horror stories of all kinds can doubtless be dug up by those sufficiently motivated to excavate. However, it is equally naive, and considerably more doctrinaire, to pretend that solid and durable scientific results arise from this kind of opportunism. The purpose of suborning scientists in order to create phantom science is certainly not to produce knowledge, nor even to produce technology, per se. Rather, it is a specialized category of public relations or advertising or legal gamesmanship. It is the sort of thing that happens when a firm is eager to get approval for a new product, to escape responsibility for a catastrophe, or to peddle a costly new gadget to the Pentagon. When the motive is actually to deliver a workable end product, there's not much point in putting one's thumb on the epistemological scales. On the contrary, in such cases it is crucial to have science that accurately describes the objective world. To put it crudely, the military wants bombs that go "boom!" In any given instance they may be conned or lobbied or sweet-talked into accepting bombs that don't go "boom," but there is a limit, even under the most venal regime, to how much of this can go on. At the end of the day, at least some of the bombs have to go "boom," and requisite to their boomworthiness is an accurate, objective science.[26]

As social institutions, science and technology undoubtedly interpenetrate each other and often enough seem to be wearing the same hat—or lab coat. But to confound the two is to mistake superficial factors—institutional affiliation, for instance—for deep ones. The way science develops over time is a very different matter from the evolution of technology, despite the fact that the two institutions often seem to be marching side by side. I am not here suggesting a reversion to the heroic image of pure science, where every practitioner is a Galileo. Nor am I clinging to the myth that the history of scientific discovery is inscribed in the formal textbook account of mature scientific theories. The logic that cements formal exposition is very different from the logic of scientific investigation at the frontier. But the latter, in turn, is quite distinct from the logic under which technological innovation unfolds as an element of social and economic history. Indeed, the key difference is that science—again meaning the systematic investigation of how the one and only physical world works—is indeed logical, as a *process* as well as in its completed and refined theoretical structures. Here, "logical" is meant as a term of high praise. Whether technology is logical, in this sense, is a rather different question; indeed, I suspect that the answer is largely negative.

In its developmental structure, science is basically dendritic, that is to say, new strands of knowledge are continually branching off from old, though all are at some deep level united. Specialties and subspecialties proliferate and subdivide with amazing rapidity. Sometimes a given branch stalls or stagnates in its growth, though in no sense does it die off. Occasionally—with increasing frequency these days, it seems—seemingly far-distant branches reunite to produce new syntheses. (Cognitive neuroscience and structural biology provide important examples of this.) Moreover, there is a feedback process wherein developments at the frontier not only augment knowledge, but alter our understanding of the knowledge already won, sometimes rather drastically. There is, of course, no way of predicting in detail how this vast organism will grow, and extrapolation of current trends beyond a few years seems futile. The process is largely self-governing. Much of its logic is modeled on the structure of the reality it seeks to understand.

This isn't to claim that science is insulated from social and economic factors—such extreme "purity" is an unrealizable dream, and not a particularly desirable one. Some branches of science are sites of enormous effort and financial commitment simply because they are vital to further technological development in certain areas. (Medicine, from my point of view, is to be viewed as a subdivision of technology.) Molecular biology constitutes one obvious example. Research into high-temperature superconductors is another, though at this point dreams of a technological payoff are still speculative. On the other hand, vast quantities of money and talent are sometimes invested in areas of science whose eventual value to technology is quite dubious. For instance, the long-lasting boom in postwar experimental physics (recently truncated, alas, by the demise of the superconducting supercollider) was centered on high-energy particle theory. This field is essential to our understanding of the basic structure of the universe, but it is hopelessly remote from "applied science" as far as anyone can see at this time.[27] Comparable investment has gone into astronomy and cosmology. The enormously costly Hubble telescope is but the flagship of an army of terrestrial and space-borne tools. Here, too, it is difficult to envision how anything of direct practical utility might emerge.

What has to be borne in mind is that, despite the global unpredictability of the overall growth pattern of science, at the local level it largely lives up to its reputation as a deeply rational enterprise. Points of growth and nodes where new branches of inquiry split off represent, when we cast aside the metaphor, the actions of scientists as individuals and groups. It is the reflections and decisions, as well as the deeds, of these people that constitute the actual practice of science. To a very great degree such thinking is commendably accurate and rational. This does not mean intuitions and hunches are absent, nor does it exclude the possibility of wrong guesses, blind alleys, and dead ends, which are certainly frequent enough even for the finest of scientists. It does not exclude pigheadedness and blind prejudice, either (since we are dealing with human beings), although I do think that the worst kind of bias is infrequent and, indeed, rather remarkable when it occurs. I merely assert that at the microlevel, scrupulous histories of ongoing research science typically reveal a series of intelligent decisions made by conscientious people. They are able, on the whole, to filter out extraneous considerations when theorizing or interpreting results, and they consistently do so. They are anything but naive about the possibility of weak methodology leading to unsound conclusions, and they are eager to guard against it. Thus, whatever is adventitious or serendipitous in the accretion of scientific knowledge, the major theme is the pursuit of objectively plausible ideas for objectively sound reasons. It may be true that reality does not dictate how science proceeds because human beings do. But to deny, as do many current advocates of the "social" analysis of science, that reality constrains successful science to proceed within very narrow channels is to pursue a willful illusion.[28]

Technology, by contrast, is very different in the mechanics of its exfoliation. The people who function at the growth points and nodes are a mixed lot, hard to characterize in any simple way. The key players may be scientists or engineers, but they may also be executives, bureaucrats, bankers, generals, or politicians. The events by which a technology advances, recedes, stalls, or branches off don't seem to adhere to any generalizable logic. To illustrate the difference, consider semiconductor physics. Once the decision has been made to investigate the topic, there is (*pace* the relativists) only one theory of semiconductors that will emerge. By contrast, consider the enormous body of extant technology that derives from our knowledge of semiconductor physics. It is anything but clear that this was the inevitable result of the physics, that it was all implicit in the science itself. That is the point of the examples with which I began this section. It might well have happened that personal computers never went beyond the stage of hobbyists' toys—mainframes might still be king. Similarly, the Internet was hardly an inevitability; it might have been quashed early on by any number of chance developments. The list could well be expanded. Tiny deviations from the actual historical pattern two or three decades ago might have bequeathed us an entirely different world. For instance, we might all be driving around in fuel-cell powered electric autos by now. Or, perhaps, had the Roosevelt administration been slightly more reluctant to commit itself to atomic research, the enormously expensive effort to isolate fissionable materials would never have been undertaken during the war and would have been found too speculative and costly for a postwar America. Thus, nuclear weapons would never have come into existence.

Alternative histories of the world are of little use, save for entertainment value, but I offer them in this case merely to emphasize the quirkiness of technology which, to say it once more, is a social and historical process, not an epistemological program. What drives, stalls, or deflects it is difficult to say, but it is hard to believe that dumb luck never takes a hand. Even at the finest scale, the way in which proferred technologies interact with the social organism, prospering or failing in the process, is not governed by anything like scientific logic. It is hard to explicate in terms of any plausible logic whatsoever. If I knew far more about condensed matter physics than I do, I might have a better idea of whether high-temperature superconductors can be developed to the point of commercial feasibility. I still wouldn't have a clue as to whether it might be a good idea to invest in the firms trying to bring such technology to the market. Bad ideas sometimes prosper and good ideas sometimes languish.[29] Beyond that, there seems little one can say.

Notwithstanding its equivocal relation with science, or the frequent irrationality of its social behavior, technology is not the great villain of our time. The recent history of our culture has included a long gambling session with technology, and there have been some nasty surprises. Yet, on the whole, I think we have come out winners. Perhaps this, too, is dumb luck but, all things considered, I'm too optimistic to accept that view. I don't think that we can really afford to quit while we're ahead, so far as technology is concerned. Attractive as this alternative has been made to seem by its most eloquent proponents, it is not only delusory but potentially disastrous.

This is not to endorse past or present anarchy or foolhardiness in the deployment of technology. One needn't agree with the full array of antitechnological indictments in order to recognize the enormous human cost that can be imposed when greed for power or money is amplified by scientific ingenuity. The paradoxical spectacle of the deepest intelligence of which our species is capable meekly rendering tribute to the grossest stupidity and cruelty is an image that brutally shatters easy optimism. That humanity is (as always) in serious trouble, and that technology has been directly instrumental in deepening our dilemma, is so clearly true that only obliviousness or insanity can question it. Yet it is equally true, if not quite so universally appreciated, that only through technology do we stand a chance of climbing out of the abyss.

The central sin of technology, if there is one, has been to ratchet up the killing power of the military to the point that warfare in which one or both protagonists have access to modern weapons is merely industrialized slaughter. Perhaps the most horrifying fact is that the greatest threat humanity has ever known is also the most antiseptic. It is now easily possible to obliterate millions of lives at the flip of a switch without a moment's worth of fury or bloodlust having passed through the souls of those who do the flipping. Indeed, the greatest ingenuity of our military planners has gone into making sure that this is precisely what will happen at a nod from those who sit atop the governmental pyramid. The most democratically chosen leadership the world has ever known holds arbitrary power over life and death to a degree that no Caesar or Khan would have been able to comprehend, let alone emulate. Whatever else technology has to offer, it has certainly endowed the world with an ample stock of macabre paradoxes.

Beyond this, there is the grim progress of industry and its appetites through the natural landscape. Vast stretches of forest and countryside are still being ripped apart in the quenchless search for resources, and, horribly often, ancient societies are left for dead in the aftermath. Physical extermination of tribal peoples is a real threat in certain places (parts of South America and western New Guinea, for instance). Even where it is not, technology, in collusion with a cash economy and an ethos of impersonal market exchange, eviscerates fragile local cultures and cuts their members adrift in a world they little understand. Yet in the very heartland of the technocracy itself (another paradox) discontent abounds. There is no accusation against technology that fails to find a receptive audience among the educated of the industrialized world. Frequently, this is a species of social hysteria that tells us more about the demons of our own culture than about the actual effects of the technology in question. Often enough, however, the accused is guilty as charged, or at least partially so. The jury is still out on the most apocalyptic threats that currently unnerve us—global warming from an enhanced greenhouse effect due to fossil-fuel combustion, the attenuation of the ozone layer by anthropogenic chemicals in the upper atmosphere—but the indictments are sufficiently plausible that to discount them is foolhardy. In sum, we are still not clever enough to avoid being snared by the devices we are clever enough to invent.

All of this argues for substantial adjustments in our political, diplomatic, and juridical institutions in order to get a few steps ahead of the unanticipated consequences of technological change, rather than trailing a mile behind, as seems to have been the case ever since the Industrial Revolution. Of course, I say this in the face of the patent fact that our political theories are hopelessly primitive compared to our understanding of the natural world. Our best insights into the workings of the political organism still come from the hoariest of sources—Aristotle, Confucius, Machiavelli, Hobbes, Burke— rather than from a mature and active science of human nature.[30] Meaning no disrespect to my friends who work in the field, I find the term "political science" to be a misnomer, at least to the extent that it suggests that the discipline might soon give rise to an efficacious "political technology." We have only the faintest glimmerings of what kind of institutional changes might be useful in avoiding the worst disasters technology can create and in maximizing the potential benefits. It is sobering to reflect that the greatest overall success of post–World War II politics—the avoidance of nuclear war—owes little to the development of humane political systems, still less to the wide dispersal of wisdom and moderation throughout our species. The chief instrument seems to have been the mindless acceleration of technology itself, which raised the consequences of all-out warfare to such a suicidal pitch that even the densest of brass-hatted heads finally got the message.

As historian Edward Tenner says, "In the real world, few trends emerge without ambiguity, beyond a reasonable doubt, before precious time is lost."[31] The intellectual difficulty of accurately extrapolating technological tendencies and the socioeconomic ripples they cause is enormously daunting. Yet it seems rather modest in comparison with the chore of devising a political agenda that will make it possible to act in accordance with such predictive wisdom as we are able to develop. To say that thinking on these questions has been rudimentary, up to this point, is almost to give it too much credit. Moreover, most of the ideas thus far offered for bringing technology under the control of humane values have been intertwined with other agendas. They have been closely linked with schemes for racial justice, socialism, feminism, animal rights, and deep ecology. How much this vitiates or enhances the analysis varies from case to case. But all too often, utopian daydreams are being spun when the first order of business is to divest ourselves of as many illusions, pro-technology or con, as is reasonably possible. Technology is not a god that will save us independent of our ability to use it with foresight, nor is it a demon that must be wholly exorcised if we are to survive. It is a servant we can't afford to lose, but a mercurial and potentially dangerous one.

The vexations of technology have inspired cogent and eloquent attacks against science from a wide variety of thinkers and ideological positions. At the same time, technology is really the first line of defense for science against the misconceptions of a scientifically ignorant population. Intellectuals, alarmed at what industrialization has done to mar the sweetness of the world, tend, since they are intellectuals, to locate the root of the evil in the toxicity of ideas. Many of them are therefore tempted to abjure the scientific mode of thought. Within the West, this has been a recurrent philosophical theme, from Herder to Blake, to Nietzsche, to Heidegger, to Foucault. On the other hand, the experience of most people in their everyday lives breeds fierce loyalty to the technological cornucopia. Ordinary citizens are not about to give up automobiles, cable television, or air conditioners under any regime, no matter how deeply tinged with green. What is more, they are almost as beguiled by the promise of future technology as by the comforts and pleasures of existing gadgets. This loyalty is even further enhanced by the common (and correct) assumption that substantial advances in medicine are only possible through the increasingly elaborate use of advanced technology. Given that people are perfectly aware that science is the indispensable condition of technological fecundity (however they may be mystified by the scientific details), there is little chance that an atmosphere of unconstrained hostility toward science will ever rule in our culture. The joys that technology gives access to may be frivolous and fleeting (at least so far as the professionally gloomy are concerned), but they will be defended with last-ditch ferocity, and that defense, perforce, includes science within its perimeter. Hence we have another paradox: the ignorant (or at any rate the unintellectual) are effectively more committed to sustaining the scientific way of thinking than a large class of intellectuals.

This is not to claim that uneasiness about technology is solely limited to professorial factions. Misgivings, not only about particular technologies, but about the idea of technology per se, infuse our culture, even at the level of popular cliché. "It's not nice to fool Mother Nature!" as one well-known advertising slogan used to put it. Numerous commentators cite the precipitous decline in "technological optimism" in our society since the feel-good days of the postwar period.[32] No doubt this reflects a genuine increase in the fear of the effects of technology, often on a rational basis. Certainly the untoward consequences of technology, intended and unintended, real or imaginary, have been a constant theme in the popular media from which most of us infer what's going on in the world. This media slant is presumably reflected in polls and surveys cited as evidence for increasing technopessimism.

This trend, however, may be at least partially misleading. The inclination to declare oneself suspicious and discontented with respect to technology may reflect, as much as anything else, a more general alteration in the tone of the ambient culture, a shift in the direction of skepticism, and pessimism overall. As Louis Menand, a sophisticated analyst of social and intellectual trends, remarks, "Academic skepticism is only a little in advance of a popular skepticism that has by now pervaded almost every area of American life."[33] Let us posit, then, that our social norms embody an idealized personality type to which many people make at least some effort to conform (at least when being interrogated by strangers). Assume further that, over forty years or so, this model, in one crucial dimension, has devolved from optimism, cheerfulness, and trust to cynicism, suspicion, and knowingness. It would then follow that people in general put on a gloomy and distrustful demeanor in response to a spectrum of perceived issues, including, of course, the promise or threat of technology. They tend to express doubts and misgivings out of proportion to any despair actually felt, simply in order to present the kind of "public face" that contemporary fashion approves of. This, I think, is a more-than-plausible scenario. There has indeed been such a mood shift, indicated by changing attitudes toward politics, government, and universities, among other things.[34] Even more important, there has been a veritable celebration of the "trust no one" attitude. For instance, a peculiar skepticism has greeted new successes in the American space program. Some cynics think the *Pathfinder* Mars mission is fakery, a simulation through Hollywood-style special effects techniques. Others believe that the overt purpose—studying Martian geology—is a cover story for something involving extraterrestrial civilizations.[35] "Technopessimism"—the popular version at least—is yet another consequence of the ongoing enthusiasm for all-purpose paranoia.

Even if the foregoing explanation is too facile, it remains the case that, for most people, enthusiasm for technology and fear of technology are united in the same breast. Even the most articulate of academic technopessimists tends to disseminate his opinions through the Internet, after all. As a philosophy of society, technopessimism involves people in too many gross contradictions with their experiences and desires to be sustained easily. Principled anti-technologism is hardly a popular movement at all. Rather, outright undiluted technophobia is almost wholly the province of a declining commonwealth of academic malcontents. It is part of an intellectual mood that prevails for reasons that may have nothing much to do with its intellectual merits. Pessimism seems to recur among intellectuals from time to time on various grounds—or pretexts. During the height of the Enlightenment, that supposed *locus classicus* of optimism, a wave of despair overtook European thinkers when the great Lisbon earthquake annihilated tens of thousands, undermining the comforting notion of a benign and rational Providence. Just a few years later, even more unaccountably, hundreds of young men did away with themselves, driven to that desperate extreme by nothing more than a reading of Goethe's *Werther*. These instances, of course, are not evidence that technopessimism is ill-founded, but personally I find the doctrine unconvincing. It seems to me yet another instance of the tendency of intellectuals to study the world by looking intently in the mirror.

I am not, to say it once more, advocating an uncritical technophilia as a desirable alternative. Technology is too heedless of human values, and the history of technology is too saturated with unintended consequences, for that to be either morally or intellectually tenable.[36] I am constrained by the realization that the globe now contains more than five billion members of our species—large, aggressive, wily mammals who are extremely irritable when hungry, and who frequently resort to violence under far less extreme deprivation. I am further constrained in that, unlike some "deep ecologists," I am ethically unable to accept a "Malthusian" solution to the problems posed by these huge numbers. Consequently, I'm driven by both hopefulness and realism to the notion that an intense commitment to technology, along with a degree of faith that its nasty side can be mastered, is the only possible path to a decent human future. To put it another way, any serious move to abandon technology or to constrict it seriously is absolutely certain to bring about disasters as severe as those that haunt the nightmares of technophobes.

Indeed, the technophobic imagination is continually inflamed by tales of disasters that actually result not from high-tech, but from low-tech or even no-tech. The noise and filth of cities, the spread of new diseases in Africa and Asia, the accumulation of toxic waste in landfills and streams, are all problems that arise from primitive or clumsy technologies. It is impossible to achieve long-term solutions without allowing sophisticated technology to flourish. Many critics of contemporary technology ascribe its evil power to the corporate elites and plutocracies who sponsor it. I am instinctively sympathetic; I'm still too much a socialist to feel much admiration for megacorporations or the megarich. Yet I'm driven against the grain to reflect that the most reliable defense we have against the dangers of Luddite populism is the implacable greed of corporations.

For example, there is at present widespread opposition to genetic modification of plants, animals, and bacteria for various agricultural, industrial, and medical purposes.[37] On the whole, this opposition is irrational to the point of foolishness and deserves to be classed with the nineteenth-century superstition that proclaimed train travel at thirty miles per hour a cause of insanity. Yet in some current cases, it seems to have the upper hand. If this continues in the long run, the interests of society and its members will be greatly harmed. I doubt that the opposition will prevail, however, simply because corporations and investors stand to gain too much from transgenic technology in its various forms. The technocorporate elite will triumph by fair means or foul—probably foul—over strongly felt popular opposition. Yet paradoxically they will be on the side of the angels, if only in the context of this single issue.

In general, neither technophobia nor technophilia is a reliable guide to how to deal with technology in relation to our social needs and our collective future. Both points of view incarnate too many fantasies and conflate issues that ought to be kept apart. Neither of them is much more reliable than the Psychic Hotline. Our problem is to tame technological innovation without choking it to death, to accept the quirkiness and unpredictability that seem to be corollaries of innovation without allowing social consequences to swing wildly out of control. We are faced with the exacting task of developing a "science of technology," a deep understanding of this interface between scientific comprehension of the world and the imperative for human control over the world. We need, moreover, the power to predict, not just to concoct post hoc explanations. Unfortunately, academic study of this issue seems, for the moment, more likely to confuse than to clarify. Many of the would-be experts in the area are ensnared in a tangle of postmodern sophistications—that is to say, delusions—of their own making.[38] Still, as the gravity and urgency of the issue become clearer, we can hope that able thinkers—scientists, engineers, historians, economists, and others—will begin to build the necessary foundations. My attitude, then, is on balance optimistic, though it is a battered and weary optimism, and far from brimming with certainty about the outcome. As Edward Tenner puts it:

> Technological optimism means in practice the ability to recognize bad surprises early enough to do something about them. And that demands constant moni-

toring of the globe, for everything from changes in mean temperatures and particulates to traffic in bacteria and viruses. It also requires a second level of vigilance at increasingly porous national borders against the world exchange of problems.[39]

A tall order indeed, but one with which we have little choice but to comply.

SIX

Nature

"Nature" is a prepotent word. It cuts a wide rhetorical swath. It amplifies the force of doubtful arguments while diminishing the cogency of arguments largely sound. People are apt to be highly forgiving of errors in logic, gaps in evidence, lapses in coherence, if only one can convince them that one speaks on behalf of what is natural. The Western world today is one that has largely rejected the idea of divine authority. At least Western society no longer constitutes a coherent doctrinal community uniformly responsive to a single mode of invoking heaven's mandate. Though there are sects large and small that recognize a special notion of divinity as the ground for moral and ethical judgment, none, not even the largest and best organized, can command wide obedience beyond its confessional borders. And even within these, assent rests largely on sustained individual commitment, which may be terminated instantly if the individual chooses to change faiths or abandon religion altogether. It is a different story with the concept—one might even say the deeply resonant myth—of "nature."

To call an act or a predisposition "unnatural" is to place it beyond salvage or redemption. On the other hand, to call something "natural" is to ascribe virtue, efficacy, and benignity. Physicians, as well as their patients, talk approvingly of "letting nature take its course," whether speaking of healing or of the death of the terminally ill. The unspoken assumption is that if we can somehow bring about a situation where mere human cunning, in the form of technology or otherwise, can be left aside, where everything is put in nature's benevolent hands, then we will have secured the happiest of all possible results. In a practical sense, there may be many situations where this premise is close to the truth, on a human scale, although, as the old joke has it, "Death is nature's way of telling you to slow down." But these locutions also encode an ideology, a conceit that the natural order partakes of the divine, and can be communed with only through the renunciation of human cleverness.

The hypocrisies of our marketing strategists constantly pay tribute to this prejudice. Cosmetics, laxatives, and soporific drugs are all marketed with devout tributes to their naturalness. Natural fabrics have a cachet denied to even the most convenient and attractive synthetics. In our day, "natural foods" have soared in popularity to claim a major share of the market. The "natural" is the virtuous opposite of the degraded manifestations of humanity's fallen state. To appeal to nature, in many contexts, is to do an end run around the misgivings or doubts that taint conventional religion. "Nature" is the code word for the way things were meant to be, rather than the way they are. In that sense, the concept of nature fully embodies our most deeply engrained teleological prejudices.

The transition from "God" to "Nature's God" to "Nature" as the chief icon and guarantor of virtue and wholesomeness is the central intellectual and moral trajectory of the last five hundred years of Western history. As Mary Douglas and Aaron Wildavsky put it, "In a secular civilization, nature plays the role of general arbiter of human designs more plausibly than God."[1] The concept of "naturalness" now largely does the cultural work formerly carried forth by "godliness." Blasphemy, in the strict sense, hardly exists any more as a meaningful category of perceived evil. Instead, the sense of wickedness that used to be attached to it is now ascribed to practices or ideas deemed to affront nature.

The quasi-deification of nature is usually associated with the Enlightenment and, subsequent to that, to the progress of naturalistic understanding and explanation that came about with the rise and intensification of the sciences—that is, the "natural sciences." Thus, this shift is usually regarded as representing an increase in the overall rationality of thought and life. The enthronement of the natural, in that it exiled or diminished the raw power of theological fiat, brought society and humankind out of the fog of arbitrary sectarian conceits and outright superstitions—or so it is generally supposed. By redefining the court of ultimate appeal as something that might, in principle, be observed, investigated, and verified by our minds and senses, by disestablishing the inbred hocus-pocus of religious discourse as a significant arbiter, our civilization presumably freed itself from an ancient shadow. This is reasonably common wisdom and, as common wisdom goes, reasonably wise. I don't intend to contest it in any general sense. Rather, my main purpose is to point out how vexed, incoherent, and contradictory the notion of "nature" is, especially as we invoke it in our social discourse. Correspondingly, there is frequently a gap, indeed an antipathy, between what is commonly viewed as natural and what emerges from the thought and labor of the natural sciences. Nature, conceptually, is a many-headed hydra, and more than a few of those heads snarl and snap at science. The veneration of nature has grown into a potent and pervasive cult (albeit one with many sects) and, like all such cults, it can sometimes demand frightful sacrifices. All too often, what is to be sacrificed is the measured, calculating intelligence with which science approaches the world.

It is plain that "naturalness" is an indistinct category, rife with obscurity and contradiction. "Nature" is invoked to all sorts of purposes, and flagrant inconsistency is the inevitable result. Those with a traditional view of sexual morality—not all of them traditionally religious, by any means—denounce homosexuality as an impermissible detour from the natural course of sexuality. Indeed, the ancient notion of "the abominable and detestable crime against nature" has retained much of its emotional force, and may even be said to have provided the inner logic behind the Supreme Court's decision to uphold state antisodomy statutes. On the other hand, sexual libertarians insist on the naturalness of homosexuality, pointing to analogous behavior in many mammals, as well as to the wide range of human cultures that tolerate, or even encourage, homosexual behavior. In either case, the reason for making "naturalness" (or "unnaturalness") the central rhetorical issue is the moral effect of having it believed that nature is on one's side. In like manner, defenders of untrammeled capitalism have, since Herbert Spenser, proclaimed it the natural order of things for the strong to prosper and the weak to fall by the wayside. Against that, socialists and other anticapitalists since Prince Peter Kropotkin have decried that same capitalism as a perversion of the natural propensity of human beings to live communally in mutual sympathy and solidarity. Generals see warfare as a natural function of humankind, while pacifists see it as a hideously unnatural warping of the innately irenic propensities of our species. Feminists see sexual egalitarianism as the state of nature,[2] critics see male dominance deeply embedded in evolution and physiology.[3] The point here is not to judge the legitimacy of any of these appeals to "nature," nor to set up any canons for evaluating them. I merely point out the seemingly irresistible lure of this rhetorical ploy, whether justifiable or not, in a host of contexts. Many people who argue in this mode insist that in so doing they are strictly respecting the separation of "is" from "ought." Often they pronounce their conclusions concerning what nature supposedly tells us about the human condition with a regretful sigh, claiming they wish it were otherwise. Nonetheless, even in those cases where the barriers between "is" and "ought" are seemingly high and firm, one catches the sense that a faint nimbus of "ought" surrounds the proclaimed "is." The lurking insinuation is that if nature has it so, it is wrong, or foolish, or dangerous, or unjust, to want to have it otherwise.

Almost inevitably, whoever resorts to this mode of argument immediately strays into danger of grave philosophical error. The root problem here is the attempt to divide the universe into the distinct spheres of the "natural" and the "unnatural." However meticulously one draws the boundary lines, however scrupulously one respects them, the error persists and haunts subsequent arguments. In responding to the solemn ontological question, "What is there?" the noted philosopher W.V.O. Quine once quipped, "Everything."[4] From the monistic viewpoint, on which I insist, a similar remark is appropriate about "nature": What is nature? The only answer that can be consistently defended, is, in essence, "Everything!" What, therefore, is unnatural? Nothing! The seamless web of nature is one which comprehends humanity and its works, not in the vanished past or the utopian future, but in the here and now—humanity, with its flaws, perversities and frank stupidity. The natural world is simply the world. One cannot conceptually prise away the complex phenomenon of humanity—what our species is and does—from the rest of the phenomenological cosmos, at least not without introducing some form of dualism through the back door.[5] The Interstate Highway System is just as natural as the body's network of blood vessels. The Grand Canyon of the Colorado is no more, if no less, a part of nature than the canyons of Manhattan. All are products of the unfolding of natural processes operating under uniform laws. Anyone who wishes is entitled to make moral distinctions among these, in whatever direction is most pleasing, but doing so inevitably intrudes a huge dollop of subjectivity into the question. Someone else is free to discern an alternative moral order in the same phenomena; nothing factual favors one view over the other. The point is that *sub specie aeternitatis* human life and all its works cannot be wrested away from "nature" or "biology" without arbitrariness and logical inconsistency.[6] Needless to say, dualistic thinking is nonetheless commonplace. In Western society, until recently, the overriding tendency was to accord a moral dignity to the "human" which elevated it above the merely natural. Of course this was done by associating humanity with an even higher realm inhabited by God, the angels, the saints, and so forth. In our day a similar distinction is current, but with an inversion of values which now ascribes virtue to the "natural" end of the dichotomy, and depravity to the human. Contemporary fashion, at least among the intellectuals who crowd the canyons of Manhattan, is to prefer the landscape "where every Prospect pleases, and only [Western] Man is Vile."[7] This cultural predilection is central to the phenomena I shall subsequently try to describe. It has weighty consequences for the way in which science is perceived and even for how it is done. Nonetheless, it is a habit without warrant in accurate ontological thinking.

It is possible, perhaps, to salvage some logical consistency for the use of the natural-versus-unnatural dichotomy if the "natural" is opposed to the antithetical notion of the "artificial" or, simply, the "human." The Andromeda Nebula clearly stands on one side of the boundary, the Sistine Chapel on the other. Nonetheless, confusions and paradoxes haunt even this narrower usage. The presence of humankind has been working changes on the landscape for countless millennia, exterminating hunted species, introducing plants and animals to new environments (both intentionally and accidentally), drastically revising the underlying ecosystem by deforestation, slash-and-burn agriculture or (as in the case of aboriginal Australians) massive brushfires intended to steer game within range of hunters. To call a given piece of landscape "wild" is therefore often to ignore the centuries of human history that went into reshaping it. A strain of ecosentimentalist thought, uncomfortably aware of this contradiction, tries to fidget its way out of the dilemma by regarding some peoples and societies as inherently more "natural" than our own industrial Moloch. But this is a move that once again opens the door to arbitrariness and rampant subjectivity in the erection of standards by which societies are judged worthy of the accolade "uncivilized." Plainly, if we are to be consistent in thinking about "nature" and insightful in studying how the word is deployed to hortatory effect by various schools of opinion, we will have to abandon the notion of a natural sphere wreathed in virtue and splendor. This doesn't automatically preclude anyone so inclined from despising shopping malls and supertankers. It does, however, foreclose the all-too-facile option of merely denouncing them as transgressions against the natural order.

All in all, thinking clearly about a host of problems that are commonly addressed in nature-saturated rhetoric requires that, as much as possible, we renounce this all-too-convenient category and the comforts of the moral certitude that apparently dwells within it. As I noted earlier, this is a task of surpassing difficulty, since it goes against the grain of deep and unquestioned cultural habits.[8] But recourse to the "natural" is, in fact, a subtle anthropomorphism. We may take courage from the fact that the practice of science has taught us, over the decades, that with enough hard work and imagination it is possible to steer clear of anthropomorphisms.

Stephen R. Kellert, a passionate environmentalist and advocate of biophilia, announces in his book *Kinship to Mastery* that "a world devoid of natural symbolism would be a world of emotionally and mentally stunted people. A society reliant wholly on artificial creation would strike us as not only odd but oppressive."[9] I confess to sharing instinctually most of Kellert's attitudes in regard to many specific environmental questions. (In fact, I confess to having fled the canyons of Manhattan for a little country house in New Hampshire overlooking the Connecticut Valley in order to get this chapter written.) I am deeply persuaded by most of the arguments for the importance of biodiversity advanced brilliantly by Edward O. Wilson.[10] Tree huggers have first call on my sympathies, even—or especially—when they are of the direct-action monkey-wrenching breed who interpose their bodies and their ingenuity between the voracity of lumber companies and the sanctity of virgin forests.[11]

Nonetheless, I think Kellert's proposition deserves close, even skeptical, scrutiny because it is, in effect, an argument from human nature. It asserts that it is in the "nature" of human beings to love "nature." In the current cultural climate, this has Rousseauian overtones. It suggests that humankind, in its uncorrupted state at least, inclines automatically to biophilic virtue. These days, the great majority of thoughtful, intelligent people would be horrified at the prospect of a world where a more or less unsullied "nature" is no longer available for meditation and delectation. But, like all such arguments, Kellert's thesis deserves to be measured against what we know or can reasonably guess about humanity on the basis of concrete historical experience. It is by no means unprecedented that peoples of one culture have found the mores of another to be "odd and oppressive" and the members of that odd culture to be "emotionally and mentally stunted." It is quite likely that a medieval European would decry our own culture in such terms for its failure to venerate the Blessed Virgin with sufficient fervor. Indeed, a well-bred Aztec would doubtless find us odd and emotionally stunted in our deplorable lack of enthusiasm for massive orgies of human sacrifice. We must therefore ask, in good cultural relativist fashion, whether Kellert is merely glorifying his own intense prejudices by describing them as human universals.

The contrary evidence is slender, because to date every culture has been embedded in an environment in which recourse to natural symbolism, in Kellert's sense, is automatic. No society has been wholly "reliant on artificial creation." Indeed, the whole notion would have been inconceivable until this century. Nonetheless, it is a fact at least slightly subversive of Kellert's argument that the "natural world" in which one culture finds beauty and meaning, from which it extracts its myths and its theodicy, would be regarded by a member of a sufficiently foreign culture (imagine a Maori transported to Iceland!) as bizarre, incoherent, ugly, and frightening—in short, hell on earth. The fad for finding the sublime everywhere—in the mountains, tundra, desert, rain forest, tidal wetlands, the endless snowfields of Antarctica—is a recent, very Western, habit. How and when it arose is an interesting study for the cultural historian, no doubt, but surely it cannot be cited as a human universal. Furthermore, we do have some very limited experience with life within environments that are an entirely "artificial creation." Crews of nuclear submarines or, at an even greater extreme, space stations, typically spend months in surroundings from which every visible trace of the nonhuman biotic realm is excluded. Crew members are carefully screened and highly motivated; but they are also under significant psychological and even physical stress. No doubt, they are delighted to see trees, birds, and meadowlands when their tours are over. Nonetheless, few of them seem to have been put off stride by the lack of such things while sequestered below the sea or above the atmosphere.

By way of mental experiment, let us extrapolate and imagine an entire culture similarly immurred in a wholly artificial environment. This is a well-known staple of science-fiction writers, who have imagined colonies dwelling in huge satellites,[12] and vast interstellar spaceships that house many generations of human travelers in the course of centuries-long voyages. There have even been attempts to create prototypes of such cultures on earth.[13] Assuming, *arguendo*, that the technical difficulties of building a self-sustaining environment vast enough to contain an entire society might be overcome, we may ask whether such a society would fall victim to the sorrows predicted by Kellert. Would it necessarily incubate massive neurosis, ennui, and depravity? Note that this is a very different question from "Would I like to live there?" For my part, I, like Kellert, would *not* like to live there. On the other hand, I strongly doubt whether I would want to live in fifth-century B.C. Athens, fifteenth-century A.D. Florence, or Moghul India, despite the fact that all of these cultures created splendid art, philosophy, and science.

Might it be that our hypothetical artificial environment could also generate transcendent art, philosophy, and science? Might it, just possibly, turn out to be a society in which virtue, courtesy, and the art of living reach new heights? Who is to say that it could not be vigorous, gracious, and free? The supposition that forests, mountains, and plenteous wildlife constitute a necessary condition for all these virtues is just that—a supposition. It is far from self-evident. Kellert (echoing his sometime collaborator Edward O. Wilson) makes certain arguments for the idea, grounded in the hypothesis that our entire evolutionary history predisposes us to biophilia: "Over the millennia, humanity's affiliation with life and natural processes conferred distinctive advantages in the human struggle to persist, adapt, and thrive as a species."[14] Like many arguments that rely on an evolutionary just-so story, this one achieves a degree of plausibility without being conclusive. Kellert could be right, but his arguments alone are not sharp enough to settle the outcome of our little thought experiment. Perhaps one day we shall discover the truth of the matter if ever serious attempts to build space colonies are undertaken. Until then, Kellert's doctrine requires a considerable leap of faith.

I don't advocate any such extravagant experiment. I am not particularly interested in refuting Kellert in this or any other way. My gut feelings are almost as biophilic as his own. Yet I am reluctant to endow Kellert's prejudices—and mine—with transcendent authority, or to believe that Nature itself speaks through them. Kellert's position, it seems to me, can be a way station en route to an extreme version of biophilia that slips into outright misanthropy.[15] This kind of extremism is typified by the sentiments of another passionate biophiliac. In a recent article in a faddish ecophilosophical journal, philosopher Keekok Lee contemplates the possibility (still very much at the outer edge of technological speculation) that contemporary industrial methods will be largely supplanted in the future by "nanotechnology."[16] This means that all sorts of manufacturing and assembly procedures will be carried out by micromachines, synthetic molecules programmed to process raw materials into finished goods while generating minimal noxious effluent or none at all.[17] For the sake of argument, Lee is willing to concede the fondest hopes of the techno-optimists: nanotechnologies are not only possible but will vastly reduce the environmental toxicity of our productive processes and allow highly efficient recycling of trash or excess material.[18] In short, she accepts as a given what most people would tend to think of as the utopian promise—utopian in an environmental as well as an economic sense—of nanotechnology. Yet she remains unhappy: "If domination is to be understood in terms of a relationship between two parties with extremely unequal powers, then humans in possession of nanotechnology are in a position to systematically replace nature with synthetic artefacts. Such a situation justifies the political image of domination with which modern science has been associated."[19]

What perturbs her, apparently, is the notion that with sufficient ingenuity, humankind will be able, in a sense, to circumvent nature, and thus to "dominate" it (although exactly what is to be dominated is far from clear). Perhaps for Lee "domination" is the sinful state into which we fall when we try to escape from an appropriately humble and reverential dependence upon nature. This seems to be the direction in which her thinking is headed as she elaborates upon her misgivings:

> The collective project of humanizing nature by systematically transforming the natural into the artificial is analogous to the absolute pursuit of money, political power, or prestige in the case of an individual person. Just as such an individual project necessarily involves trampling on the legitimate interests and ethical demands of other people, the collective project tramples on the legitimate ethical demands of nonhuman others. This is tantamount to collective egomania.[20]

Lee cannot even accept a world where human prosperity reigns without putting serious pressure on the species and ecosystems that are now endangered by our current relatively primitive technology. Apparently, it is impossible to distance ourselves from "nature" benignly; to declare independence is still to insult and dominate. Only humility, gratitude, and vulnerability are acceptable. Note, too, that in Lee's jeremiad, the spiritual disaster that nanotechnology will inflict on us is not to be measured in anything that human beings actually feel or suffer. Rather, it is the ontological shadowland of "nonhuman others" that must be vindicated. In this quasi-theology, to enter the state of domination/damnation we needn't act to evil effect, deliberately or otherwise. Our "knowledge" is sufficient to condemn us. "Modern science" itself is the source of the evil taint. To *know* is to be guilty of dominating what is known.

I don't wish to tar Kellert with the biophilic extremism that Lee and many others express. Kellert is a careful and insightful thinker on many environmental issues, and I believe it would be the better part of wisdom to take his concrete advice on most of these. Nonetheless, even in Kellert's work there are hints of the tendency to reify "Nature" and to set it up as a quasi-divine being. This tendency, greatly amplified, characterizes many of the more uncompromising biophilic thinkers who, in trying to construe love of nature as an ethical, rather than merely an esthetic, imperative, become ensnared in the constraints of their own subjectivity.[21] These they mistake for sacred law. The most damning thing about this brand of nature worship is that it leads to outright contempt for scientific rationality. Moreover, the ill-effects of such extremism go far beyond mere wallowing in philosophical error. On a number of levels, the ultras are, in effect, enemies of a sane and effective environmental movement. Their chief defect is that their environmental philosophy is essentially a call to worship which will be heard only by a small, eccentric minority. Even worse, their open misanthropy, as well as their doctrinal absolutism, makes them convenient foils of those who are indifferent or hostile toward environmental causes. They exemplify the fact that environmentalism is a crusade which sometimes seems to need saving from its most fervent partisans more than from its enemies.[22]

At the heart of the delusional aspect of nature worship is a great piece of hubris masquerading as humility. It consists in the false assumption that an ideology so deeply felt and compelling must arise from an extrapersonal, and indeed, a transhuman source. As with more conventional religion, there is a failure to come to grips with the fact that every choice is a choice predicated on human values, in all their historical contingency, not on values that exist in "nature" external to us. Appeal to a transcendent realm (whether "divine" or "natural") in which values supposedly inhere is really a way of cheating on behalf of one very human set of values as against others. The conventionally religious ethicist can at least appeal to an ontological claim, the existence of a divine realm transcending nature and the material universe, as warrant for the moral judgments propounded. By contrast, those who try to ground their ethical position on "nature" or the "natural" fly in the face of the monism they at least implicitly embrace. This is as true for judgments concerning our duty toward nature as for any other ethical theses. We are the source, and the only source, of the values we embrace and advocate, even when the intended beneficiaries are nonhuman entities. Whatever rules we come to live by, we have created—not received—them. To put it another way, no whale ever tried to "Save the Whales." No whale ever cared in the least whether the whales are saved or not. We humans— some of us—are the only ones who care. In denying or evading this, the eco-left paradoxically joins hands with the religious right.

It is perhaps just as well that none of us can be truly consistent in preferring the natural to the artificial. Sharks are more "natural" than life rafts— at least according to vulgar understanding—but, if forced to dive off a sinking ship and swim for my life, I'd rather meet a life raft than a shark! I daresay that radical environmentalists would share this view if the choice were actually forced upon them.[23]

However, in less exigent circumstances, the widespread truisms disparaging our own ingenuity in favor of what is ascribed to "nature" can have considerable force. Consider two cauliflowers, say. Specimen A comes from the toil-hardened hands of an organic farmer who indignantly eschews the synthetic pesticides and fertilizers that the chemical industry belches forth. Specimen B comes from the toil-hardened hands of migrant workers enslaved to an industrial-scale agribusiness which gleefully douses its crops with the whole spectrum of chemical gimmicks. Which is more wholesome? Every instinct—or, rather, every culturally implanted truism—confidently answers "A!" Of course, the answer may well be "A." But then, it may not! The point is that the answer cannot be inferred as most of us would instinctively try to infer it, by reasoning that since A is the more "natural" product, it is the more benign. There is, of course, an intelligent argument that approximates this kind of logic: the organically grown item probably does not contain any chemical that did not exist in the environment in which the human race evolved; the nonorganic vegetable, on the other hand, may well bear at least traces of chemicals that never saw the light of day until industrial chemistry began to be applied to agriculture. On this ground alone, one might well surmise that consumption of B entails greater risk. But this surmise has to be tentative; the investigation simply can't stop here. Other possibilities arise. Countless substances that have existed "naturally" since time immemorial are decidedly incompatible with human health, as dozens of overadventurous mushroom gatherers discover every year. Perhaps the organic cauliflower harbors a toxin exuded by a mold or fungus that the chemical arsenal lavished on its inorganic cousin has squelched. Or perhaps another evolutionary argument (advanced by the biologist Bruce Ames)[24] is correct: organic gardening is a synonym for practices that select, in a Darwinian sense, those plants best equipped to flourish and mature in the presence of a spectrum of predators and parasites. Thus the organic cauliflower reaches our table looking firm, white, and healthy because it has produced exceptionally large quantities of the "natural" pesticides that protect it against these enemies. It follows therefore that A may well be loaded with such pesticides, none of which has been tested for its possible ill-effects on human beings. On the other hand, since corporations are wary of lawsuits, the artificial pesticides used on B have been chosen for their relative harmlessness to humans. Furthermore, B has been designed to come to market after its pesticide dose has been obliterated by time and the elements.

I don't advance these arguments to champion standard supermarket produce over what is available at the health-food store. I simply point out that the question of which is truly healthier may be tricky and delicate, impossible to answer without a careful and exhaustive assay of the kind that no one is actually likely to lavish on a couple of cauliflowers.

What it all comes down to is how certain ingested molecules interact with human body chemistry. Which of them are present in which plant, and in what quantities? Molecules, I think most of us would willingly concede, behave without reference to the social history of how they came to be incorporated into a particular cauliflower. To infer, a priori, that the more "natural" cauliflower is by definition more healthful than the "artificial" one is no more sustainable philosophically than to infer that A is better than B because organic farmers are nicer fellows than the corporate sharks who run agribusiness. The trouble is, however, that fallacies of this sort are rife in our culture, even among scientists. The "natural" is a category that is continually invoked, explicitly and implicitly, and almost always under the assumption that the natural trumps the artificial at every turn.

"Nature" is a term that is hard to put into play without an implicit appeal to commonly received notions of transcendence. It invokes a hierarchy of values that equates virtue with a hypothetical (and often fictitious) "natural" state of affairs. The impulse to prefer the "natural" product to the "artificial" without further investigation or analysis clearly illustrates this. This propensity is shared by most strata of contemporary society, including the great majority who, for one reason or another, usually ingest the morally suspect B just as it comes from the supermarket, rather than the virtuous A, which the Organic Health Food Shoppe would have been happy to supply at a premium price. Since such preferences rest on implicit ideas of transcendence, they are epistemologically deficient. But they nonetheless reveal a widespread cultural logic that infuses current mores. The great danger of such logic is that it short-circuits the empiricism, the attention to concrete and particular circumstance, on which accurate science depends. In its way, it replicates the intellectual dangers of revealed religion.

How does the idea of "nature" play out in the important environmental questions that confront us? In one sense, for all its vulnerability to philosophical critique, the widespread veneration of nature, the strong cultural propensity to endow the natural with transcendent virtue, is a good thing. It obliges millions of laypeople, whose grasp of the scientific questions involved is at best vague and shallow, to take seriously some of the potential threats discerned by environmental scientists. In that sense it is pious fraud, justifiable not by its innate truth, but by the healthy effects it has on the popular temperament. Environmentalism would never have become such a powerful political force, such a pervasive aspect of mass culture, and such an inextricable aspect of our social ethic, had it not been able to appeal to the feeling that environmental threats are affronts to nature itself and therefore essentially sacrilegious. This is especially true in the case of issues that do not involve an immediate threat to human health and safety, issues such as the preservation of biodiversity, protection of endangered species, and the defense of the tropical rain forest.

Humanism and rationalism, unaided by invocation of the transcendent, are sufficient, in the abstract, to justify all these causes. It is worth noting, for instance, that Edward O. Wilson, the most passionate, prolific, and influential advocate of biodiversity, eschews the rhetoric of transcendent nature in his own writing, although his eloquence produces similar moral effects. In the conclusion to his great book, *The Diversity of Life*, Wilson justifies his intense biophilia by appeal to any number of important human values:[25] human life on earth is sustained by an intricate web of ecological relationships whose dazzling complexity is only minimally understood; future investigation of organisms and ecosystems will unearth all sorts of information, as well as particular substances, vital to human health and well-being; the study of even the most obscure species is a source of endless fascination and intellectual wonderment, and may be the key that unlocks an even broader understanding of the world; depriving our planet of species and habitats will impoverish us intellectually; beyond that, it will impoverish us esthetically.

Nonetheless, through all his earnest advocacy, Wilson refuses to plunge into outright sacralization or deification of nature. He is too much the rationalist and scientific humanist. As he says elsewhere, in defending the "intrinsic" value of even the most obscure species, "I don't mean metaphysically intrinsic, I mean that because it is so unique. It has such a vast and ancient history. And because it has so much potential value in many dimensions to humanity."[26] In the end, Wilson is "anthropocentric" rather than "biocentric," although he would probably reject that dichotomy as unnecessary, unhelpful, and distasteful. Realism about human beings nonetheless compels us to recognize that logical rigor and philosophical soundness carry little weight in civic discourse. They simply cannot stir up sufficient fervor, at least not among a large enough population, to drive mass movements. These movements, even the most virtuous, apparently need a dose of irrationalism to keep going.

Environmentalism draws upon the nature worship that unites many disparate ideological strands in our culture, binding together members of many religions, as well as many nominal agnostics and atheists. Yet it would be absurdly priggish to insist that environmental ideology completely purge itself of thinking and rhetoric that, explicitly or by subtle suggestion, reinforces the tendency to think of nature as sacred. If Wilson, along with other equally committed defenders of biodiversity, clings to an essentially humanistic outlook, many of those who are influenced and inspired by him simply cannot be so philosophically abstemious. The people who are most likely to take ecological concerns seriously usually find it hard to take an anthropocentric view of the moral questions involved. They can make a wholehearted commitment to environmental causes only because they invest nature with sacred qualities. If environmental activism were the sole effect of their piety, even hardheaded reductionists might forgive their seduction by theoretically defective philosophy. Toleration of quasi-theological language is a small price to pay for the political muscle to achieve meaningful environmental goals. Unfortunately, that is not the end of the story. The amalgam of ideas, assumptions, and values that informs the opinions and actions of most environmental activists often leads to serious miscalculation on specific issues.

The problem is that a concept of nature that exalts it as a transcendent realm tends to relapse into its own, decidedly nonscientific, brand of reductionism. Conformity to what is labeled "natural" becomes the touchstone of all value judgments. Consequently, any departure from this standard takes on the tincture of sin. This is a Manichean view of the world, albeit one aligned on a different axis of values than traditional notions of virtue and evil. How does this work in practice? Consider the question of whether the low-frequency electromagnetic radiation emitted by ordinary power lines and transformers poses any danger to human health. The notion that there is indeed a danger—particularly in connection with juvenile leukemia—has gained great currency within the past five or six years. The jeremiads of activist Paul Brodeur have had wide influence, and millions of people now cast fearful eyes on the commonplace cables and devices that link their homes to the power grid.[27]

For various reasons, these fears have seemed greatly overblown from the first.[28] Exhaustive epidemiological research has since confirmed this skepticism. Low-frequency radiation seems to have no pathological effects on children or adults, and no biological effects in general. These extensive studies—undertaken to disconfirm what was from the first an inherently improbable hypothesis—have cost us, at a very conservative estimate, hundreds of millions of dollars. Added to this has been the cost of lawsuits, construction delays, unnecessary rerouting of power lines, and so forth. Finally, we must add in the nonpecuniary effects wrought by fear, anxiety, and guilt, which have afflicted thousands, if not millions, of people who were panicked by Brodeur and his allies. Even today, notwithstanding the well-publicized debunking of the supposed dangers of low-frequency radiation, many of these people remain anxious and suspicious. Nothing that science might say can neutralize their misgivings.

Why did the low-frequency radiation scare take root so swiftly, and why is it so hard to quell? Part of the reason is the authoritative manner of Brodeur and others who first raised the issue, which conveyed the image of competence allied to a militant concern for justice. It was also very significant that young children were thought to be at greatest risk from the supposed threat. This made it far easier to arouse concern. The extensive scientific ignorance of most of the population also played a role. It disarmed the skepticism that should have greeted the initial claims and greatly amplified the effect of weak anecdotal evidence and *post hoc, ergo propter hoc* logic. It obscured the difference between flawed studies of the problem and methodologically sound ones, especially where issues of statistics and their interpretation were involved.

But beyond this there were what one might call "metaphysical" reasons. These arose from the conviction that something unnatural had been turned loose in the land, and that disease and death would be the inevitable consequence. What seems to have been at work is the axiom, rarely made articulate, but deeply held by many people, that affronted nature will always avenge itself on guilty humanity. It played a crucial role in the low-frequency radiation scare. To many people, the very word "radiation" suggested atomic weapons, nuclear power plant disasters, radioactive fallout, and so forth. Of course, low-frequency power-line radiation is far milder stuff than what comes out of a light bulb, let alone an atomic blast, but the onus nonetheless settled on transformers and electric blankets simply because what they emit was given a technical, and therefore ominous-sounding, label. "Electromagnetic radiation" *sounds* like something that intrudes on the natural order of things. Nature, then, must have its revenge for such a transgression, here in the dire form of children struck down by cancer.

This also accounts, in part, for the stubborn persistence of the scare. Where does the evidence that absolves low-frequency electromagnetic radiation come from? From those very same necromancers—scientists!—who conjured up the evil stuff in the first place. Once the tag "unnatural" has been firmly affixed, it is very difficult to shed. The odor of brimstone persists, and the strongest objective evidence cannot dispel it entirely. Even worldly and, by conventional standards, well-educated people are susceptible to such delusions. Part of the difficulty is that the entire controversy has been inappropriately shifted into the moral realm. The factors involved have been assigned moral status and have been given roles in a myth about moral struggle. Power lines offend against the natural order; therefore they are demons; they spew poison. Unsophisticated as it sounds, this tale has been naturalized in the folklore of a culture that is always on the lookout for transgressions against nature. It will not disappear easily.

How does this emotional atmosphere affect genuine environmental problems of the kind that draw the grave concern of most scientists? If popular panic over nonexistent dangers is deplorable, might not popular panic over real dangers be at least somewhat useful, even if it is grounded in very imperfect understanding of the issues? Perhaps, but the matter is by no means simple. Surges of strong popular feeling are not necessarily the healthiest currents to be caught up in when trying to work through complicated problems where both theory and evidence are in a fragmentary state. Presumably, such issues deserve the most penetrating and accurate scientific investigation possible. However, it is all too obvious that scientific work can be obstructed, derailed, or subverted by a number of factors. These include the inertia of bureaucracies as well as the selfishness of what, in a more naive but probably more perceptive time, used to be called the moneyed interests. They also include popular opinion or, what is worse, popular opinion inflamed by fear and premature certitude. Even more, there is the possibility that any of these factors may subtly infect the attitudes of competent and reputable scientists who, in their own minds, are honestly trying to put forth their best efforts.

A crucial case in point is that of the global warming controversy. Let me admit from the start that I find this issue confusing and frustrating to deal with. Climatology is nowhere near my area of special competence, but I am familiar with the general issues and with the methodological difficulties involved, particularly when it comes to the travails of constructing reliable and useful mathematical models. At the moment, all I can say is that I am quite flummoxed; I have no clear idea of where the truth lies. I don't know whether anthropogenic carbon dioxide, along with other greenhouse gases, will actually cause a significant rise in average global temperature over the next few decades. I'm even more ignorant of what effect such a rise will have on weather patterns, on ecological communities, on economics, or on human welfare in general. Since I am conservative (with a very small "c"), I instinctively side with the pessimists—those, like Stephen Schneider, who think that significant global warming is all but certain unless CO_2 emissions are drastically curtailed, and that it will negatively impinge on the human community. [29] This is not to say that I believe that Schneider and his allies are irrefutably right, but rather that it is far too risky to gamble that they are wrong.[30] Dozens of scientists of the highest distinction have concurred: "Increasing levels of gases in the atmosphere from human activities, including carbon dioxide released from fossil fuel burning and from deforestation, may alter climate on a global scale. Predictions of global warming are still uncertain—with projected effects ranging from tolerable to very severe—but the potential risks are very great."[31]

Since the signatories include such stellar figures as Walter Alvarez, Michael Atiyah, David Baltimore, Hans Bethe, Robert Gallo, Murray Gell-Mann, Sheldon Glashow, Stephen Hawking, Dudley Herschbach, Roald Hoffman, Leon Lederman, Roger Penrose, Carlo Rubbia, Carl Sagan, Glen Seaborg, James Watson, and Edward O. Wilson, I am hardly ashamed to share their uncertainty—or their feeling that the wisest course is to play it safe by assuming the worst and acting accordingly.[32] This means, at the very least, preparing the ground for a technolgy in which pure hydrogen and other energy sources that do not generate CO_2 replace hydrocarbon fuels as much as possible. Most of the devices that utilize the latter—principally gasoline and diesel engines, oil- and coal-fired generators, and kerosene-burning jets—are based on old ideas, and, even apart from the danger of global warming, produce unpleasant side effects in the form of noise and noxious effluent. Therefore, there is every reason to replace them with new ones. Of course, "free" hydrogen is not really free for the taking—it must be produced by electrolysis of water. Therefore, we shall have to increase our electric-generating capacity enormously by adding plants specifically dedicated to hydrogen production. These will have to be driven by power sources that don't employ combustion—solar (including indirect forms like wind and hydroelectric), nuclear fission, and, if at all feasible, nuclear fusion. We shall also have to improve the technology of our power grid to allow concentration of primary power plants in areas where these resources are most accessible and where, in the case of nuclear plants, there is small danger to population centers. Perhaps this will be facilitated by the emerging technology of high-temperature superconductors, which may allow long-distance, low-loss power transmission.

Beyond this, it is quite probable that if the models predicting serious global warming are even approximately correct, we have already crossed a threshold in terms of the greenhouse gases thus far pumped into the atmosphere. Significant global warming might be inevitable, therefore, irrespective of whatever emission-reducing measures we now take. It follows that novel technologies and economic arrangements will have to be put in place to adjust to a wide array of consequences.[33] Conceivably, this will prove to be an even more massive task than eliminating carbon-based fuels from our energy base. Making all these changes—both to limit emissions of CO_2 and to deal with climate changes that might nonetheless develop—involves extensively retooling the global economy. It is a vast and daunting undertaking, but one which is quite feasible in theory. The sooner we get started on it, the better.

Unfortunately, one of the serious impediments to necessary change and innovation is the environmental movement itself as it now exists. Merely by mentioning "nuclear" as one of the primary power sources supporting a hydrogen-based economy, I run the risk of triggering loud and eloquent tirades from the activist community. For a number of reasons, it will be exceedingly difficult to purge "nuclear power" from the demonology of most environmentalists. Yet the alternatives do not seem compellingly preferable. Large-scale hydroelectric projects—unavoidably they would have to be large-scale—are almost as hateful as nuclear plants to the eco-devout, and probably for better reasons, all things considered.[34] It is unclear that pure solar energy and wind farms can take up the slack. Yes, there are good reasons for being wary of nuclear energy. The danger of additional Chernobyl-type disasters is foremost. But it has to be remembered that the Chernobyl explosion resulted from a combination of extremely imprudent design and incredible stupidity on the part of the plant's operators. Moreover, Chernobyl, even though it came close to being a "worst-case" nuclear plant catastrophe, does not rank, on the scale of human death and suffering, anywhere near the top of the list of historical misfortunes. One can confidently predict

that within the next five years, an earthquake, volcano, or typhoon somewhere in the world will have vastly crueler consequences than Chernobyl.
Given the strong possibility of creating fission plants which, by design and
siting, will be virtually immune to anything resembling the Chernobyl melt-
down, it would be deeply irresponsible to rule them out in advance.[35] True,
there are other serious problems with fission power. The main by-product
of fission is a host of highly lethal radioactive isotopes that cannot be neu-
tralized, and therefore must be safely stored for the indefinite future. In addition, nuclear power necessitates a large traffic in fissionable materials that
could conceivably be purloined by a rogue state or terrorist faction bent on
acquiring nuclear weapons.[36] However, nothing I have seen convinces me
that these difficulties are insuperable. The time may have come to let nuclear
power back into our good graces, at least provisionally. The more certain
global warming becomes, the more we shall need to accommodate this
possibility.

The universal revulsion, among environmentalists, against nuclear power
gives considerable insight into the convoluted ideology that now drives environmentalism as a social movement. I have cited the good, concrete reasons for approaching the whole question of fission reactors with great
caution. But when one contemplates the fervor and absolutism with which
atomic energy is deplored by the environmental faithful, it becomes clear
that there are factors that go far beyond scientific skepticism. The situation,
alas, is not all that different from what we find in the low-frequency radiation controversy. It involves a great deal of ignorance, irrationalism, and
immunity to relevant evidence. Part of what condemns nuclear power, in
the minds of its most irreconcilable foes, is its mere association with nuclear
weapons. It conjures up images of Hiroshima, Nagasaki, the Cold War arms
race, and so forth, even though civilian power plants have little connection
with these things aside from symbolic resonance. Even beyond this, however, we find a widespread feeling that atom splitting in its own right constitutes a primordial crime against the natural order. From this point of view,
atomic power can never be beneficent.

The reluctance of environmentalists to plan for the global warming they
themselves predict is another instance of the same mentality. As Gregg
Easterbrook points out, "Many environmentalists consider the mere mention of adaptation [to climate change] heretical because it implies that
humanity can overcome global warming rather than be overcome by it, thus
shifting the emphasis away from green guilt."[37] Perhaps this is an overstate-
ment, but it captures an important psychological truth about environmentalism: that in part it is an act of ritual abasement before a personified
"nature" rather than a program of practicable measures for dealing with concrete environmental dangers. In all of these issues, activists hold before us
the numinous image of a perfectly good and wise nature, and we are caught

up in a system of values and a canon of judgment that reject careful analysis of risks and benefits. In this atmosphere, it is difficult to weigh choices in the light of the facts that scrupulous science provides.

An example far removed from the global warming controversy displays the psychological mechanisms involved. Consider the medical technique of magnetic resonance imaging (MRI), which complements the CAT scan as a noninvasive method of getting a detailed cross-sectional picture of what is happening in the human body. Recall that MRI was originally known as NMR (for "nuclear magnetic resonance"). Why was the original terminology abandoned? Nothing about the technology itself changed. But physicians realized that many of their patients, influenced by the environmental and anti-nuclear-power movements, were put off, even frightened by the very word "nuclear," though NMR has nothing whatever to do with fission, radioactivity, atomic weapons, or the like.[38] Similarly, much of the opposition to the technique of preserving foods by exposing them to radioactive cobalt has been evoked not because of the actual scientific issues involved, but because people are terrified by the fantasy that they are about to be poisoned by a radioactive diet.[39]

It would be comforting to believe that this kind of logic represents only the fringe of the environmental movement. The problem, I fear, goes much deeper than that. Environmentalism has no credentialing process. Scientific acumen is not a requisite for participation or even leadership, and considerable prominence has been given to figures whose scientific competence is nearly nonexistent.[40] Moreover, environmentalism harbors a strong edenic strain, the desire for the whole of humanity to revert to a purportedly "natural" lifestyle.[41] Sectarian ideologies flow into this. Many "deep ecologists" regard hunter-gatherer societies as the ideal to which we ought to return. Eco-feminists envision a matriarchal, goddess-worshiping, vegetarian paradise from which all traces of male dominance have been purged. Individually, these factions amount to little, even within the internal politics of environmentalism. But they reflect, in particularly pointed form, the overarching "theological" presuppositions of the movement, the incessant appeal to "Nature" and the morality of the "natural."

The practical implications of this propensity are serious and unsettling. Consider once more the changes necessary to meet the global-warming threat. It is unlikely that these will be accomplished if we insist at the same time that human values worldwide have to be made over in the image of the ecological ideal. Irrespective of what ecotopians prefer to believe, increasing numbers of people in those countries on the verge of industrial maturity (most notably China) will eventually insist on electric appliances and air conditioning for their homes, automobiles for their personal use, and access to long-distance air travel. These desires will have to be accommodated. A bucolic existence built around organic subsistence farming and the virtues

of the simple life is not a real alternative, despite what environmental ide-
ologists say. In fact, a strong case can be made that the more our human
population can be concentrated in modern, well-designed cities, the less will
be the burden on the landscape and its wild species.[42] If environmentalists
insist on pursuing a maximalist program, where concrete issues are continually linked to demands for the moral reconstruction of their fellow humans,
they are likely to retard rather than advance the most essential environmentalist causes.

It would be pleasant to believe that the most scientifically sophisticated
environmental advocates invariably avoid the intellectual traps that entangle
their more naive comrades. Sadly, this doesn't seem to be the case, to judge
by some of the recent, well-publicized environmental advocacy literature. A
minor but puzzling instance occurs in Stephen H. Schneider's eloquent *Laboratory Earth*. The book is a concise but thorough exposition of the case for
regarding substantial global warming as a likely and disastrous consequence
of the heavy use of fossil fuels. The sketch of the relevant atmospheric science is clear and compelling. But one turns with particular eagerness to the
final chapter, which promises to outline the policy options available to us
to lessen or counteract greenhouse-gas emissions.[43] Many useful things are
said there about the technical, economic, and demographic aspects of the
problem. Conservation and energy efficiency are especially stressed. Yet there
is a curious lack of attention to alternative energy sources and the possibility of supplanting high-carbon fuels. Since this is a central issue, one gropes
for an explanation of this strange omission. Might it be that Schneider, writing principally for an environmentally committed readership, is loath to tread
on the sensitive feelings of antinuclear activists, opponents of hydroelectric
dams, and the like? To my direct knowledge, Schneider has, in the past, been
an advocate of nuclear power as part of the strategy of fossil-fuel substitu-
tion.[44] Perhaps he has since learned to be very discreet about this issue, for
fear of alienating the environmentalist rank and file. In any case, the crucial portion of his book has been gravely diminished by lack of attention to
the energy question.

A more disconcerting example is provided by Paul and Ann Ehrlich's *The
Betrayal of Science and Reason*.[45] The Ehrlichs have done Herculean work over
the past twenty-five years in bringing environmental issues, particularly
population growth, to the attention of the public. Their newest book is a
denunciation of what they term "brownlash," the emergence of a cadre of
scientists and journalists who scoff at many environmentalist pieties. In the
Ehrlichs' view, many of these people defy sound scientific consensus. Therefore, they are scientifically marginal, or merely incompetent. The wide publicity given to their views is the alleged result of a cynical campaign
orchestrated by the corporations and industries that balk at the cost and in-
convenience visited on them by prudent environmental policy.

My heart is with the Ehrlichs on many of these issues, particularly when it comes to their ongoing war of words with the late economist Julian Simon, whose "something will turn up" attitude toward the problem of population growth I find childishly irresponsible.[46] Yet I am discomfited to find that their treatment of a number of important issues is desultory and question-begging. It is instructive to read what they have to say on certain questions in comparison with the writings of journalist Michael Fumento, particularly his *Science under Siege*. The Ehrlichs briefly mention Fumento, tagging him as one of the brownlash mercenaries. Initially one accepts their estimate at face value. Fumento is a man of the right and publishes to a large extent in conservative organs like *National Review* and the *Wall Street Journal*, which does nothing to endear him to most serious environmentalists—or to me. The Ehrlichs are well-respected professional scientists, while Fumento, by contrast, is merely a journalist with a law degree—surely an ideal provenance for someone whose task is "to make the better case appear the worse." It is surprising, then, that a close comparison of their books goes a long way toward undoing that initial judgment. In particular, looking at the respective treatments of the dioxin question, I am struck by the thoroughness of Fumento's research and exposition.[47] It becomes absolutely clear that the persistent environmentalist claim that dioxin is "the most toxic substance known" is wildly false. The evidence is conclusive, because it comes from the health histories of thousands of people who have undergone moderate to severe exposure to dioxin. The damage to their health has been very slight, and may be fairly said to be undetectable, except in cases of ridiculously heavy contamination. Even there, health problems appear to have been relatively minor. True, dioxin is quite deadly to a number of nonhuman species and therefore must be regarded as a serious environmental threat. But that is not the issue. The enormous fear of dioxin that has been built up in the public mind comes from its image as something whose slightest trace will inflict death or grave illness on human beings. As Fumento shows, this is nonsense.

By contrast, the Ehrlichs fudge and fumble at the issue, doing little more than to quote briefly from some medical journals without providing adequate context or perspective.[48] They do nothing to address the evidence and arguments that Fumento marshalls. It cannot be that they are unfamiliar with them, for Fumento's book is among their references. What seems to be at work is a phenomenon all too familiar within environmentalism. Once an "environmental risk" has been denounced and dragged to the bar, there is no way of subsequently acquitting it.[49] The Ehrlichs seem desperately unwilling to let go of the myth of dioxin's diabolical toxicity. They are unable to concede the obvious truth that dioxin has been grotesquely oversold as a threat to human health. It is as if they fear that to back away from any well-known environmentalist claim is to destroy the movement utterly. The result is that they propound "science" as flawed as some of the brownlash junk science they want to denounce.

The dioxin affair is not a minor one. Hundreds of millions of dollars have been spent at Love Canal, Times Beach, and elsewhere to seal off the threat of "the most toxic substance known." Hundreds of people have been forced to relocate, and in the process, many of them have been frightened out of their wits. Similar fear haunts thousands of Vietnam War veterans who have been told that their exposure to dioxin via military defoliants like Agent Orange puts them (and their children) at perpetual risk of bizarre ailments. It now appears that all this was largely a false alarm. This is embarrassing for the environmental movement, which was dioxin's prosecutor-in-chief, but the initial misjudgment was understandable, if unfortunate. It would be truly disastrous, however, if dogmatism and intellectual rigidity prevent activists from conceding their mistakes. In that case, environmentalism will rapidly forfeit credibility and influence, and the brownlashers so detested by the Ehrlichs will have easy going, even in cases where purported threats are real and grave.

Paradoxically, the environmental movement would be better served if it studiously read Fumento's book, rather than the Ehrlichs', even though this would obviously have to be done through gritted teeth. The Ehrlichs identify many real villains. For the most part, these are classically heartless plutocrats or their servile minions. It is well to have such cynical scoundrels exposed and denounced. Yet the Ehrlichs apparently cannot bring themselves to admit that environmentalism has ever cried wolf, even in situations more clear-cut than the dioxin controversy. They take no note whatever of the follies of the low-frequency radiation scare, or of the related brouhaha over cellular phone system antennas. They close their eyes to the substantial fraction of the activist community that embraces New Age irrationality in various forms. Consequently, they ignore the degree to which environmentalism subverts itself. Fumento, by contrast, holds a mirror to environmentalism, one that reveals the flaws that righteous indignation papers over. Whatever his purpose, he has diagnosed dangerous propensities that cry out for immediate correction. A movement that cannot react with anything but chagrin to the news that a purportedly dire threat is nonexistent, or that a potential danger has been grossly overstated, is in deep trouble in the long run. Radical environmentalism barely conceals its secret wish that humanity *should* suffer, since it has so grievously offended Nature. A trace, perhaps more than a trace, of that attitude can be found in the Ehrlichs' manifesto. Ultimately, it represents the kind of emotional self-indulgence that the movement can ill afford.

Natural science has a hard time with nature, or rather, with "Nature," the myth that abides in our culture. The admiration for science strongly present in surveys of public opinion is certainly sincere enough, especially when it is directed specifically to the technological fruitfulness of scientific knowledge. Yet a dangerous undertow lurks beneath this admiration, a strong current of suspicion and misgiving. As much as science is praised, the countermyth that science is never to be trusted undoes much of that praise. With part of its soul, at least, our society regards science as the great enemy of nature, and in that guise fears it. Science dissects nature; nature is the continual victim of the scientist's sacrificial blade. The scientific quest for deep knowledge of nature is an invasion and a violation. Since nature always avenges itself, the sacrilegious act of probing too deeply brings disaster down on humanity's brow.

How much of this is inarticulate feeling, how much explicit doctrine? In some circles—specifically, the feminist faction that asserts woman's special closeness to and empathy for nature—it is an emphatic and fully wrought theme. But the explicit version usually doesn't get heard beyond tight sectarian confines. What is more generally in play is a vague, sporadic sensibility rather than a doctrine. It echoes in commonplace maxims like "Science doesn't know everything," or "There are some things man was not meant to know." Atavistic as it may seem, old fears rule in many hearts, fears of offending mightier beings than ourselves by appearing to know too much about them. Mythology and scripture resound with this theme, and it has not lost its power.[50]

The popular response to the achievements of molecular biology is particularly telling. Denunciations flow whenever some new achievement is announced in the news media. Most of these go far beyond sober analysis of the personal, social, and economic effects. Their rhetoric strives toward the apocalyptic; symbol is more important than reality. Although these jeremiads flow from a wide spectrum of religious and political viewpoints, they are united by the common theme of hubris and retribution. Eminent intellectuals are not immune from these anxieties. Consider the following response to the story of Dolly, the cloned sheep: "But we—we humans, that is—should be haunted by Dolly and all the Dollies to come and by the prospect others will appear on this earth as the progeny of our omnipresent striving, our yearning to create without pausing to reflect on what we are simultaneously destroying."[51] It is an old story revisited. The Faustian ambition of Western man has once again invaded a realm it is forbidden to enter, and will be hurled to deserved destruction! But the story notably lacks concreteness. It never makes clear who is actually to suffer for the overweening pride of science, or how, or why. The crime is singularly victimless. Or rather, the victim is an abstract ethos of restraint and pious refusal of knowledge. Yet the whole idea of cloning touches a sensitive cultural nerve.

The passage quoted above is the work of Jean Bethke Elshtain, one of our most astute social philosophers and moral critics.[52] Since I think it is greatly overblown, I am surprised and disappointed.[53] How deeply the Frankenstein myth has sunk into the contemporary psyche![54] Similar sermons pour forth in response to transgenic mice, or patenting of genotypes, or the discovery of genetic correlates of homosexuality.[55] Most of these strike me as silly.[56] Perhaps I am being a naive techno-optimist, but it seems to me that within the next fifty years humanity will turn out to have benefited enormously from all this biotechnological inventiveness.[57] A recent "Declaration" by associates of the International Academy of Humanism is on the mark: "We believe that reason is humanity's most powerful tool for untangling the problems that it encounters. But reasoned argument has been a scarce commodity in the recent flock of attacks on cloning."[58] The agitated public reaction to biotech and to new discoveries about human genetics is deeply disconcerting.[59] It may turn out to be unavailing and ultimately irrelevant, but it is very revealing of the subterranean structure of our common cultural assumptions about nature and knowledge.

It has not come to the point where angry civilians storm the labs to forestall blasphemous research. (It may come to that, so far as biology is concerned, although physics is probably safe enough.) Yet the image of science as the cunning, duplicitous assailant of nature has a caustic effect over the years. Some of the eventually triumphant opposition to the superconducting supercollider was stimulated by the hubris of the physicists themselves, expressed in their claims that they were on the track of a "final theory" or even "the theory of everything." Outbursts of similar resentment continually recur, triggered by developments like the computer Deep Blue's triumph over chessmaster Gary Kasparov, as well as the cloning of Dolly the sheep. Misgivings about the scope of science and the breadth of its claims stir the blood of even the sophisticated and knowledgeable. The great anthropologist Mary Douglas demonstrates, in her book *Purity and Danger*, just how ineradicably we humans harbor the notion that categories and bounds, the "natural order of things," must be upheld and kept inviolate: "[A]ll margins are dangerous. If they are pulled this way or that the shape of fundamental experience is altered. Any structure of ideas is vulnerable at the margins."[60] To transgress boundaries, even symbolically, is to threaten the stability of the cosmos, and it is therefore dangerous and abominable. In modern times, however, science has continually boasted of its refusal to recognize boundaries and its determination to smash through the barriers that stand between us and the knowledge we seek. It is not surprising, then, that the psychosocial mechanisms that Douglas posits come into play at times to besmear science as transgressive and blasphemous. It is interesting that Douglas, in a subsequent book, with Aaron Wildavsky, subjected the environmental movement to a similar analysis and discerned in it a ritual rather than a practical purpose, the task of symbolically maintaining the ordered categories of nature and culture that contemporary industrial society declines to respect.[61] In her view, this accounts for much of the irrationalism, hyperbole, and lack of proportionality between means and ends that she discerned in militant environmentalism. Much of its actions are ritual actions, contrived to ward off ritual pollution. "The rank and file are kept together by choosing the widest possible sectarian appeal—man's natural goodness, his corruption by big hierarchical organizations, and his redemption through a return to the natural, undifferentiated order of things. The common danger to nature is viewed as coming from a technological attack on nature."[62] The deep fallacy in this mode of addressing problems is clear enough. It consists in invoking the numinous to settle empirical questions. As a method of ritually cleansing believers of symbolic pollution, it might, for all I know, work well enough. But it is not a very promising way of dealing with real, physical pollution.

Science has no choice but to live with such primal anxieties. Scientists are mistaken if they think that because the society that surrounds them revels in the plenty that science bestows, it has developed a frame of mind that reflects the scientific ethos. More ancient tendencies are in play in the collective imagination. As ever, they impel human beings to fear and to propitiate the superhuman forces that seem to undergird the world of appearances. In today's culture, it is largely the image of an idealized nature that has assumed the superhuman role, and science, in that it appears variously as the interpreter, protector, and tormentor of nature, stands in an equivocal position, or worse. No matter how eloquent its apologists, science will continue to be the focus of suspicion and resentment. In a deep sense, it really is the most transgressive aspect of the social organism. Part of the duty of institutional science, therefore, is to maintain, in the face of all these atavistic fancies, its own firm independence as well as the sovereignty of the intellectual outlook that makes scientific thinking possible. Scientists should do so as tactfully as possible, but they must do so resolutely.

SEVEN

Ethnicity

American society is remarkable for successfully integrating into its mainstream people whose ancestors came from a wide variety of European cultures. At the same time, it has been dreadfully unsuccessful in finding a comparable place for descendants of the aboriginal peoples of this continent, or for the descendants of black slaves. Similarly, the members of the "latino" cultures of the Southwest and Puerto Rico, who became de facto "Americans" when the expansionist policies of the United States encroached on the territories in which they were living, have not been easily or happily absorbed. From its founding as a political entity, America has defined its culture, without challenge or question, as the culture of a people who look to the British Isles for their linguistic and ethnic roots. Its laws grow out of the bedrock of English common law, notwithstanding the elaborate edifice of constitutional law that interpenetrates it. Its religions—and religious conflicts—have developed historically from the spectrum of faiths that abided, and quarreled, under the British Crown, with Anglicanism flanked by dissenting Protestantism on one side and Roman Catholicism on the other. Until the end of the nineteenth century, the white population sustained this British character, modified only slightly by the presence of French and Dutch culture in some marginal areas and the influx of large numbers of Germans (and, of course, Irish, who were at least equivocally "British"). Since then, the arrival in quantity of members of other white European nationalities has tinctured American culture somewhat without seriously altering it. Two or three generations have sufficed to anchor these populations in the common Anglo-American ethos, so that they identify with their ancestral societies only in terms of vague sentiment.

Recently, the underlying demographic picture has begun to change so swiftly and drastically that many thoughtful people wonder if the traditional "old stock American" culture can retain its predominance. For the past three decades, since immigration law was reformed in the mid 1960s, large numbers of new immigrants have been arriving in this country. But comparatively few come from Europe and very few from western Europe. On the other hand, the black population has been augmented by new arrivals from the

English and French-speaking Caribbean and from sub-Saharan Africa itself. Meanwhile the latino population (meaning, roughly, Spanish-speaking but at most partially European in ancestry) grows even faster proportionately, thanks to the huge number of newcomers from Central and South America and the Spanish-speaking Caribbean. Concurrently, immigration from Taiwan and China itself adds substantially to a Chinese-American community that has been steadily growing, via natural increase, for more than a century. Ethnic Vietnamese and Koreans are also now well represented, and, even more so, Indians and Pakistanis. Combined with new arrivals from the Arab countries and Iran, Pakistanis and Indian Moslems have established Islam in this country as a major faith. As a religious community, Islam has also been bolstered by the substantial number of American blacks who have abandoned the Bible for the Koran.

The demographic effect of all this new immigration has been uneven. Major cities on the east and west coasts are now polyglot communities, where Korean neighborhoods crowd against Dominican and Indian. The suburbs currently accommodate large numbers of the more prosperous and professionally skilled of these groups. Rural America, especially in the prairie states and the mountain west, remains relatively unaffected. Yet it is clear that a general sense of unease has been growing among the core white population whose ancestors came to this country before the last decades of the nineteenth century. This disquiet is obvious not only in magazine articles and op-ed pieces warning against the dilution of traditional American culture by the newly arrived millions, but equally in the rise of new and often disquieting social movements. Ominously, the emergence of so-called militias attests to the growth of nativist organizations with a strongly racial ideology and a taste for violence. These groups rail against a "New World Order," a code word for the weakening of a traditional sense of American national identity. The militias are far outside the mainstream, but the anxiety to which they give such extravagant form haunts many far less volatile citizens. Even the recent moves to unify the mainstream Protestant denominations can be seen to express a subtle nativism.[1] These gestures reflect a desire to circle the wagons, not only against the less decorous religiosity of the evangelical and fundamentalist denominations, but also against the cultural threat of ethnic multiplicity, which puts in question the central position of "white Protestant Christianity" as the hegemonic social formation. Optimistic commentators tend to stress the enormous power of American culture to absorb new ethnicities and to convert them a bit at a time to mildly colorful variants of standard "Americanism." Perhaps the optimists are correct, but the case is by no means clear-cut, and equally eloquent and persuasive analysts argue for a much gloomier and more troublesome scenario.

All this is more than familiar to anyone who keeps even a casual eye on current events. The major political and social questions at issue go far beyond the purview of this book, whose central concern is the cultural status of science. Yet it is worth pondering the effects of the current highly variegated ethnic mix on the position of science and scientists within contemporary American culture. We must consider not only the attitudes of the varied elements of the ethnic spectrum toward science, but also the ways in which friction, rivalry, and competition among these groups skew such attitudes. Fortunately, the question of ethnic rivalries within American science need not be raised. Whatever other troubles beset science, the tendency to factionalize along ethnic lines has not been one of them, at least not within recent decades.

This will seem a blithe, even irresponsible, statement to those people who are seriously concerned about the severe underrepresentation of American blacks in the scientific community. The reflexive response of those most earnestly concerned with racial justice to such an evident disparity is to assert that systematic racism within the institution in question must have at least something to do with it, and probably plays a major role. After all, why should laboratories and academic science faculties be any different from police departments, trade unions, and executive suites? Given that American history is so much a history of the dirt done to blacks by countless formal laws and informal social codes, how is it possible to acquit science in the face of the apparently damning absence of blacks from its ranks?

The answer, I think, lies in the enormous ethnic and racial mix that has come to characterize science, even without any systematic program for fostering that diversity. It is no secret that an inordinate proportion of American scientists are of Jewish background. Their numbers, both relative and absolute, grew to something like their current level at a time when everyday social anti-Semitism raised few eyebrows. Even more recently, comparable numbers of East and South Asians—here I link immigrants with native-born Americans of Asian ancestry—have flooded into scientific institutions of every kind, despite a degree of prejudice against them in the ambient society that, at certain times, has matched hatred of blacks in its viciousness.

The durability of anti-black attitudes obviously has a great deal to do with the exclusion of blacks from science, but it seems to me that racism has operated, in this instance, not through the institutions of science as such, but rather through a host of other mechanisms that are relevant to the education of young people, and thus to their chances of preparing for scientific careers. Obviously, foremost is the inadequacy of the underfunded, demoralized school systems through which most black children have to pass. We also have to take into account civil and political institutions that convey all kinds of negative messages to these youngsters. At the same time, however, some of the blame goes to codes of behavior and sets of expectations that are internal to black culture itself. These are not embodied in formal organizations or explicit rules, but in tacit value systems and unstated assumptions that reflect both demoralization and deep alienation from the axioms of white society. They are thus very hard to pinpoint. Even to bring the subject up courts charges of racism in today's touchy atmosphere. Consequently, they are especially difficult to address through systematic reform. The device most frequently employed is the favorable publicity given to black people who have achieved intellectual distinction.

When it comes to science in particular, the emphasis is of course on scientists, engineers, astronauts, and the like—people who can be put on view as "role models." Often this is augmented by ploys involving films and television shows, where the roles of scientists and physicians can be filled by black actors. These tactics are, unfortunately, of limited effect, not least because the cultural attitudes they try to address and reform are marked by cynicism about the very middle-class values that role models incarnate. Moreover, the role-model strategy, if it is pursued honestly, suffers from a serious shortage of raw material because there are so few black scientists, and very few of these are authentic scientific celebrities. That is the very problem these methods hope to address. Hence they are vitiated by an all too familiar vicious circle: lack of role models keeps blacks from pursuing scientific careers, while lack of blacks with successful scientific careers makes role models scarce. Unfortunately, things are made worse by the less than honest methods that have been promoted by some zealots and endorsed by some educators. These will be discussed in some detail below.

Finally, candor compels us to take account of an assumption that has always permeated our society and which has lost little of its force over the centuries: the doctrine that for hereditary reasons blacks are simply not smart enough to achieve intellectual distinction—certainly not in the most abstruse and demanding areas of science. This obviously prejudices whites (and, more generally, non-blacks) against the notion that blacks can appropriately aspire to scientific careers. Even worse, it implants the same bias in blacks themselves, despite outward shows of denial. Most liberals and egalitarians behave as though the whole issue could be swept aside by sufficiently vigorous denial, repeated sufficiently often. Unfortunately, it will only be swept aside when a substantial body of black scientists of the first rank emerges under circumstances that immunize their reputations against charges of favoritism or hyperbole. These scientists have not yet appeared, nor is it anything more than a pious hope that they will appear within another generation. Until they do appear, the dogma of black intellectual inferiority will remain well entrenched, even within the hearts of people who would never publicly admit to such ungenerous views. Abstract arguments for the intellectual parity of all "races" of humankind are of limited effect because they are at best sketchy, plagued by gaps that must be plugged by hope in place of hard evidence.[2] Thus, to stipulate that there are no group differences in innate intellectual ability as though this were a settled fact is to give a hostage to uncertainty. Decency, however, compels it. There is no other way to eradicate the legacy of a deeply shameful past than to assume that the doctrines that sustained so many infamous practices are simply false. This assumption is in some degree prayerful. Even so, we must posit it as a given at this point in history, since the consequences of not doing so are so plainly horrific.

An omnipresent consideration in any study of the relationship of science to ethnic difference is the historical fact that science, as a cultural institution and a mode of thought, is essentially a Western invention, largely created in western Europe in the course of the seventeenth century. Before the 1600s there was a body of knowledge that incorporated some partial truths about the natural world in a vaguely systematic way. Occasionally this was amplified or extended by mathematical analysis and logical deduction. To call this mode of knowledge "science" is not totally inaccurate, but the usage begs questions and involves a certain amount of charity. Methodologically, as well as substantively, this kind of science was weak and vulnerable to displacement or submersion by notions in which we now would recognize no scientific truth. There was no methodological consensus, even among the tiny cadre of individuals who might meaningfully be described as "scientists," no core of reliable texts beyond some classical tracts on mathematics and astronomy, augmented by a few more recent mathematical works. The idea of the laboratory as the proving ground for new knowledge had not taken form. Observational astronomy, in its tools, methods, and purposes, was not all that different from what had been practiced by the Chaldeans and the Mayans.

By 1700 the situation had changed enormously. A complex of linked ideas had coalesced into a consistent method for acquiring and affirming scientific knowledge. Observation, experiment, and calculation had established mutually fruitful relationships, and had fostered an outlook that has distinctly modern contours. The authority of the ancients and of religious dogma in scientific matters was already irretrievably damaged, not least because of the huge disparity in empirical and explanatory power between what the new methods had revealed and what was claimed in the ancient texts. Scientific instruments had progressed spectacularly in sophistication, accuracy, and usefulness. Even more important, the understanding of how they were to be integrated into a systematic search for new knowledge had matured. A consensus had emerged about how such new knowledge was to be tested and judged, mathematically as well as physically, as the mathematical machinery appropriate for formulating ideas in physics and astronomy reached an unprecedented degree of conceptual power. All this constituted an epistemological revolution, as well as an enormous broadening of what was actually known about the universe. Moreover, it was at last clearly understood that science has to be an open-ended enterprise, not a circumscribed and final body of doctrine. The idea of "research" as a continuing stream of new results and syntheses had taken form. The continuity of the natural world, the essential unity between terrestrial and celestial nature, had come to be accepted, at least among the most astute. Prototypes for the institutional machinery that was to turn science into a profession with clear goals and standards were in place. The connection between the progress of "pure" science and the generation of powerful technology had not become fully evident, but it was grasped intuitively, at least in some special cases, by a few prescient souls. Beyond that, the concept of science as a self-nurturing structure, constantly expanding and continually converting new achievements into the foundations for ever newer achievements, had become infused into the culture.

All this had been done despite the deep religious, dynastic, and cultural antagonisms that had riven Europe despite the Thirty Years' War and the Puritan Revolution. One may argue that, in their grim way, the confessional struggles of the seventeenth century had contributed to the frame of mind that made the birth of modern science possible. Even devout souls, contemplating the endless, unavailing carnage that the various sects were all too willing to inflict on one another, began to understand that reasoned colloquy must impose no confessional conditions on its participants, and that the intrusion of religion into every area of dispute and discussion not only raised tempers, but hopelessly clouded issues. Perhaps the greatest foundational contribution of England's Royal Society was not the fragments of scientific theory that found their way into the early numbers of its journals, but the decision to ban theological matters from its debates as idle, unproductive, futile, and disruptive. This surprisingly hardheaded determination to isolate science from the arena of the religious imagination was the fountainhead of a long tradition that distinguishes empirical knowledge from numinous speculation. This tradition is a necessary condition for modern science to exist at all. Isaac Newton's notoriously divided soul is itself a wonderful metaphor for the way in which science forged its own path out of the obscurantism and sectarian bloody-mindedness that had savaged Europe for centuries. Newton, in the oddity and passion of his personal religious doctrines, was completely a man of his time, or rather, of an even earlier generation. Yet, in the methods he used and in the canons of scientific knowledge that he advocated, he seems to leap over the centuries. No wonder he is Western civilization's greatest culture hero.

In recent years, contemporary scholasticism—that is, the brooding cleresy that stalks the dim corridors of postmodernism—has challenged the idea that what I have just described really occurred.[3] It denies the existence of the scientific revolution, or at least questions whether it is to be regarded in an unambiguously positive light. From this perspective, revolutions in the mode and content of knowledge are said to be an endlessly recurrent historical theme. Each era has its characteristic "episteme," and change occurs as the mode of social existence changes, one "way of knowing" succeeding another as one "form of life" succeeds another. The key point, apparently, is that none of this represents progress toward a more accurate apprehension of objective truth. On the contrary, what any society claims to know merely codifies how authority flows within that society. Knowledge is not power; rather, power is "knowledge."[4]

This view strikes me as more than faintly ridiculous. It is sustained by careless, impressionistic argument, woolly philosophy, and doubtful history. But I don't propose to refute it here.[5] Suffice it to say that I am loyal to the traditional, if currently unfashionable "Whig interpretation" of science history, which seems to me not only edifying, but quite accurate, in general.[6] It is high time that Whiggery climbed back into the saddle. What interests me, however, is why the nihilistic posturings of the postmodernists gained such currency in the first place. The answer, I think, lies in the simple fact that the triumph of the scientific revolution, and the commanding position in the modern world of the institutions to which it gave rise, is very much a "Western" triumph. The scholarly subculture that is so intent on debunking this story as philosophically, if not historically, unsustainable is driven by a mood that plagues progressive intellectuals generally. With some justification, these thinkers regard the ascendency of the West as a disaster for the rest of humankind. They deplore Western expansionism, colonialism, the slave trade, the extirpation of the native cultures of the Americas, the economic hegemony of the capitalist system, and the ongoing pauperization of the undeveloped world. Such things are well worth deploring. But this stance, whatever its moral satisfactions, leads inevitably to frustration.

The commanding position of the "West," along with its economic assumptions, its social and political institutions, and its popular culture, has never seemed more secure. There is little that scholars, nestled in their more or less pleasant cloisters, can do to reconfigure the constellation of power and wealth that prevails in the world. The manager of any modest mutual fund has far more power at his disposal than any professor can possibly deploy, *qua* professor. So far as tenderhearted intellectuals are concerned, the upshot is resentment, bitterness, and a steady determination to savage the reputation of whatever Western institution comes within rhetorical reach. For some, Western science is the target of choice, not least because until recently it has been the most unambiguously admired of our cultural institutions, its noble reputation seemingly eternally assured. Without this strong undertow of revulsion against the spectacle of triumphant Western capitalism, it is doubtful whether postmodernism in general or its idiosyncratic approach to science in particular would have made many inroads. All the hostile strains in "science studies" are driven by it.[7] To put the matter reductively but truthfully, the urge to debunk the "Whig myth" of the ascent of science grows out of the desire to see the West displaced, if only symbolically, as the progenitor of a uniquely powerful and reliable mode of knowledge. In effect, the debunkers are acting as agents of ethnic resentment, even if they are only surrogates rather than proper representatives of the injured groups.

Facts, alas, trump noble intentions. The central epistemological truth remains that birth of science gave humanity an unprecedented power to see into the workings of reality. The central historical truth remains that it was in Western society, and none other, that, all within a few decades, the scientific ethic, emboldened by crucial factual and mathematical discoveries and canny empirical methodologies, solidified and began to manifest its unique power. This is not to deny the contributions of non-European cultures, including the Babylonian, Greek (which was, let us remember, decidedly "non-European"), Arab-Islamic, and Indian, as well as the Chinese. Indian mathematics, for instance, not only gives us the concept of zero, but was, as late as the seventeenth century, at least as advanced as European. Yet it never quite reached the critical take-off point. It failed to become the self-sustaining intellectual engine that appeared in Europe in the era of Descartes, Fermat, Huygens, Newton, Leibniz, and the Bernoullis. Chinese technology, in the preindustrial era, was easily a match for that of the West in terms of ingenuity and utility. Yet it failed to generate a tradition of abstract systematization and speculative inquiry that, in its turn, might have spawned true science.

Why was European culture of the seventeenth century able to avoid the blind alleys that might have kept its nascent science stagnant or severely limited its potential for development? This question has been on the historian's table for generations, and while many answers have been confidently offered, none has won general assent. The complementary question has been equally knotty: Why was the manifest intellectual power of Chinese and Indian civilization constricted in ways that prevented them from synthesizing their insights into a general and autonomous scientific method? The prospect for obtaining for definitive answers to either of these questions does not appear particularly bright. But there is no honest way of avoiding their underlying premise—the close historical correlation of modern science with the culture and intellectual tradition of western Europe.

This leaves the postmodernist clan with a limited set of choices if it is to avoid the indigestible conclusion that something uniquely admirable is to be credited to Western culture. One is simply to take refuge in a perverse or whimsical philosophy that denies that science possesses the virtues traditionally ascribed to it, or simply refuses to regard them as virtues. The anthropologist Marvin Harris has noted that "post-modernists replace science and reason with emotion, feeling, introspection, intuition, autonomy, creativity, imagination, fantasy and contemplation. They favor the heart over the head, the spiritual over the mechanical, the personal over the impersonal. For post-modernists, there are no privileged paradigms. Science gets no closer to truth than any other 'reading' of an unknowable and undecidable world."[8]

Postmodernism here erects a misleading dichotomy. The practice of science is in fact pervaded by introspection, intuition, autonomy, creativity, imagination, contemplation, and even fantasy.[9] But it also has the virtues that stereotype assigns to it: intellectual discipline, endless self-scrutiny, an ethical commitment to objectivity, and a deep respect for the clinching argument, even when it punctures appealing daydreams. It is precisely this meld of intuition and discipline that makes it possible for science to get much closer to truth than other "readings." The childish solipsism that postmodernists embrace cannot be taken seriously as argument, but it does disclose the powerful currents of resentment, distrust, and frustration which lead some tenderhearted souls to adopt it in the first place. "The central ethical piety of postmodernism," notes Harris, "is the belief that science, reason, and objectivity have done nothing to solve the major problems of the twentieth century such as homelessness, the pauperization of minorities, the threat of nuclear terrorism, famine, and war."[10]

Of course, studious application of "science, reason, and objectivity" to this article of faith demonstrates that their removal from contemporary discourse would do nothing to solve any of these vexing difficulties. It would, indeed, render them all the more menacing. Alas, those who have already taken leave of science, reason, and objectivity are not moved by such an argument! But one cannot deny that these folk are overflowing with sympathy for the oppressed, particularly those who are easily identified as victims of white, Western, capitalist culture, and that this sympathy has been the crucial motive for endorsing the postmodernist stance! It is no wonder, then, that of the disciplines that might be termed in any way "scientific," cultural anthropology has provided one of the few examples where postmodernism and explicit antiscience have made inroads.[11]

The logic that has won out is that to endorse science as a "way of knowing" more privileged than traditional non-Western knowledge systems is to concede the moral superiority of victimizers to their victims. "Following the lead of Clifford Geertz and under the direct influence of post-modern philosophers and literary critics such as Paul deMan, Jacques Derrida, and Michel Foucault, cultural anthropologists have adopted an increasingly apodictic and intolerant rhetoric aimed at ridding cultural anthropology of all vestiges of scientific 'totalizing' paradigms."[12] The charges against science grow more damning as the postmodernist mood becomes more and more truculent: "Science's critics also assert more than guilt by association, and they level the charge that science is actually the storm-trooper of Western Enlightenment colonial capitalism whose aim is to search and destroy all ideologies that may threaten to breach the walls of this Western edifice."[13] Antagonism toward science is well on its way to becoming the ideological litmus test for aspiring cultural anthropologists. "Graduate students seem relieved to be told that all knowledge is gender or class or culture relative . . . and that science as we know it is a male white European enterprise, and so tainted and of no more value than magical incantations."[14] "Science as generalizing taxonomy," proclaims one would-be scholar nurtured in this tradition, "is itself bound up with the construction of those who wield it as white."[15] The fervor with which this kind of doctrine has been embraced prompted one bemused observer to note:

> Returning recently from an extended period of fieldwork, I was surprised to find that the rejection of empirical science is still gaining popularity among anthropologists as critics make ever bolder declarations. Labeled the "romantic rebellion . . . , more and more anthropologists are rejecting the empirical basis, logical methods, and explanatory goals of the so-called "natural" sciences as being inappropriate for the study of human affairs.[16]

Fortunately for science, and even more fortunately for the various peoples in whose name postmodernist scholars ritually denounce science, the conceit that scientific knowledge is the implacable enemy of the Third World is limited, for the most part, to the cloisters of First World universities. Historically, a number of non-Western cultures have been confronted by Western science and have come away from the encounter, not in despair over their own epistemological shortcomings, but with a passionate resolve to assimilate the scientific viewpoint into their cultural outlook. They have done this in full awareness that it is *their* traditional assumptions, rather than the canons of science, that have to be modified when there is conflict.

The first example is the West's very own internal "Other," the Jews. Until the end of the eighteenth century, the European Jewish community was the very model for the persistence of unthinking faith, suffocating ritual, and rationally unjustifiable cultural habit. Voltaire mocked Judaism (rather too ferociously) as the ultimate bastion of obscurantism, superstition, and resistance to Enlightenment clarity of thought—the last place, in short, where the scientific ethos might be expected to take root. Yet within little more than a century, science and rationalism had been so far adopted as reigning ideals by most Jewish intellectuals that Voltaire's stereotype was completely turned on its head. Not only were Jews enrolled in the scientific community in numbers ridiculously disproportionate to their percentage of the population, but the very notion of "Jewish" ideology posited a secular, materialistic philosophy inspired by scientific discovery and exalting scientific rationality above all things.[17]

The example of the great Asian cultures—China, India, and Japan—is also quite pertinent. The miraculous modernization of Japan following the Meiji Restoration is a tale universally known. It involved the amazingly swift assimilation of many Western ways, including the establishment of a fully mature scientific and technological infrastructure (all too evident at the Battle of Tsushima).[18] The key point is that this occurred precisely when national and cultural pride were at a zenith. The fanatical Japanese chauvinism of the period proved remarkably consistent with the determination to absorb a thoroughly alien system of knowledge. The Japanese, furthermore, did not find it necessary to construct a retroactive myth that would contradict or modify the story of science as a Western invention. The aim of Japan was simply to match, and then outdo, Western scientific achievements, not to rewrite history. The same story is essentially true of China, although with a delay of several decades that is in part the result of Japan's extraordinary technological—and military—success. India, too, has reached parity with the Western world in the competence and prestige of its scientists.[19] Ironically, the stresses of Indian politics and ethnic rivalries have bred a brand of Hindu fundamentalism that denounces science as a foreign intrusion. Chauvinist hotheads now insist that "Vedantic" science be installed in its place. The fact that rationalizations for these demands are sometimes echoed by a few Indian "intellectuals" trained up in the postmodern faith is an illustration of the instincts common to both the avant-garde Western "left" and old-fashioned reactionary obscurantism.[20]

There has been an enormous quantity of postmodern pontification about "incommensurability," the impenetrable barrier that is held to segregate the epistemology of one culture from that of another.[21] Dogma holds that this not only makes it impermissible for one "way of knowing" to criticize another, but in fact even prevents one culture from apprehending how another comes to know and understand the world.[22] But the examples I have just cited demonstrate that "incommensurability" does not prevent "Western" scientific thought from penetrating the supposedly inconsistent epistemological regimes of "non-Western" cultures. These examples ought to have stifled this postmodern cliché before it gained much currency. That they haven't done so is damning evidence of the leaky quality of postmodern logic. It is also true, and lamentable, that the adoption of scientific standards for investigating the material world does not automatically entail the adoption of Enlightenment attitudes toward human dignity and social equity. But then, even in the West, science has flourished in situations where these ethical imperatives have been treated with contempt.

At any rate, so far as the core epistemological question is concerned, the decidedly nonpostmodernist anthropologist Robin Fox comes near the truth when he observes: "[T]he world we have is one in which this particular tradition of scientific inquiry—arising miraculously in one part of the world in one historical period—is all we have by way of a system of knowledge that will indeed transcend local cognitive boundaries. See how eagerly those who do not have it embrace it. And why? Because they are not stupid; they see it pays off."[23] I would quarrel with this only to the extent of partially challenging the assertion that those who do not have science always "eagerly embrace it." That embrace occurs, it seems to me, when the culture that finds itself confronting science is genuinely secure in its own values and confident of the worth of its own achievements—criteria that certainly hold in the case of Japan and China.

It is cultures whose self-confidence and sense of accomplishment are shaky that seem most eager to find some pretext for deriding Western science. These cultures are not those distant from the West, but on the contrary, those that the West has engulfed. Examples include the aboriginal societies of Western industrial countries like the United States, Canada, Australia, and New Zealand. The most noteworthy instance is America's black subculture, whose disaffection and anger with the white mainstream is the stuff of daily headlines. These groups, which live in the shadow of the great achievements of the West (as well as its great crimes), have the least effective psychological defenses against the dreadful sense of permanent dispossession. They are aliens on their own home ground and must be content with the leavings of a wealthy civilization whose defining institutions were developed by people who held them in low esteem. It is these "internal colonies" that have most consistently adopted the program set out for them by postmodern academics. They have declared war on a science that consigns their own ancestral traditions to the status of myths and legends rather than hard knowledge.

A notable instance is the recent book, *Red Earth, White Lies,* by the Native American activist Vine Deloria. Deloria's curious volume embodies the bitter discontent of the professional militant. It is hard to condemn his many grudges. The poverty and political weakness of most American Indians warrant unbounded anger, especially when they are joined to the memory of the violence and broken promises that created this situation. But Deloria's fury carries him over the edge into fantasy. He maintains that anthropologists and archaeologists have nothing valid to say about the origin and history of native peoples. The scientific methods these researchers employ to determine dates, durations, and ancestral affiliations are worthless because (at least to a Native American) Western science is itself worthless. According to Deloria and many other American Indian activists, the only valid way of determining the history of tribal peoples is to consult the tribal traditions that encompass this information. On this view—and it is shared by other academics of Indian background[24]—"science" is nothing but the tribal tradition of a new, and obnoxious, group which has no traditional ties to this land, and therefore can say nothing about how the First Peoples came to be. It is interesting that Deloria, despite all his heavy scorn for "Western" science, has no scruples about borrowing a bundle of conceits from that other flock of Western intellectuals, the postmodernists. Perhaps, as a full-fledged professor, he sees it as a good career move to combine the emotionally compelling aura of the activist with the academically hip cachet of the postmodern dialectician. However that may be, Deloria has lost no opportunity to swell the circle of the faithful through book tours, lectures, and the usual machinery of publicity. His converts are not limited to other Indians. Academic trend-mongers respond to his flattery of their doctrines by accepting his.[25]

This in itself would be sad, as most crank cults are sad, whatever their political pretensions. But what makes it serious as well is that it comes in the midst of a major quarrel that pits North American archaeologists and ethnographers against the most militant factions of Native American rights groups. Under pressure from the latter, Congress has enacted a law giving tribal peoples custody of human remains and artifacts that have historical ties with those tribes. Whatever the merits of this measure, in practice it has been stretched beyond all sensible limits.[26] The weakest, most implausible claims of kinship have been honored. Hardliners do not hesitate to claim title to all ancient materials that can be said to have come from their "historical" territories. This includes items thousands of years old that cannot reasonably be associated with *any* extant Indian group. Nonetheless, thanks to the pusillanimity of various officials, museum research collections have been stripped of invaluable material in order to allow Indian claimants to rebury it in secret locations. Field archaeologists have had years of work stolen from them by ethnic chauvinists who have prevented them from studying and publishing their own completely legal finds. As one glum researcher remarks: "It is easy to be pessimistic about the future of American archaeology. The rift between myth and science, between emotionalism and rationalism, seems so great, so fundamental, that it would appear that there is very little common ground possible on which both Indians and archaeologists can stand together."[27]

There is an implicit political dimension to all this appalling nonsense: Native American activists have a stake in making it appear that their claims to particular tracts of land are anchored in the trackless depths of time. The complicated truth—that this continent, over the millennia, has seen repeated waves of migration, conquest, displacement, and even extirpation of one ancient group by another, long before the Europeans came—spoils the neat moral effect of that tall tale. Hence the archaeologists, whose professional duty is to investigate and report the disagreeable truth, must be robbed and silenced. The public's tendency to idealize the Indian as its chosen embodiment of ancient ecological wisdom has aided in this thuggery.[28] (Note how easy it is to do this kind of symbolic cheerleading, rather than support the concrete reforms necessary to end poverty, disease, and social disintegration on the reservations.) Archaeologists, by contrast, are thought of as monomaniacal eggheads and pettifoggers. Consequently, public and academic officials have been far more solicitous of Indian complaints than archaeologists' protests. Furthermore, the archaeologists have been betrayed by the enthusiasm of some of their own colleagues for the relativistic science bashing promoted by postmodernism. It is hard to make a case for the unimpeded pursuit of scientific truth when even some supposed peers declare that one "truth" is as good as another.

This is not the only instance where activists or sympathetic intellectuals decide that the most effective thing they can do for the self-esteem of the dispossessed is to anoint their folk wisdom as fully epistemologically equivalent to standard science. "Western science," says one such thinker, "is only one way of describing reality, nature, and the way things work—a very effective way, certainly, for the production of goods and profits, but unsatisfactory in most other respects."[29] The "ethnoscience" of many other *ethnoi*, therefore, must be raised to equal, if not superior, esteem. In New Zealand, it is the Maori who get the accolade. Activists demand that Maori science should be taught alongside the conventional version. Notes one distressed observer, "The idea of Maori science seems to make sense at first hearing, partly because of a vernacular but inaccurate definition of science as 'a body of knowledge,' and partly because it appeals to the fairness of teachers, who genuinely want different perspectives to tell both sides of the story. The latter appeal is misleading, and echoes creationist requests for 'equal time' for their story."[30] In Australia, it is the Aboriginal cultures that require redress. Naturally, the postmodernist faithful who dominate the field of "science studies" are eager to supply it. A report on a recent conference announces:

> This is to remake the debate about the boundary between science and non-science, so beloved of old-time philosophers. "We" show that "our" knowledge is on the science side of the boundary. At our workshop there was very little of this endeavor. Yet many were in the business of drawing boundaries: science is to be kept as "the other." A neat inversion of that former enterprise. It is understandable for those who have suffered at the hands of science—and many Aboriginal communities have suffered terribly in many ways, under a regimen of scientific treatment—to want to distance themselves from what they see as the polluting effect of science. And there is no doubt that having science as "other" is useful. It provides protection; possibilities for establishing a community of knowers united in elaborating a form of knowledge that is "not science."[31]

It is notable that this kind of blithe, self-infatuated nonsense passes without critical comment among those academics who list their disciplinary affiliation as "science studies."

In the United States, the most troublesome cultural conflicts over science are likely to emerge in the larger context of racial antagonism between black and white. At the base of the problem is the severe self-doubt that permeates the black community and which all too often expresses itself through its opposite, a vainglorious celebration of every real or imagined achievement or virtue that can be labeled "African." This is emotionally understandable, but politically disastrous. What has happened is that the most "American" of all peoples, one whose history is deeply, tragically, and, most important, nobly entangled with the history of American civilization, has largely endorsed a mode of cultural display that ferociously asserts its irreconcilable alienness.[32] The myth fails utterly on the level of logic. Most of its common themes are absurd. American blacks are West African in ancestry (partially; most have European ancestors as well), yet it has become an article of faith that they are the heirs of ancient Egypt. Egyptian culture had no historic connection with West Africa, and is "racially" unaffiliated as well.[33] Likewise, the growing enthusiasm for Islam ironically endorses a history of Arab incursion into black Africa which was hardly more humanitarian than the West's colonialism. In fact, it initiated the export trade in black slaves that provided the precedent and model for predatory Westerners. Finally, even those black nationalists who only embrace pure West African cultural motifs are not immune to paradox. The most powerful and widely known black cultures were, in most cases, eager partners in the slave trade. Their warriors incessantly raided weaker groups to seize captives who could be dragged in chains to European coastal outposts and sold to white slavers.

To embrace a fiction in support of one's sense of cultural identity is not necessarily idiotic or debilitating. Most groups do it to some extent—think of the generations of American schoolchildren who were weened on Parson Weems's tall tales about George Washington. In the case of American blacks, however, the net effect on the country as a whole is to persuade many whites (even those who profess publicly to the contrary) that the black population is hopelessly, irremediably foreign, even more so than recent immigrants from Korea or Iran. Scratch an enthusiastic multiculturalist, one who beams complacently upon kente-cloth apparel and Kwanza celebrations, and you will usually find someone who believes, at bottom, that the gap between black and white is irreconcilable.

Because modern science is one of the chief props to European cultural prestige, it is an area in which black cultural nationalism feels it must offer countermyths. Many of these involve overestimation, sometimes mounting to grotesque exaggeration, of the supposed scientific and mathematical accomplishments of ancient Egypt. Specimens of this genre appear in Ivan van Sertima's peculiar book, *Blacks in Science, Ancient and Modern*. Even more extreme is the bizarre *Science Baseline Essay*, by Hunter H. Adams, which not only claims that the Egyptians understood quantum mechanics, but that their science was linked as well to deep psychic powers. Since I have commented on these works elsewhere I shall not rehearse their absurdities in detail here.[34] However, as a university teacher I have run into a number of black students who take these writings perfectly seriously and react with voluble fury when they are challenged.[35] Despite frequent and public refutations of this kind of literature, it prospers and even grows.

Another distressing example has just emerged from a university press under the title *Ethnomathematics*.[36] The editors firmly announce, in their introduction, that "The Eurocentric myth is tenacious, pernicious and silencing, distorting perspective and inducing myopic vision."[37] Whether they proffer reasonable evidence for this view is another question altogether. "The ethnomathematics developed by different groups are likely to be more efficient at solving problems related to their cultures than academic mathematics," declares one author, although no examples are provided to help us judge this rather remarkable proposition.[38] Of course, supposedly progessive intellectuals, with whose relativistic view of knowledge this claim so nicely dovetails, will probably accept it on faith, notwithstanding that the author elsewhere affirms that his view of the human condition derives from that of the notorious Nazi philosopher, Martin Heidegger.[39] Another contributor, even more fervent, enjoins us to remember that

(a) People of color were the original founders and innovators of mathematics and science.
(b) Europe was never isolated from Third World (actually First World) mathematical and scientific achievements.
(c) European capitalism developed because of Europe's incorporation of the mathematical and scientific ideas and techniques of the First World into their capitalist superstructure.
(d) Europe dominated, enslaved, and colonized Africa, Asia, and the Americas and thereby stopped and/or reversed most, but not all, forms of first world intellectual, mathematical, scientific, and technological activity.[40]

Here, at least, we merely have fiction rather than the epistemic relativism characteristic of so much academic writing. Like many other Afrocentrists, the author wishes us to know that Europe's proudest achievements were looted from older, darker-skinned peoples. This has the virtue of conceding that these were genuine achievements and worthy of pride, not merely socially constructed delusions. Still, it is hard to read such stuff without feeling contempt and pity—contempt for the shallow chauvinistic bombast that spits in the face of plain fact, pity for the pain that drives essentially well-meaning people to the anodyne of delusion. The listed "rules" aim to create a classroom atmosphere where black children (principally) will be encouraged to learn mathematics. They offer cultures and peoples, rather than individuals, as role models. But, beyond the harangues about stolen legacies, the author seems to have given little thought to either the mechanics or the pedagogical psychology of mathematics teaching. Ethnic pride is the be-all and end-all. The objection to such a methodology is clear. Even conceding that this approach might initially inspire hard work and successful learning, its beneficiaries, if they pursue the study of mathematics at all intensely, will eventually come to realize that they've been sold a bill of goods. Just how much damage such a discovery will do is hard to see, but clearly there will be no good effects.

It is noteworthy that *Ethnomathematics* also contains a piece by the quirky Martin Bernal, whose recent books on the supposed non-European foundations of classical Greek civilization were widely acclaimed by many who did not take the trouble to weigh their plausibility very carefully.[41] In the present instance, Bernal makes claims for the depth and scope of Egyptian mathematics that far outrun any evidence that has come to light.[42] As a mathematician, I'm moved to reflect that Bernal has now added mathematics to the subjects on which he has confidently spoken out without bothering to learn much about them.

The publication of books like this under the protection of a full-fledged academic imprimatur is still a fairly infrequent phenomenon, but not quite as infrequent as one might hope.[43] Even an institution as august as the American Museum of Natural History has been known to indulge in similar acts of "epistemic charity" toward non-European cultures in order to ransom their images from the taint of irrationality and superstition.[44] The Museum's Hall of African Peoples, for instance, contains a display case labeled "Science and Divination," intended, no doubt, to assure young black American visitors that their black African forebears pursued and applied systematic scientific knowledge. Unfortunately, the contents of the display are indeed reminders of the superstitions that prevailed in these tribal societies. The artifacts shown were employed in decidedly unscientific divination, magical control of natural processes, and so forth.[45] It is likely that the prevalent wish for "positive role models" induced the curators to label this "science." Possibly, the relativism and the disdain for orthodox science which have savaged the intellectual integrity of cultural anthropology had a hand in the episode as well. Whatever the case, neither the general public nor, in the long run, the black population will benefit from such pious fraud.

One of the stories that organs of science education—and this includes museums—have the duty to tell is the difficulty and agonizing slowness with which the scientific worldview developed. In the telling, one of the things that must be made clear is that all sorts of historic and contemporary "knowledges" do *not* merit being called science, intriguing as they may be from other points of view. It is the educator's duty to insist that "superstition" is, indeed, often the name for such things. The purpose is not to insult or patronize other cultures, but to clarify how remarkable it is that the scientific ethic did, at last, succeed in emerging from a stew of human confusion. Educators must also emphasize how fragile, in a societal sense, the scientific ethic yet remains. It is at least disquieting that an institution that would certainly resist the threats and blandishments of conventional creationism will act of its own accord to efface the boundary between science and its negation in order to soothe a tender political conscience.

Crank theories are hardly new or particularly striking phenomena. That academic and scholarly institutions now seem to be willing to serve them up, suitably rewarmed, is nonetheless disconcerting.[46] The motives of those who do so are usually blameless, even if their judgment isn't. Who wouldn't want to encourage minority children to learn science and pursue scientific careers? But merging this worthy intention with the hallucinations of embittered identity politics seems likely to make things worse overall. A generation of young people steeped in crackpot doctrine while being shortchanged on scientific substance is not likely to get very far in science, engineering, or medicine.[47] It's much more probable that the pattern of exclusion, resentment, and compensatory fantasy will simply repeat itself indefinitely. The irony is that a generation of university-nurtured intellectuals, acting out of laudable motives—and an entirely misplaced confidence in their own judgment—will have contributed to perpetuating the gross injustice they claim to want to abolish.

The problem, finally, is an ancient one: the propensity to confound justice with vengeance. Organs of "progressive" academic opinion continue to vomit a stream of invective against "the West," all it has been, and all it has done. Simultaneously, they vindicate the oppressed and excluded by heaping inordinate praise on their "incommensurable" ways of knowing.[48] Western civilization will, most likely, survive the onslaught. The real penalty will fall on the heads of the intended beneficiaries.

How much additional trouble lies in wait for science because of the hostility or misunderstanding of blacks and other minorities? The new crypto-cult of science studies sometimes slips into a rhetoric simultaneously accusatory and triumphalist:

> [T]he pursuit of a scientific realist agenda amounts to an attempt to arrest the process of essence-construction as an ever-larger share of the world's population appropriates the science that is suitable for its needs. If I am correct in this diagnosis, then to discover the essence of tomorrow's science we should look to the ways in which recently enfranchised citizens of the republic of science—women and especially people of color from all over the world—separate the wheat from the chaff in the West's scientific legacy.[49]

This kind of posturing, however, is an arid game and the academics who insist on playing it find in the end that their only real allies are racial and gender chauvinists of the crudest stripe, augmented from time to time by fervent believers in New Age fads or even more ancient superstitions.

Having mentioned gender, it is appropriate to address briefly the question of sexual bias in science and the criticism it has drawn from feminist theorists. As the quotation above indicates, the campus-radical indictment of science as a Western institution almost always, in the very same breath, denounces it as well for being a patriarchal, sexist institution. The politics of ethnic resentment in this arena rarely loses touch with the politics of radical feminist resentment. Nonetheless, the objective situation of women, particularly with respect to science, is vastly different from that of black Americans, despite what one hears in the anthems of identity politics. For this reason, a chapter on gender and science is absent from this book.

The hard feelings of feminists toward science are understandable given the history of science as a bastion of male exclusivity. This bitterness, however, may well have outlived its usefulness and relevance. The simple fact is that in terms of prestige as well as numbers, the situation of women in science has been improving swiftly, largely without the aid of formal "affirmative action" mechanisms or the intervention of the brand of "gender feminism" current within most women's studies programs.[50] "Equity" feminism seems to have done the job perfectly well.[51] The most significant fact is that the culture of science seems to have been left largely unaltered by the recruitment of large numbers of women (just as it has been left unaltered by the influx of east and south Asians).[52] Certainly, in the crucial areas of methodology and epistemology, nothing much seems to have changed. If there have been any consequent alterations in the direction of research, they are subtle and difficult to detect.[53]

This has not prevented the emergence of feminist science critique as a major campus industry. Each year, the claims for the relevance and power of supposed feminist insight grow more and more extravagant. "Feminist epistemology," paying particular attention to the flaws of "gendered" science, has proved a fast track for a number of scholarly careers. In terms of volume of publication, this genre completely swamps work on science and ethnicity. Yet in the long run—and the not-too-long run—it is unlikely that all this ringing rhetoric will have a lasting effect. It is merely part of the history of academic gamesmanship and will have little influence on science, or on the apprehension of science by intelligent nonscientists. It will fade rapidly after its little season, not least because of the changes that are now underway in the sexual demographics of science. It is academic makework of the dreariest kind. Even now, it is little known or regarded outside the sectarian circle of academic feminists and those who feel politically obliged to heap praise on them. It may nonetheless be responsible for one concrete effect: to dissuade bright, but unduly susceptible, young women from pursuing scientific careers.[54] Yet the cultural and institutional reach of gender feminism is small, and science is therefore unlikely to be much affected or afflicted by it.

By contrast, "racial" theories of science, though far less articulate and far less congruent with the current mannerist affectations of the academy, are likely to have greater staying power. They answer to something real and disturbing in the wider culture. They are even likely to have some effect on how science is organized and funded, because the absence of blacks in science continues to weigh heavily on the consciences of funders and administrators and the politicians to whom they answer. The academy is obsessed with race, the more so because it is unable to accomplish much of substantive value in resolving the country's racial dilemma. Science, by contrast with other disciplines, is largely indifferent to race and ethnicity. At this point in its history it is as thoroughly "open to talents" as is possible for any human institution. The paradoxical result, however, is that from the black point of view, science seems to be the most exclusive of institutions. A corollary of openness to talent is lack of indulgence toward paucity of accomplishment.

For reasons cited previously, the black community has not produced very many accomplished scientists. Even those scientists who count themselves as emphatically liberal on racial politics tend to draw back when it comes to compromising the meritocratic standards of research science. One may strongly support affirmative action on the level of undergraduate admissions without being willing to compromise traditional standards when it comes to promotion, tenure, recognition, and awards for scientists. Consequently, blacks continue to be almost invisible in professional science. From one point of view, this bodes well for eventual racial reconciliation. When large numbers of blacks find their way into scientific careers (with, presumably, a proportionate amount of distinction), it will be understood that favoritism had little to do with it, and that competence and merit played by far the major role. However, that hypothetical development is bleak consolation for those who are understandably dismayed by the current statistics. The most likely prospect is that black intellectuals will view science with increasing suspicion, and that their attitudes will propagate into the black population at large. Whether this will have much effect on young black people who are thinking of science as a career is hard to say. Their numbers have been steadily, if modestly, climbing, and it would be tragic if that trend were reversed.

My own experience—very limited and not rising above the merely anecdotal—suggests that at the college level, black students pursuing science seriously are not only growing numerically, but are coming into the field with a perfectly serious demeanor not visibly affected by black chauvinist propaganda. At least I have never seen a black student in a mathematics class respond to a passing remark about Greek mathematics by rising to declare that the Greeks really stole all their ideas from Africa. For whatever it's worth, this is encouraging. Nonetheless, the prospect is that for all the reasons cited, the black population as a whole will retain and develop increasingly dubious ideas about science. Perhaps we caught a whiff of this in the "Kemron" controversy, where large numbers of black AIDS victims sought a dubious treatment solely on the basis of its supposedly African provenance.

Of course, the American black population is alienated from mainstream white opinion on a host of issues. The institutional reach of black discontent is limited, and is not likely to stymie scientific work in many cases.[55] Nonetheless, the hostility of any large segment of the population is not something that science can afford to regard with equanimity in the present climate. Black discontent need not be an explicit ally of the other social forces inimical to science in order to form a dangerous de facto partnership with them. The lamentable fate of the superconducting supercollider testifies to the fact that an antiscience coalition can be composed of a number of mutually antagonistic ideological strands. Beyond this, there is the patent injustice of a situation where, whatever the mechanism, skin color keeps large numbers of people from sharing the joy and wonderment of science and bars them from participating in the greatest intellectual adventures of which our species is capable.

EIGHT

Education

By any reasonable measure, the United States is by far the world leader in science as a whole, and in all major branches of science. More than a third of the world's recent scientific literature—including engineering and medicine—was written by American researchers. Even more impressively, almost half the citations that occur in the literature as a whole are to American papers. In virtually every field, the U.S. leads the world in total output. Although by some measures, average researcher "quality" is higher in other nations (particularly in Scandinavia), America is among the top five in all areas even by this standard.[1] The naive expectation might be, therefore, that American scientific success must be linked to a superior system of scientific education. Virtually no one believes this, however, unless the evaluation is restricted to professional education at the graduate level. Instead, one hears a constant litany of criticism of American science education.

> The paradox is that we, here in the United States today, have the finest scientists in the world, and we also have the worst science education in the world, or at least in the industrialized world. . . . American scientists, trained in American graduate schools, produced more Nobel Prizes, more scientific citations, more of just about anything you care to measure than any other country in the world, maybe more than the rest of the world combined. Yet, students in American schools consistently rank at the bottom of all those from advanced nations in tests of scientific knowledge, and furthermore roughly 95% of the American public is consistently found to be scientifically illiterate by any rational standard.[2]

This is supported by countless surveys and statistical comparisons showing that at least through the secondary level, and possibly beyond, science education in this country lags behind that in many other industrialized countries in scope, thoroughness, and rigor. American schools, on the whole, are said to offer uncoordinated, diluted, superficial accounts of science, often taught by instructors whose own grasp of the subject is at best hazy. In a key area like mathematics, there is an overgenerous ratio of rewards (in the form of grades) to manifest skill. For its part, biology labors under a continual drumfire of criticism from religious zealots who, more than a century after Darwin, remain unreconciled to the most profound insight in the subject— the theory of evolution.[3]

This is not a new situation. The problem has been as bad—and sometimes worse—for decades. Yet during that period, American science has been at least reasonably successful in recruiting sufficient talent to its ranks to maintain its predominance. How may we explain this apparent paradox? One important factor is usually called the "brain drain." Foreign scientists at all levels, from beginning graduate students to distinguished senior researchers, continually flow into American scientific institutions. This is relatively easy to account for. Obviously, given that the United States is the "world leader" in so many fields, it is the best place to make one's name in those areas, since being in this country will bring one close to the most important ongoing research activity and will also provide a platform for displaying one's own accomplishments. Moreover, the strictures against outsiders that plague would-be immigrants in so many industrialized countries are virtually absent from institutional science in America.[4] Newcomers are neither exploited, nor stigmatized, and, whatever their status in the eyes of the INS, they instantly become full-fledged scientific citizens.

Even beyond the university or the laboratory, America, for all its ethnic tensions, is still uniquely welcoming to immigrants, especially those who are comfortable with "middle-class" values and notions of achievement.[5] This has worked to the benefit not only of research scientists, but of the tens of thousands of foreign-born, and often foreign-trained, physicians, whose ubiquitous presence may be noted at virtually any large hospital. These immigrants have become a major presence in the medical community at large. The United States is still a paragon of material prosperity compared to most of the world, especially when we discount the enclaves of urban and rural poverty. The comfortable lifestyle of the American upper-middle class, easily available to most scientists who have settled into fruitful careers, remains seductive to educated people around the world, especially because it is psychologically far less confining than the conformity imposed by cultures in which traditional notions of authority and decorum reign. To many scientists, freedom from social rigidity and conventional piety seems especially appealing, and America offers it unreservedly. For all these reasons, the pool of American scientific talent has been continually fed by foreign sources, with considerable mutual enrichment.

Beyond this, American science still feels the benefits of the shrewd and fertile governmental policies that were in place in the 1950s and '60s. These days, the term "sputnik" is probably not in the vocabulary of most people under the age of forty, but a few decades ago it had awesome symbolic resonance in this country. Recall that it was the nickname for artificial satellites of the type that the Soviet Union began to loft into orbit in 1956. The fact that the Russians were the first to launch such a device provoked a degree of shock and distress in this country that is hard to imagine nowadays. The trauma was compounded by the fact that what seemed like endless months passed before the fledgling U.S. space program matched this achievement. It did so only after a few painful, very public failures, and even then, only with a device that seemed puny compared to the ponderous (180-pound!) sputnik.

For several years, the news was full of gloomy talk suggesting that the United States was about to be eclipsed by the Soviet Union as the world leader in technology. The apparent feebleness of American rockets, compared to the mighty Russian boosters, was widely appropriated as a metaphor for this nation's perilous position. The political fallout from the Soviet Union's spectacular public-relations coup continued to roil public discourse well into the 1960s, and certainly had an effect on the 1960 presidential election. John F. Kennedy's cries of alarm at the so-called missile gap were a key factor in building his narrow margin of victory over Richard Nixon.

That all this was more or less a mirage became clear a decade later, after American success in military rocketry as well as the *Apollo* moon-landing program exposed the severe limitations of the relatively primitive and inflexible Soviet technology. But while the sputnik scare lasted, it had marked and enduring effects on American science policy. These effects certainly included heavy governmental subsidy of scientific research, much of it directed to military or space-related projects, but also with a strong "basic research" component benefiting "curiosity driven" science. A further corollary was a crash program to improve science education.

Whatever the general limitations of the crash-program methodology, it must be recognized that the postsputnik effort to upgrade education had a swift and spectacular payoff. It was centered around the recruitment of talented young people while they were still in high school. They were provided with various enrichment programs, including special accelerated courses during the summer break. Those who went on to study science in college and found their way into graduate programs were provided with generous financial support through direct fellowships, as well as budgetary allotments on the research grants of senior scientists. After receiving advanced degrees, many of them were provided with jobs through academic expansion financed by government subsidy.

Personally, I am typical of the generation of scientists that rode this wave of governmental enthusiasm and generosity. In my case, the National Science Foundation played fairy godmother. While in high school at the end of the 1950s, I spent a summer at an NSF Summer Institute in solid-state physics. As a graduate student in mathematics, I was supported, for the most part, by an NSF graduate fellowship, as well as benefiting from summertime study programs run with NSF dollars. When I came to my current institution in 1969, it was to take a line created by an NSF "Centers of Excellence" grant, which paid the salaries of newly recruited faculty for several years.[6] Starting in the 1970s, this halcyon period for science—and science education—began to wind down for various political and cultural reasons. But the scientists nurtured and encouraged by governmental programs—and by private foundations that have since become much less generous toward science—now form the backbone of the American scientific establishment (if we must call it that) and have been responsible, over the last thirty years, for developing and maintaining the unprecedented excellence of American science. The sputnik panic may seem silly in retrospect—nowadays the endless problems afflicting the Russian MIR space station provide an ongoing cliff-hanger and stand as a metaphor for the fecklessness of Soviet-style technology. But, however much pious fraud about America's immanent danger was involved in the process, sputnik provided the pretext for programs that had astonishing results and which are sorely missed.[7]

The sputnik-inspired policies were intended to produce a corps of professional scientists. Science education has another, parallel goal—to impart a minimal understanding to the public at large, irrespective of profession. I shall consider this aspect presently, but first I want to stress the recruitment and training of active researchers. Frankly, this order reflects what I consider to be the relative importance of the two functions. The lack of what is usually called "scientific literacy" among the population at large—which extends, indeed, to most nonscientific intellectuals—is a grave and vexing problem. However bad it is or may become, it is nonetheless secondary to the question of maintaining the quantitative and qualitative strengths of professional science. A scientifically illiterate public is a headache; a serious shortfall in the numbers and competence of scientists would be an enormous disaster. Indeed, a prime reason for taking the scientific literacy problem seriously is that it threatens, in some ways, to limit the capability of the scientific community to sustain and renew itself. Though it is emphatically unfashionable to say so these days (especially within the pedagogical establishment), the nurturance of new generations of scientists is the most important duty of science education, and has first call on our attention and concern.

One aspect of the science-recruitment policies of the sputnik era that deserves frank acknowledgment was their unabashedly elitist character. The methodology, admitted openly and without embarrassment at the time, was to identify promising talent at a young age, and to make special provisions for its development. These provisions included a great degree of educational segregation. Bright young people were conspicuously set apart from their contemporaries. They were placed in special, enriched classes and sent into extracurricular programs that only admitted those of comparable talent. It was assumed that, as they moved through the educational stream, they would maintain a rather close-knit formation, associating primarily with their intellectual peers and with mentors drawn from the ranks of accomplished scientists.[8] The operating hypothesis was that talent, whatever its origins in genetic endowment or early childhood experience, is a relatively rare resource, that positive measures are necessary to seek it out, and that once found it requires and deserves special treatment.

In my view, this hypothesis is largely correct. Even if it is not, we nonetheless lack, at present, the pedagogical insight that might make it irrelevant. The plain fact is that we cannot hope to turn average students into scientists of professional caliber, even if we could somehow induce them to aim for scientific careers. Science is an elitist calling, and it draws upon abilities that are manifest in only a small segment of the population, irrespective of whether they might be latent, if only we knew how to tap them, in a far larger spectrum of people. The first concern of a pedagogy that aims at turning out scientists, therefore, is to identify this aptitude and at the same time to insulate it, in certain ways, from those personal and cultural factors that might discourage its maturation. This means that talented young people must be singled out for praise, encouragement, and close attention, even at the risk of provoking a degree of snobbery in their attitudes. In our present situation, intellectual snobbery is not the major affliction of our culture. Much more dangerous is the prevalent anti-snobbery that scoffs at intellectual distinction and at the hard work and deferral of immediate gratification that are so important in achieving it.

Youth culture, with its values and attitudes, has recently drawn the admiring attention of many "cultural theorists," not least because they see in it (probably through rose-colored glasses) a "site of resistance" to the hegemonic values of capitalism. The profound, and worsening, anti-intellectualism of this ethic seems not to be of much concern to such observers, since intellectualism (at least, of the sort that doesn't sit at the feet of French post-structuralists) is held to be a snare and a delusion. However, from the point of view of someone who wants to see science continually renewed by fresh and enthusiastic young talent, present-day youth culture constitutes a singularly corrosive environment.

These days it is increasingly difficult to maintain an elitist point of view in educational circles. It goes against the ethic in which so many of today's young academics have been reared, one which challenges the worth of what were formerly lauded as great cultural achievements, and which insists on praising the demotic and the quotidian, thus bringing the "margins" to the "center." Scientists, even those who are strongly egalitarian in their general political views, have not been persuaded, on the whole, to abandon meritocratic standards when it comes to science. However, scientists seem to have become more and more irrelevant to educational policymaking, even where science education is concerned. More typical of modish educational philosophy is the following passage:

The challenges to empiricist science and positivism that have been launched during the past generation have, we believe, been insightful and productive. . . . In disclosing the value-ladenness of all claims to knowledge, and in exposing the gender-related biases in the natural sciences, for instance, these critical analyses have created space for new ways of interpreting and responding to events in the natural and social worlds.[9]

A none-too-encouraging instance of the reign of these ideas arose in connection with the attempt, a few years ago, to establish a new set of science-education standards. This was done under the sponsorship of the National Research Council of the National Academy of Sciences.[10] Here, if anywhere, one might think, the viewpoint of scientists should certainly hold sway. Such, alas, was not the case—at least not initially. The first draft was marked by pronouncements reflecting the intellectual sway of postmodernists who tend to view science with ill-disguised suspicion and distaste. "Nowhere appeared a statement that scientists seek to find regularities in nature, or to discover and explain new phenomena or laws, or to reach shareable and testable insights about the lawfulness and order of the natural world."[11] Fortunately, some alarmed scientists were able, at the last minute, to undo the mischief and substitute something sensible. But the episode is telling. It is evidence that blatant antiscience can peddle itself to the educational establishment by denouncing the unfashionable elitism that supposedly taints science.

A recent book by David E. Drew, a professor of education, illustrates the current doctrine.[12] It fulminates against the notion that talent for science might depend on innate gifts—for Drew, this is a wicked fiction of the privileged—and urges the end of elite educational programs designed to give special encouragement to the gifted. Drew's primary motive—the abolition of racial disparities among scientists—is admirable. But his methodology, which dogmatically assumes a perfect equality among humans, contradicts common sense and experience and would impose pointless handicaps on the most promising potential scientists.

Unfortunately, the reach of resentful anti-elitism is not limited to bureaucratic documents and ed-school screeds, which (at least it could be argued) are destined to repose unread and ignored.[13] The same animosity has made inroads into the training of supposed scientific professionals. Recounting the history of "reforms" of medical education in his country, the distinguished Dutch physician Wim Lammers relates:

There is a *numerus clausus* for the medical faculties [of Dutch universities], thus a selection of the candidates has to be made. When the criteria for this selection were to be decided, there was a very active group of anti-science pedagogues in Holland. They seized the opportunity and convinced our (then) egalitarian-oriented government that selection on the basis of intellectual abilities—what the medical faculty wanted—was not fair and would create unequal chances. Furthermore, they reasoned, to be a good doctor "other qualities" than intellectual talent are needed, and those qualities cannot be measured; thus a lottery is the best and fairest procedure. It was then often implied and sometimes explicitly stated that students with talent for science possessed these "other qualities" to a lesser extent than the average student; thus selection on the basis of intellectual talent was wrong. Because the then minister of education in the national government had been previously a professor of pedagogics and was one of the worst specimens of that gang, the medical faculties had not much chance. The best result these faculties could achieve was that the lottery became a weighted lottery-system in which intellectually talented candidates had (slightly) better chances.[14]

As yet, there is no counterpart to Lammer's hair-raising tale in this country, so far as I know. Even in Holland, as he further reports, a new and less doctrinaire government has modified the lottery system so that the most gifted students are exempted from it, although great damage has already been done: "Medical science has lost a number of very promising young scientists by the modish ideas of a few muddle-headed gurus."[15] In the United States, the elitist principles of medical school admissions may be compromised from time to time—for instance, to admit the offspring of wealthy contributors—but such lapses are rare.[16] The same general observation applies to graduate work in engineering and science. But this may be due to the institutional independence of American universities which, unlike those in many European countries, are not under the direct administrative control of government officials. At the bureaucratic level, there are probably more than a few idealists who would dearly love to impose the Dutch lottery model on American schools. Their lack of success is encouraging, but the atmosphere is not without ominous undercurrents.

If, for instance, we look at the National Science Foundation—the patron, as mentioned above, of some of the most science-friendly (and unabashedly elitist) programs of the 1950s and '60s, we find an organization whose commitment to first-rate research has noticeably flagged. I can speak with firsthand knowledge only in regard to mathematics, but the science press (as well as the informal grapevine) bears the grim news that the pattern is more general. A couple of decades ago, gifted and productive researchers found it relatively easy to obtain NSF support for their work. These days, such grants are increasingly rare in many areas of science (mathematics is one), and productivity and morale have begun to decline accordingly. One might remark that the grant program has thereby been rendered even more "elitist," but this will bring joy only to those who are delighted by grim ironies. Even as funding for top-flight research declines, the NSF has initiated a number of new programs explicitly hostile to traditional notions of excellence and eager to supplant them by the canons of political virtue long prevalent among postmodernists.

For example, the feminist pedagogical theorist Sue V. Rosser is currently

Senior Program Officer for Women's Programs at NSF. A sense of her ideas concerning educational reform can be garnered from a symposium she has assembled and edited called *Teaching the Majority*. Its aim is to create new forms of science instruction that will supposedly be more accessible to women, non-whites, and homosexuals. For her, science, in substance as well as organization, is deeply tainted by sexism, racism, and all the evils that throng the demonology of the academic left. Among her emphatic recommendations:

> Include problems that have not been considered worthy of scientific investigation because of the field with which they have traditionally been associated.[17]

> Undertake the investigation of problems with a more holistic, global scope and use interactive methods rather than the more reduced and limited scale problems traditionally considered.[18]

In this context, Rosser also opines that women are "better able to deal with complex problems and ambiguity" while men are "more comfortable dealing with dualisms and problems that have one correct or concrete answer."[19] Aside from resting on extremely dubious psychological speculations, such dogma encodes sexist assumptions about the incapacity of women for systematic logical thought, although, as is common in this kind of "feminist" literature, that ancient prejudice is craftily reworded to celebrate the ostensible value of non-(straight, white)male ways of knowing. Most of the papers in the book share the same assumptions and the same hortatory tone.[20] Like other work in the genre, the book rather fuzzily envisions the transformation of science into something cuddly, warmhearted, and demotic. As one contributor puts it, "It is imperative to ensure inclusive language, provide diverse role models, and offer mentoring opportunities."[21] ("Mentoring" here presumably means a process involving a fairly exact matchup of sex, race, and sexual proclivities between mentor and protegé.) The pedagogy behind such a scheme has little in common with the kind of talent scouting that drove NSF's spectacularly successful efforts of thirty years ago. I don't mean to imply that talent scouting is necessarily bound to sexist, racist, or exclusionary practices, merely that it aims primarily to nurture scientific gifts, not to cure the manifold social ills of Western culture by applying a quaint mélange of postmodernist and gender-feminist nostrums.

A vast conceptual and ideological gulf separates what Rosser and company are up to from the program that is really needed to infuse science with a steady supply of young, eager, able recruits. Expanding such a program to take special account of segments of the population that have been overlooked in the past is probably justified in some degree. But it is a minor part of the task and does not oblige us to reformulate our notions of scientific aptitude.

What needs to be celebrated is not epistemological pluralism, but the emergence of genuine talent, pure and simple, even when that talent happens to be white, male, and Episcopalian! Though I bear no personal animus toward Rosser and rather sympathize with her democratic yearnings, if not her philosophical excursions, the fact that she is in charge of an NSF program (backed, presumably, by NSF funds) while gifted and productive scientists are having their NSF grants pulled out from under them is outrageous. NSF's mission is to keep the natural sciences healthy and thriving. It ought to be prudent about engaging in social engineering of any kind. This is especially true when that engineering is based on the more vaporous and banal varieties of academic theorizing.

Rosser's position with NSF is not an isolated anomaly. It reflects an attitude that has crept into that agency as its mission has broadened to include general, nonspecialist science education. Here, for instance is the "Guiding Principle of NSF's Systemic Initiatives":

All children can and all children must learn rigorous science, mathematics, and technology.
- This guiding principle represents a fundamental change in the way educators and society view children and the way they view science, mathematics, and technology education.
- Misconceptions:
  — Only the brightest and most talented students will succeed in learning science and mathematics, or, put another way, these subjects are too difficult and inaccessible for most to understand. Therefore,
  — Not everyone needs to understand science and mathematics to succeed and to contribute to society.
- This guiding principle, that all children can and all children must learn rigorous science, mathematics, and technology, represents the powerful economic and social mandate of NSF's systematic approach to reform.[22]

These words resound splendidly—but they are wishful nonsense. No doubt they are inspired by noble motives, but taken in anything like a literal sense, they are wildly false. The precepts dismissively labeled "misconceptions," while somewhat too sweeping, are much nearer the truth of the matter. To start with, the idea of "children," in the everyday sense of the term, learning "rigorous science and mathematics" is palpably absurd.[23] In the course of history a few actual children have managed the trick—Pascal, Gauss, and Galois come to mind—but such talent is as rare as that of Mozart or Mendelssohn. Even Newton was unacquainted with rigorous science and mathematics before the age of twenty! The most charitable thing that can be said about this declaration is that it is hyperbole in the service of the idea that "scientific literacy" should be commonplace in the educated population.

I shall visit the problematical idea of scientific literacy shortly, but for the moment, I want to consider what the attitudes embodied in this manifesto imply for the recruitment and training of professional scientists. The most salient feature of the statement is its categorical rejection of the idea that there is any wide disparity among people in their innate ability to master science. A blithe egalitarianism reigns. Any attempt to confront this pipe dream with an "elitist" view of the spectrum of talent would be rejected with pious horror. In contrast to the successful policies of the post-sputnik era, there is no support for the idea that aptitude ought to be singled out and encouraged through enriched educational programs and the tutelage of experts. On the contrary, the notion that talent is universally and equally distributed is fervently endorsed. As in Lewis Carroll's Caucus Race, "All have won and all must have prizes." Or, as Mr. Rogers used to sing to toddlers, "Everybody's fancy, everybody's fine." This is a benign dictum for nursery school, but not likely to be of much use in developing the next generation of quantum field theorists.

In fact, this country has had a pretty good preview of what happens when the doctrine that rigorous scientific learning is for everyone is adopted as the ground of educational practice. A few decades ago, the "New Math" was all the rage among pedagogical theorists. The idea was to ground elementary and secondary school mathematics education in the formalistic, abstract approach that characterizes contemporary research mathematics. Notions from set theory, modern algebra, and symbolic logic were introduced into the curriculum, to the bewilderment of a generation of parents who found themselves unable to comprehend what their fourth-graders' math homework was supposed to be about. Unfortunately, the children themselves were equally bewildered. Even worse, most teachers were little better off. After a few years of unavailing struggle, New Math was written off as a colossal error, and even today, mention of the topic causes retrospective shudders in many people who were run through that particular gauntlet.

However, there is more to the story. I was one of the first students to be exposed to the New Math when it was still in an experimental phase.[24] At that point, the program was limited to markedly talented high-school students. Teachers were drawn from the ranks of those who had done some serious graduate work in mathematics. From my perspective, and that of most of my fellow students, the results were, in fact, superb. We got a valuable head start in the abstract, rigorous language of mathematics, as well as the attitude of mind that goes along with it. We learned the importance of systematization and generalization, and were naturalized into the ethic of deductive formal demonstration. Even within the tiny circle of my teenage acquaintances, the experience produced a significant number of people who went on to careers in advanced mathematics and physics. I would even guess that those who went on to careers in medicine, law, or nonscientific fields benefited substantially from the experience, if only because they learned the value of sustained, precise reasoning.

The lesson is fairly clear, if not particularly comforting to knee-jerk egalitarians. The elitist style of science education is quite wonderful when targeted to an embryonic scientific elite. However, when it is arbitrarily foisted upon young people of modest ability (and their typical instructors), disaster ensues. This is nothing more than common sense might have suggested, but unfortunately it had to be learned through painful and wasteful experience.

Notwithstanding the fine words of the NSF's "Guiding Principle" in respect to "rigor," we are not likely to see a recurrence of the New Math approach in mathematics or any other branch of science education. What is far more likely is a widespread "dumbing down" of teaching practice, at least as regards the most talented and promising students. The more one examines the practical applications of current philosophies as they reach into actual classrooms, the more one realizes that what ensues is a weakening of science education for those young people with real potential for scientific careers. This is hinted at in the rationale that the authors of the "Guiding Principle" offer by way of justification: "This view of learning is reflected in the professional standards of the National Council of Teachers of Mathematics [NCTM], the American Association for the Advancement of Science [AAAS], and the National Research Council of the National Academy of Sciences [NRC]."[25] This is true at some level, but misleading. These endorsements come not from the scientific community, but from educational theorists who have co-opted its prestige. In particular, the NCTM's so-called professional standards, far from being uncontentious, have provoked considerable controversy. In the course of that debate, most of the professional research mathematicians who have looked into the question have become quite critical of the NCTM approach.[26]

Before discussing the NCTM and its foibles, it pays to consider in some generality the philosophical and ideological strands that are drawn together in various programs of science education reform, including that of the NSF. Until one has spent some time navigating these murky waters, the situation can seem contradictory and confusing. What, for instance, does the postmodernist scorn for the idea that science reports truths about the natural world have to do with the rhetoric of the "Guiding Principle," which, whatever its other sins, seems endearingly enthusiastic about the value of scientific knowledge? What does the social constructivist dogma lying at the heart of many of the pronouncements of "science studies" have to do with the blithe, if misguided, assumption that scientific expertise is the heritage of each and every schoolchild? In fact, these elements are correlated in what turn out to be rather pernicious ways. Together, they form the rickety underpinning for ideas about science education which, however wrongheaded, are currently quite fashionable. They have become the basis for educational practice in many schools.

Let me outline a few thematic concepts that pervade the discourse of current attempts to "reform" science teaching:

1. Knowledge of science is "constructed" by the student rather than "learned." The process is not a passive one, but rather engages a repertoire of cognitive processes that go far beyond rote memory.
2. Science, like all knowledge, is "socially constructed." It is determined by the contours of social institutions, the ideology that undergirds these institutions, and the interests they pursue and defend.
3. All "knowledge" is really discursive practice, the verbal and linguistic result of the interaction of available "texts." These include not only conventional books, papers, and the like, but also the habits of description, categorization, and conceptualization that are historically accessible to us. Thus it is an idle metaphysical conceit to think of knowledge as a faithful representation of a reality independent of discourse.
4. Every culture (and subculture) has its own distinctive mode of apprehending the world, its own "way of knowing." There is no overarching master discourse which stands above these. All have equal epistemic legitimacy and dignity.
5. All individuals are comparably gifted in their innate ability to acquire knowledge. Apparent disparities are always the result of inequitable social arrangements and the uneven allotment of privilege.

The first precept is the most valid, and also the most benign.[27] We might well call it "psychological constructivism." It merely acknowledges that the human mind is not a mere receptacle or Lockean tabula rasa. In acquiring a body of knowledge, particularly a highly structured and systematic knowledge structure such as science, we continually engage in an active process of building and revising intellectual categories, creating cognitive repertoires for manipulating and utilizing new concepts, comparing and consolidating our new knowledge with what we knew previously, and readjusting our relationship with the phenomenal world on the basis of what we have just learned. It is reasonably useful as a reminder of the severe limitations of the "memorize and regurgitate" style of teaching, especially in science classes. But by itself, it has very little to say about how science really ought to be taught. However, in schools of education and elsewhere it has often been advanced as a pretext for what is now known as "constructivist" learning. The idea here is basically that an instructor ought not to function as an authority figure who imparts and explains what is "known" to a presumably ignorant body of students. Rather, the duty of the teacher is to create situations and environments where the students, in dialogue with one another, will reason through to the correct general principles by the dialectic of hypothesis and experience. "Instruction" principally consists of throwing out hints and suggestions from time to time. The success of the instructional process is then to be measured by the degree to which the emergent explanatory schemes create consensus, rather than by the degree to which they match the picture offered by standard science. A corollary of this philosophy is that it is not particularly important for a teacher to understand science well. The teacher, after all, is but a coparticipant in the process of constructing "knowledge." (The deemphasis on teacher expertise is probably just as well. As physicist Alan Cromer has pointed out, the training of teachers in constructivist methodology is marvelously suited to creating confusion about the scientific points at issue.)[28]

The second principle, "social" as opposed to "psychological" constructivism, both rationalizes and compounds the delinquencies of "constructivist" science education. As doctrine it is neither valid nor benign.[29] In theory, it is quite distinct from psychological constructivism and is in no sense entailed by it. Yet for influential proponents of constructivist education practice it is an important prop. The underlying reasoning is that "science," like any other knowledge, merely represents the consensus of a given community as to what meets its particular epistemological desiderata. Hence, the "science" constructed by the classroom community as a result of appropriate dialogue and negotiation is fully deserving of that designation. Since all communities are epistemologically autonomous, there is no point in allowing standard textbook science to come in as a *deus ex machina* at the end of the process in order to correct or revise what the students have constructed as a community of "knowers." "One prominent constructivist text," M. R. Matthews notes, "advises seeking harmony between scientific and children's conceptions only up to that point where continued teaching bears adversely upon a 'child's self-esteem and their [*sic*] feeling for what constitutes a sensible explanation.'"[30] The adequacy of an explanatory mode is communally defined, and the "real world" is a red herring, especially when conventional science is offered as an authority on it.

Obviously, this point of view is further reinforced by the third of my listed doctrines, which might simply be called "postmodernism": we cannot really "know" anything; we are merely conscripted into one discursive universe or another, and it is illusory to appeal to a spurious ideal like "objectivity" as an arbiter among such perspectives. Clearly, postmodernism and social constructivism run in tandem. Each appeals to the other in constructing defenses against common ideological enemies. Their mutuality of interest is clear to anyone who keeps an eye on current academic life. The notorious initial draft of the NRC's science education standards explicitly yoked the two as authorities for a constructivist doctrine of science education.[31]

The fourth point of view cited might simply be labeled multiculturalism.[32] In the context of science and science education, it insists that there are perspectives, ways of knowing, that have been—unjustly—excluded from the investigative and expository practices of Western science. These perspectives are those of heretofore despised and reviled communities, especially those of non-white people such as Africans and Native Americans, but also including a supposed "women's community." Social constructivist and postmodernist articles of faith—that no community of knowers enjoys a privileged epistemological position above any other—obviously stokes the confidence of multiculturalism's advocates when they insist that these "Otherly" perspectives be brought into today's science classroom, despite their poor fit with standard science.[33] "Science is a way of knowing and generating reliable knowledge about phenomena. Other cultures have generated reliable knowledge about natural phenomena, therefore reason invites exploration of the possibility that other cultures may have different sciences."[34] This of course involves a rather generous standard in evaluating what constitutes "reliable" knowledge. It reflects the same attitude that led the American Museum of Natural History's curators to affix the label "science" to African divinatory practices. But it is a common attitude among those factions proposing to reform science education in the name of equity and justice for historically oppressed ethnic and racial groups.[35] Sadly, it has led in practice to a good deal of grotesque nonsense, the sort of thing well represented in *Ethnomathematics* or, even worse, the notorious *Portland Baseline Essay in Science*.[36] The latter, bizarre as it is, has been adopted by a number of urban school districts as an official text.

We come, finally, to the fifth principle: absolute and unadorned egalitarianism. This is anti-elitism raised to an absurd pitch. It is, so to speak, a solipsistic elaboration of the doctrine that all communities of knowers are of equal epistemological legitimacy. All individuals are equally competent as well! This is not, however, an endorsement of unbounded individualism. Since this is a doctrine with a left-wing provenance, it holds that this presumed parity of competence must be realized in collective or communal modes of knowing, where consensus and reconciliation of all knowers is the ideal. In practice, what it endorses is "collective learning," in which students work together in groups, building a shared understanding of the scientific or mathematical issues at hand. As a practical matter, this is not necessarily an unsound notion. I believe that students in science and mathematics often *do* get a more thorough understanding of the subject by working together (at least some of the time) on homework and in reviewing the content of textbooks and lectures. At this level, there is no novelty, let alone radicalism, of doctrine. Things become dicier, however, when it is suggested that evaluation, grading, and the like be done on a collective basis. Further, doctrinaire insistence that all students are essentially at par with one another in terms of competence—an attitude certainly reflected in the NSF "Guiding Principle," for instance—precludes the recognition and encouragement of notably gifted students. In fact, it denies that they even exist! This automatically denies them any chance of rapid advancement in scientific learning, since the fictitious assumption of equal competence mandates that the pace of learning will be dictated by the most recalcitrant learners. It is, how-ever, a principle that is enticing to multiculturalists because it promotes a classroom situation where the disparities between differently socialized sub-populations that might show up are made to vanish by doctrinal fiat. To put it more concretely—and bluntly—if the system is rigged to make sure all students come out looking the same, then, by definition, there is no dan-ger that black students will come out looking comparatively bad. Note that these ideas also have the ancillary effect of disparaging the apparent talent of accomplished scientists as nothing out of the ordinary, and certainly un-related to any superior insight into physical reality. This devaluation of tal-ent and authority greatly appeals to the postmodern sensibility:

> If learning is the process of knowledge construction, should not that process provide the most appropriate goal? And who better can evaluate knowledge construction than the constructor? Evaluation from a constructivist perspec-tive should be less of a reinforcement and/or behavior control and more of a self-analysis and metacognitive tool. Constructivist learning is not supposed to mirror reality, but rather to construct meaningful interpretations. Constructivist evaluation is the mirror for reviewing the construction process.[37]

A recent announcement from a science education publication, the *Journal of Research in Science Teaching*, illustrates the way in which postmodern themes are united with an appropriately hortatory political tone:

> In this special issue of *JRST*, the connection between marginalized discourses (i.e., progressive, critical, feminist and poststructural theories) and our under-standing of liberatory and democratic science education for all will be explored through the following questions: What are the implications of marginalized discourses of pedagogy for science education? How does the lens of pedagogy that one employs influence work in science education? How do the multiple approaches and practices that fall under the umbrella of pedagogy influence science education and societal change? The studies presented in this issue should draw from a number of debates concerning schooling and the need for liberatory education, the social construction of science and of identity, and systems of race, class and gender oppression and domination.[38]

Whatever else might be said about such rhetoric, it certainly indicates two things. First, schools of education—even their science education depart-ments—are desperately eager to ape the silly, self-righteous, mannerisms of recent literary "theorists," trendy cultural anthropologists, and so forth. Even worse, the idea that science educators have a special duty to identify and nurture the next generation of professional scientists has fallen by the way-side. Rather, in the view of radical science pedagogy, the highest responsi-bility is to struggle against science inasmuch as science is an "authoritarian form of knowledge."[39]

> Students share with teachers the position of general subordination to the hegemonic knowledge of the natural sciences. . . . Students are also subordinated to the teacher and the teacher represents the authority of the natural sciences. It is from this position of double subordination that students must appropriate (i.e., take over) the knowledge of the natural sciences.[40]

Science education ought, by this doctrine, to consist of "appropriation" of knowledge by students, especially by those socially disadvantaged. "Appropriation," insofar as I can make it out, means accepting the science one finds comfortable, while rejecting what seems hopelessly "alien" and replacing it by myths of one's own devising—a natural enough process if one has no hope of "mirroring reality" but wishes merely to construct a subjectively meaningful "interpretation." Despite its leftist provenance, this seems to conform quite closely to the creationist ideal of how science instruction ought to proceed. Indeed, one prominent science educator, W. W. Cobern, though apparently of the progressive persuasion on most issues, exhibits extraordinary solicitude for the creationist perspective. He invokes constructivist theory as his primary justification. Biology students, he insists, ought not to be expected actually to "believe" evolutionary theory. "Belief," he writes, "is not a science education goal."[41] "Moreover," he adds, "even if it were practical to teach for belief in something like evolution, to do so would be tantamount to proselytization."[42] Constructivist educational theory, steeped in relativism and dedicated to the parity of all belief systems, merely recommends that students be taught to "comprehend" evolutionary notions without crediting them, rather like a pre-postmodern anthropologist studying the belief system of a remote tribe.

It seems a reasonable guess that those who compose and applaud the manifestos just cited would find the following problem appropriate for a high-school algebra class:

> In the year 1637, all the Pequod Indians that survived the slaughter on the Mystic River were either banished from Connecticut or sold into slavery. The square root of twice the number of survivors is equal to 1/10 that number. What was the number?

Problems of this type are characteristic of some of the more politically charged proposals for mathematics "reform"—for instance, those advocated by the book *Ethnomathematics* discussed in the last chapter.[43] Many of the best-publicized efforts to remake mathematics education are similarly drenched in visions of political virtue (if not always in so histrionic a fashion). The so-called NCTM standards are closely associated with these attitudes. Not surprisingly, California, where vociferous contention over educational issues seems to be a fixed element of the political culture, is the center of the storm (though there have been squalls elsewhere).[44] An aim of the reformers was to make mathematics instruction serve progressive causes.[45] Having won an important round through the adoption of new statewide standards in 1992, this faction soon found itself embroiled in a lengthy, contentious slanging match, which often seemed to be as much about cultural and political values as about pedagogy. This is hardly surprising since cultural and political motivations were barely disguised in the rationale for these innovations. The upshot (as of late 1997) is that the opponents seem to have won out. Yet another set of standards has been adopted, this time reflecting a very traditional mathematics curriculum.[46] Nonetheless, it is worth analyzing the rhetoric of the battle. As one of the defenders of the 1992 "progressive" Framework claimed:

> What we have now is nostalgia math. It is the mathematics that we have always had, that is good for the most part for high socio-economic status anglo males. . . . We have a great deal of research that has been done showing that women, for example, and minority groups do not learn the same way. They have the capability of learning, but they don't; the teaching strategies [to] use with them are different from those that we have been able to use in the past with young people. . . . We weren't expected to graduate a lot of people, and those who did graduate and go on to college were the anglo males.[47]

This approach embraced the egalitarian postulate that all people are equally endowed in mathematics. One critic notes, "It [the 1992 'Framework' for mathematics instruction] strongly discourages tracking, counseling that schools should 'Maximize time spent by students in heterogeneous groups. Minimize time spent in tracked or special-ability groups.'"[48] Moreover, in light of the Framework's commitment to expounding mathematics through "problems that illuminate the mathematics side of social issues," the commentator also observed that "many of the changes advocated by the frameworks seem to follow a much broader social agenda than that needed just to improve mathematics education."[49] This view was shared by informed people not committed to a "conservative" position on social questions. In discussing California's "Whole Math" (as it is called) with one such leading critic (math professor Abby Thompson of the University of California at Davis), I once suggested that the program seemed designed to brake the progress of bright kids. Thompson replied that I underestimated its anti-elitist fervor: "I don't think this is quite correct. The idea [espoused by 'Whole' Math advocates] is more precisely that there *are* no smart kids, that all children are equally talented at math, and that any differences that appear are due to societal failure, and probably racism and sexism to boot."[50]

Research mathematicians, well aware of the great disparity of talent that exists even within their extremely selective profession, find this absurd. Nonetheless, it was a passionate article of faith among the reformers. "Collaborative learning" is the order of the day, and beyond the admirable goal of inducing bright students to lend a hand to those not quite as quick, the aim is to disguise disparity of achievement, at least so far as grades and transcripts are concerned: "[I]f all students in a group receive the same grade—a grade based on the achievement of the group as a whole—often more incentive occurs for individuals to try to ensure that all group members understand the group's solution. Thus, another way to grade cooperative problem solving is to give all group members the same score."[51]

From this point of view, the idea that students might be "tracked" according to ability is not only a pedagogical error, but an outright impiety, a vicious scheme presumably inspired by racism, sexism, and imperialist greed. This flies in the face of the fact that tracking is an extremely valuable—possibly indispensable—device for producing a new generation of productive American mathematicians.[52] Perhaps the reformers were aware of how much their cherished doctrines contradict the insights of mathematical professionals. At any rate, some of the reform rhetoric seemed specifically designed to lull misgivings of research mathematicians. One pronouncement described the aim of the new standards as enabling students to

- recognize and apply deductive and inductive reasoning;
- understand and apply reasoning processes, with special attention to spatial reasoning and reasoning with proportions and graphs;
- make and evaluate mathematical conjectures and arguments;
- validate their own thinking;
- appreciate the pervasive use and power of reasoning as a part of mathematics.[53]

On its face, this sounded quite encouraging. It seemed to aim at developing in young people a good part of the repertoire of attitudes and cognitive skills that are required for the study of advanced mathematics and certainly for research work in the field. At worst, it appeared a little overambitious, since the intended beneficiaries were at an age where even bright and interested children will have a hard time living up to such intellectual demands. Looked at naively, this catalog of skills is quite in tune with the frame of mind mathematicians develop on their way to achieving professional competence. However, the truth was that statements like this were part of a bait-and-switch game.[54] What underlay the high-sounding ambition to inculcate mathematical maturity was a desire to justify a radical constructivist philosophy of mathematics education. In its pure form, this means that students are not to be taught mathematical skills in anything like the conventional sense. Rather, they are to be made to confront questions and problems, and to wrestle with them (collectively, of course) in the hope that as a "community" they will succeed in developing a body of mathematical insight—as it were, "constructing" their own mathematical knowledge. The teacher is not to instruct them in the correct way of doing things, nor even in the underlying formal concepts. Rather these must be evolved autonomously by the class as a community of knowers.

Some writers go so far as to suggest that it's not even important that the instructor be well versed in the mathematics to be learned, since the "instructor" is merely a co-creator of the hoped-for knowledge. "Students, are also sometimes urged to discover truths that took humanity many centuries to elucidate, the Pythagorean theorem, for instance," notes one critical study of the situation,[55] which points out that futility and frustration must inevitably result. What, after all, are the chances that an eighth-grade class might be even minimally successful in devising an adequate methodology for dealing with simultaneous linear equations? Undeterred by these obvious objections, California constructivist reformers insisted on trying to put such ideas into practice. As another unfriendly, but accurate, critic of "Whole Math" characterized it:

> California's standards, for example, specify that students be able to do certain calculations in their heads—a clear and reasonable recommendation. The problem is that the standards also tell the teachers how to meet that goal, directing them to stage situations in which students will discover their own "effective strategies" of figuring in their heads.
>
> Student self-discovery is a current fad in math instruction—every student his or her own Archimedes. This "whole math" approach rejects direct instruction of the student by the teacher. The California standards are rife with such self-discovery prescriptions as: "A sense of number and quantity should be fostered naturally as students interact with objects and events in their environment."[56]

It should be obvious that no real-life mathematics class can actually work this way. Even the most fervent reformer must be content in practice with a methodology which at most loosely conforms itself to the radical constructivist ideal. Teachers have to do more than throw out vague hints if students are to get anywhere; standard methods, notations and manipulative procedures must be introduced, however surreptitiously. The problems arose because of what was omitted: constant drill and practice with computational technique, careful and repeated explanations on the part of instructors. Also missing was productive interplay between generalizations and instances, for which the former must be given precise and efficient form (which the intuition of even a very good student will only vaguely suggest, at best). Serious mathematicians, in contrast to constructivist educational theorists, realize that "to understand and apply reasoning processes" and "to make and evaluate mathematical conjectures and arguments" requires mastery of computational and manipulative skills to the point where these are more or less automatic. The flight of the mathematical imagination requires very solid foundations.

For this reason, the great majority of research mathematicians, on becoming aware of the constructivist program, were disturbed or even horrified. For many (like Professor Thompson), this awareness began when their own grade-school children, subjected to the constructivist regime in math classes, came home confused, puzzled, and dispirited. A further reason was the poor quality of many of the reformist textbooks, which sometimes, in their eagerness to be "conceptual," verged on incoherence and mathematical illiteracy. The distinguished University of Wisconsin mathematician Richard Askey provides this example from an Addison-Wesley text called *Focus on Algebra*:

> In the review problems . . . there is the following one:
>
> Explain why $\sqrt{4}$ is rational while $\sqrt{5}$ is irrational.
>
> The answer given in the teacher's edition is:
>
> $\sqrt{4} = 2$, which is rational.
> $\sqrt{5}$, in its decimal form, does not terminate or repeat and therefore cannot be written as an integer over an integer.
>
> I am offering \$100 for a proof of this last claim, i.e., for a proof that $\sqrt{5}$ has a decimal expansion that does not repeat without first showing that it is irrational.
>
> I told this problem to two number theorists at a meeting at Illinois State, and they both laughed. I would too except that this is too serious to laugh.[57]

Let me assure the reader that Askey is not overstating the matter in the slightest. Askey is, in fact, a member of the American Mathematical Society's committee on undergraduate education.[58] This group has raised a number of serious questions about the NCTM Standards, reflecting many of the concerns I have raised here, including the way in which the ostensible reforms slight gifted students, deemphasize formal mathematical reasoning, and redefine teachers as "facilitators" of a vague process of "knowledge construction," rather than instructors responsible for imparting specific content on the basis of their own expertise. On the last point it bluntly notes, "The Standards have led some teachers and leaders of teacher in-service development activities to call for a dramatic reduction to the point of almost total elimination of direct instruction."[59] Wherever the peculiarities of the Standards may have arisen, it is clear that, despite some constructivist propaganda, contemporary research mathematics isn't the source.

The negative reaction to doubtful elements of constructivist-inspired reforms in California mathematics education was so broad and strong that a traditionalist new "Framework" recently swept it away. However, it is far from clear that this will settle the issue, let alone lead to vast improvement in the overall quality of mathematics instruction. On the one hand, the prospects for the traditionalist approach are at best modest. It probably will give the average student a surer grasp on routine kinds of calculation, the sort of thing needed for checkbook balancing or computing the quantity of carpeting needed in a redecorating project. But, unless truly gifted, knowledgeable, and motivated teachers can be recruited, more sensible curricular standards can have only limited positive effects. By themselves, they will not enrich the education of the truly talented. In their own way, conservatives are just as "anti-elitist" as multiculturalists. The concerns of scientists in general and mathematicians in particular are not their concerns. Nor are they particularly inclined to offer talented teachers the kind of incentives that will recruit them into the classroom and retain them there. Moreover, we should bear in mind that many "traditionalists" who rose up against the math reforms would be perfectly happy to see creationism taught as part of public school biology.[60]

On the other hand, since the philosophy motivating the movement is well entrenched in schools of education and in various educational bureaucracies, the "constructivist approach" is not likely to be permanently suppressed by its California misadventure. Fervid support for the NCTM reforms and similar programs lives on.[61] In the final analysis, it does not really flow from a conviction that such ideas promote superior teaching and learning of elementary and high-school mathematics. Rather, hypertrophied political piety lies at the root. Constructivism and its variants offer convenient pretexts for the display of self-perceived political virtue. They make it possible for well-meaning math teachers, and the well-meaning ed-school theorists under whom they study, to think of themselves as activists addressing urgent political and social problems through their educational practices.

Some critics have taken to applying sarcastic nicknames that reflect this aspect of the supposed reforms—"Rain Forest Algebra," for instance.[62] They recognize how eager reformers are to infuse math teaching with a constant stream of political homilies—like the Mystic River Massacre problem quoted above. Here, however, I must admit to having sported a bit with readers' sensibilities, whether they deplore this particular problem as imposing a "politically correct" harangue on the young captive audience of a math classroom or approve of it as injecting a needful reminder of the moral downside of American history. The Pequod Massacre Problem does, in fact, come from a genuine algebra textbook—but not a recent one, nor even one reflecting political attitudes of which current constructivist reformers would be likely to approve! In fact, the example is drawn from a text published in 1857, Daniel Harvey Hill's *Elements of Algebra*. Hill's book was based on lectures and exercises given to students at an antebellum South Carolina military academy. Like many of his other problems, the cited example is intended to drive home a heartfelt political point—but the point is not the homicidal depravity of white Christian European imperialism! Rather, Hill, secessionist "fire-eater" that he was, wanted his students to look upon Yankees as especially greedy, mercenary, two-faced, untrustworthy, rapacious and cruel—in pronounced contrast, of course, to chivalrous slave-owning Southrons.[63] (History buffs will recall that Hill went on to serve as a prominent

Confederate general in the Civil War.) The aim of my little (and, I suppose, rather lame) joke is obvious: as a forum for political indoctrination, science instruction can be made to serve many different creeds. Therefore, we ought to bear in mind the very old lesson that an outward show of virtue may well conceal something rather nasty.

Still, my principal concern is not to fulminate against the larger political aims that may motivate the NCTM reforms and similar efforts. From my point of view, these aims are in a general sense admirable, however objectionable it may be to pursue them through the ideological colonization of science and math instruction. I am not even particularly convinced that such spurious reforms threaten a severe degradation in mathematics instruction, though they assuredly do not help. Even without "political correctness" in math teaching, math education in American primary and secondary schools is dreadfully weak and spotty.[64] The self-righteous silliness of the NCTM's current line ought not to obscure the fact that real reform is desperately needed.[65] And it is most needed in precisely the area that "liberationist" reform is most determined to ignore—the identification, encouragement, and nurturance of mathematical talent.[66] The sad fact is that right-wing populism is just as likely as left-wing populism to dismiss such goals as "elitist."

To this point, I have ignored what remains the most serious, as well as the most obvious, of the threats to effective science education that abide in our culture. This, of course, is the continuing and unremitting opposition of fundamentalist Christians (and some other religious groups as well) to the teaching and learning of evolutionary theory in biology classrooms. Much could be said about it—but little needs to be. The situation is all too familiar. This in itself is a depressing reminder of the staying power of the anti-evolutionist crusade. At best, honest scientists and science educators have achieved a stalemate, one which keeps the creationists out of the classroom, in most instances, but leaves their passions intact and their support undiminished. Their constituency is at least as large as that which endorses evolution, and probably far more dedicated, on average. The situation is not, moreover, stable in itself. Scientists, philosophers, science journalists, and educators continually produce explications of evolutionary theory for a popular readership, not because there is something new to say, but in order to avoid creating the sense that the scientific community is cowed by the perennial creationist onslaught.[67] It is a treadmill situation. Year after year, creationist ideologues refurbish and reword their ancient arguments to create the impression that new evidence or novel theory has just emerged to challenge "Darwinism." Defenders of evolution and of the necessity to teach it are under constant pressure to develop effective responses. The strategy requires not only scientific accuracy but a shrewd sense of polemical cut-and-thrust. One also has to keep up with new wrinkles in the popular conception (or misconception) of science and with whatever fresh terminological chaff (Kuhnian paradigm shifts, chaos theory, or quantum mysticism, for instance) creationists mix into their ongoing snow job. This is unremitting and wearisome work, and whoever finds the time to do it is thereby giving up time that might well be spent doing something creative and intrinsically worthwhile. But it is an obligation laid on scientists and those who love science by the fears and prejudices of a demotic culture that is far from mature and, all in all, not quite sane, at least when it comes to facing up to the discomforts of the scientific worldview.

It is interesting, if dismaying, to note how far the attitudes and doctrines of the postmodern cultural left flatter the interests of creationists and similar groups on the cultural right. Perhaps I have been laying too much stress on the antics of academic leftists, though I suppose this is natural enough for an academic who continually has their dubious example before his eyes. The cultural and political forces that express themselves through creationism and other manifestations of religious traditionalism—the anti-abortion movement, for example—are incomparably more powerful, in a practical sense, than the academic left has been or ever will be. This is a society that, by and large, has little use for leftist ideas in its politics, in the way it thinks about economic issues, or in its approaches to public policy. It is amazing, therefore, that leftist intellectuals, along with a spectrum of their favored values, have obtained refuge and assistance from a number of powerful institutions—universities, academic presses, wealthy charitable foundations, professional associations, and even a few government bureaucracies. The influence of modish theory on such organizations as NSF and NRC, discussed above, is a particularly striking instance, and very difficult to account for. Whether a more traditionally minded left, primarily concerned with wages, hours, employment, trade-union organizing, and concrete checks on the power of private capital, could have found the same welcome as a "left" centered on cultural theory, identity politics, and the vaporous propositions of postmodernism is a question that I, as a traditionally minded leftist, find intriguing. To me, the survival of a community of left-wing thinkers in any form is somewhat gratifying, while at the same time the descent of the left into profound silliness is greatly depressing. But then, I suspect that the obscurantism of the postmodern left has something to do with its persistence in the good graces of the university, although the precise connection eludes me.[68]

The academic left, however, is in a precarious position in the long run. It lacks an authentic social base and, what is worse, it has shown a remarkable capacity for undermining and demoralizing the liberal-humanist consensus while demonstrating little power to generate a new consensus in its place. As an avatar of postmodernist nihilism, the current academic left has considerable corrosive power but little capacity to build. Therefore, in dwelling on the propensity of trendy-left attitudes to work mischief on science education, I have been concentrating on a transient phenomenon rather than on a permanent feature of the intellectual and educational landscape. Even the field of radical science studies, though it is by now well dug in at a number of academic strong points, is likely, I think, to fade into relative insignificance over the next decade or so. Its intellectual flabbiness, combined with its lack of a real-world constituency, will almost certainly do it in.[69] Altogether, the idiosyncrasies of the postmodern left have produced a large body of commentary (mostly inane) about science, as well as occasional challenges to sound science education, and have steered the energies behind much-needed educational reform into wasteful channels. These depredations have been well worth challenging, certainly as an intellectual exercise, and in recognition of the academic's duty to combat nonsense among fellow academics. But the effects on institutional science, and on the interaction between science and the wider public, have been minimal and will be short-lived.

Nonetheless, the hostility of the postmodern left toward science is interesting, not for what it says about "the left" but for what it says, often quite unconsciously, about a much wider culture in which left-wing ideas play hardly any role. In this respect, the left does not even represent its own traditions and perspectives; rather, it acts out some widespread social and cultural fears while dressing them in its own singular and at least faintly comical language. The presence of science hatred within the university community is indicative of how far the general level of anxiety and resentment over science that exists in the wider social world can embed itself in the corpus of intellectual life. The campus left and the religious right seem to inhabit opposite poles in terms of values, tastes, and life-style. Right-wing anti-science and left-wing anti-science use mutually unintelligible sets of passwords and countersigns and fulminate against different perceived evils (although, in the case of genetic engineering, they seem to have found, for once, a common enemy).

Yet there are closer parallels than appear on the surface. Both are deeply distressed by the philosophical monism and the resolute refusal of teleological presuppositions that the scientific tradition relentlessly drives home. In the case of the postmodernists, this fact is disguised by the absolute skepticism and relativism that they proclaim at every opportunity. But relativism is immensely difficult to sustain as a sincere philosophical position. Its much more natural role is as a shock force behind which a far more sincere credulity can move in and colonize the ideological ground once dominated by Enlightenment rationalism. This is precisely what one observes in the real-life maneuvering of postmodern academics who exploit the caustic power of postmodern nihilism to dissolve healthy skepticism, so that they can then propound emotionally gratifying dogma insulated from close interrogation.[70] Without some such intellectual strategy, it is hard for today's left-utopians to convince themselves that nature seconds their treasured assumptions and that history truly intends to vindicate them.[71] It is worth noting that some of the more wily creationists have caught on to this particular trick, which is especially useful in proselytizing college students.[72]

Beyond this, the cultural left and the cultural right are linked by profound anti-elitism. Of course, in both cases such anti-elitism is at cross-purposes with other, contradictory assumptions. The right also venerates a strongly hierarchical and plutocratic social system. The academic left, for its part, not only maintains a rigorous pecking order, where a favored stratum of superstar gurus reigns unchallenged, but often boasts of the obscurity and difficulty of its theoretical diction, accessible, it is supposed, only to initiates. Yet in the end, the left insists on its passionate egalitarianism, which we have seen at work in the various proposals for science education reform examined above.

The right, on the other hand, tends to despise an intellectual elite which, continuing the long Enlightenment tradition, smiles scornfully at its deepest assumptions, especially those grounded in religion. In both cases, scientists constitute a special focus for anti-elitist resentments. Left intellectuals (the postmodern version—not those scientists who happen to be leftists adhering to an older tradition) are put off by a complex intellectual system which is not only indifferent to their political hopes, but which also refuses to yield its secrets to their supposed theoretical sophistications.[73] The right, on the other hand, tends to admire science in its instrumental aspects—it is the source of bombs, missiles, and stealth aircraft, after all—but detests the worldview which a coherent appreciation of science entails. Obviously, this is emphatically true of the picture science offers of the history of the universe in general and the emergence of the human race, via evolutionary processes, in particular. To creationists, evolution is the too-clever intellectual toy of a rarefied elite. "Darwinists" are deaf to the accumulated wisdom of ordinary folk, as embodied in their cherished legends of faith and divine power. Almost always, creationist political efforts embody this emphatically populist tone.

One of the present ironies of campus life—not a particularly delightful one—is that paleoanthropologists who by their very vocation are on the frontline of defenders of evolutionary theory are obliged to consort day after day with some of their enemies. These aren't fundamentalist Bible-thumpers but rather the cultural anthropologists across the hall who have, in many cases, embraced the postmodernist ethos with passionate intensity. This new faith insists that, like all knowledge, evolutionary biology is a culturally specific narrative, "true" only in terms of the assumptions and values of the society that spawned it. Thus it is no more to be honored than Navajo or Maori creation myth. In logic, there is no reason for advocates of this ideology not to extend a similar courtesy to the millions of American creationists who swarm outside the university gates. That they have not done so (at least not very volubly) is not a consequence of their theories, but merely reflects ancient antagonisms between academic culture and the constituency of conservative religion.

In reality, postmodern cultural anthropology is a bare inch away from being an outright ally of the creationist movement.[74] Is it too alarmist to suggest that as postmodern doctrine shifts in coming years, a syncretic alliance might develop between principled cultural relativists and fundamentalist anti-science?[75] The fact that much of the most intense anti-Darwinist sentiment emanates from black fundamentalist churches could well facilitate the process. Indeed, it is not very hard to believe that sooner or later some of the postmodernist heavyweights, reasoning that "Paris is worth a Mass," will go over to the cultural right, where a huge popular constituency awaits, far larger than the thin legions of the left. In the twisted history of ideological quarrels, much stranger things have happened.

Science education in this country is beleaguered by a number of powerful cultural forces. The general culture itself, though without strong ideological passions or theoretical enthusiasms, is poor soil for cultivating a widespread understanding of science. It is, to use blunt words, a lazy, fatally unambitious culture, strongly resistant to demands for sustained and coherent thought on any topic (aside, perhaps, from sports, guns, and automobiles). Beyond this, cultural conservatism strongly influences the national mood, and this, while receptive to technology, is deeply hostile to the frame of mind required for a serious scientific culture. Furthermore, although conservative thought on college campuses is for the most part subdued, the static created by modish cultural leftism works its own kind of mischief. Both leftists and conservatives have, in their different ways, contributed to a slack public educational system, which is particularly feeble when it comes to science. Neither faction is likely to accept serious reform gracefully, the left because of its dogmatic anti-elitism, the right because of its own version of anti-elitism as well as its resistance to public expenditure and its fervent preference for "localism" in all things. All this constitutes a serious impediment to maintaining the educational machinery necessary to restock professional science with a continuous supply of able recruits. When it comes to the ever-elusive goal of achieving what is usually called "scientific literacy" for the general population, it is hard not to conclude that the task is hopeless.[76]

Scientific literacy has been the stated goal of science educators and sympathetic scientists for decades. There is continuing debate about precisely what it entails. How much specific science should be known, in what areas, and with what depth? How far should it aim to empower laypeople to make informed scientific judgments? How much of the overall curriculum should it occupy? Beyond this, there are questions of specific pedagogical methodologies, and of the role of elementary, secondary, and college-level education in the process. Also at issue are the ways science should be tied to professional training, humanistic education, and the study of ethical and societal issues. At this stage, nothing like a consensus has emerged over what scientific literacy should consist of, let alone how to achieve it. A few people who have spent years as scientific literacy activists have come to despair of achieving anything very ambitious. Physicist and science educator Morris H. Shamos, in his book *The Myth of Scientific Literacy*, writes, "The main thrust of this book has been to show that the common conception of scientific literacy as a goal of general education in science cannot possibly be realized, and that if science education is to have any real meaning for this large group of students, its objectives must be sharply redefined and modified."[77] On the other hand, many educators are reluctant to abandon the talismanic power of the term.[78] In their introduction to *Science Matters*, designed as a kind of crash course in scientific literacy, Robert M. Hazen and James Trefil announce: "Scientists and educators have not provided you with the background knowledge you need to cope with the world of the future. The aim of this book is to allow you to acquire that background—to fill in whatever blanks may have been left by your formal education. Our aim, in short, is to give you the information you need to become scientifically literate."[79] Nonetheless, these authors are nearer consensus than might appear from these superficially contradictory pronouncements. For Shamos:

[O]ne objective of general education in science, perhaps the most important one . . . must be to encourage such an appreciative audience, one that at least understands how and why so much needs to be spent on science and technology . . . to keep pace with the developing world. Another objective . . . must be to help the student, and society generally, feel more comfortable with new developments in science and technology. They need not so much to understand the details but to recognize the benefits—and the possible risks.[80]

Contrast Hazen and Trefil:

For us, scientific literacy constitutes the knowledge you need to understand public issues. It is a mix of facts, vocabulary, concepts, history, and philosophy. If you can understand the news of the day as it relates to science, if you can take articles with headlines about genetic engineering and the ozone hole and put them in a meaningful context—in short, if you can treat news about science in the same way you treat everything else that comes over your horizon, then as far as we are concerned, you are scientifically literate.[81]

Both recipes seem to amount to the same thing, whether or not the intended result is labeled "scientific literacy." The aim is not to have members of the general public attuned to science to the point where they can make expert scientific judgments, but rather to equip them to understand who the experts are, why they really are experts (as opposed to mere posturers), and how to seek them out when expertise is necessary. This is a big step down from more grandiose notions of scientific literacy still propounded elsewhere—for instance, the insistence of the NSF's "Guiding Principle" that "all children can and all children must learn rigorous science, mathematics, and technology." The strength of such a modest program is that it comes to terms with the limitations most people have to face, limitations on available time and energy, as well as ability. That, however, is also its political weakness; it recognizes that the gulf between the understanding it tries to foster and the scientific competence of experts will remain enormous. In return for a considerable amount of hard work, it chiefly offers the insight that one must remain a perpetual outsider. It will not be easy to accept that kind of shortfall in a democratic culture that is saturated by a public myth of universal competence, one which posits the equal weight of each person's opinion on every subject.

To be reconciled to the gap between experts and nonexperts requires a certain maturity to begin with. By "maturity," I mean the reasonably ungrudging acceptance of the propositions that science is epistemologically competent and that it is no more ethically deficient than most other human institutions—probably a good deal less so. Without this kind of goodwill and trust as a social background assumption, it is hard to see how a minimal scientific-literacy syllabus could have the desired effect. On the other hand, it seems unlikely that this largely benign view of science can be hoped for in a society that is not already scientifically literate to an appreciable extent. How are we to evade the logic of this vicious circle? Our chances of doing so seem even smaller when we recall that we are surrounded by a popular culture whose images of scientists—from *Alien* to *Jurassic Park* to *Lorenzo's Oil*—typically portray them as myopic, unimaginative moral idiots.[82]

Nonspecialist science education has to pass along a general sense of the content, methods, efficacy, and authority of science. Yet it must do so without evoking the complex of resentments and anxieties that so ominously besiege science in contemporary society. At best this would be a daunting task. In the current climate, where our various social and political mechanisms are pulling in a dozen different directions at once, it may be an impossible one. Some politicians cast everything "scientific" in terms of ridiculously narrow technological goals, others are busy assuaging the tender feelings of creationist fanatics, while still others are fervently stoking the science phobia that the environmental movement has been unwise enough to stir up. Voters periodically hack away at school budgets, while educators adopt nostrums that inappropriately aim to use science education to salve social grievances of all sorts. Universities not only nurture the quaint affectations of postmodernism, but unwisely intrude them into the study of the social aspects of science and technology, and into the philosophy of science education as well. For their part, working scientists, mindful as ever of the pressures of a research career, shuffle their feet in the devout hope that someone else will assume the task of building scientific literacy.

It is hardly surprising, then, that such efforts as have been made to address the problem systematically have been massively bungled or squandered. As we have seen, would-be reformers of primary and secondary science education have, with the blessing of trendy educational theorists, charged into the fray brandishing highly questionable philosophy and weighed down by the doctrinal demands of assorted social activists. At the college level, reform has been even more erratic. Administrators find themselves tempted to allow general education in science to devolve into the hands of "science and technology studies," a province liberally infested with crank theories as well as ideological conceits.[83] It hardly needs stating that the climate currently prevailing in the science studies community is openly hostile to anything resembling the conventional notion of scientific literacy.[84] To these partisans, what the general public should know about science is how wrongheaded and deluded scientists are and how insupportable it is for scientists to claim to know what, in fact, they really do know. For the moment, I can recommend no better antidote to the pretensions of science studies than well-informed scorn. But I cannot, alas, discount the possibility that in many colleges and universities, indoctrination in its credo will come to be regarded as meeting the "science literacy" requirement.

If there is to be any real hope to institute science literacy as a consistent outcome of the education of nonscientists, an enormous amount of sheer wrongheadedness must first be cleared away. This is an Augean undertaking, considering what has already been done to deflect or sabotage the prospect of eventually having a scientifically literate public, and taking into account a political atmosphere that generates unhelpful static on all sides. A blunt fact about the situation is that in order to disperse wrongheadedness, we shall have to clear away the wrongheaded. This is the sort of thing that academic discourse goes to great lengths to avoid saying plainly, but in this case it desperately needs to be said plainly. A substantial fraction of the people who have assumed authority over the philosophy, policy, and practice of science education are thoroughly unfitted for their positions, notwithstanding the prestige of some of the organizations that have endorsed their credentials. This means that the struggle, both for widespread scientific literacy and for an improved system to develop promising students into scientists, physicians, and engineers, will be long, unpleasant, impolite, and personally hurtful to many. I do not envy those who will carry it on, though I wish them all the luck in the world. The last word, then, should go to an educator, Joseph McInerney, who is doing his damndest to make science education into what it should be:

The science classroom, perhaps more than any other site of learning, should provide students the intellectual tools to recognize and challenge fallacious argument and false sentiment. It is here that students should develop and apply the skills of rational inquiry that will enable them to come to their own conclusions about important and complex questions without being seduced either by the weight of uninformed public opinion or by self-proclaimed moral umpires. . . . Our task as science educators is to ensure that discussions of values and ethics in science become models of rational inquiry rather than verbal free-for-alls where uninformed individuals generate more heat than light as they share mutual ignorance.[85]

NINE

Health

The soulful face of Deepak Chopra stares out at me from the cover of *Newsweek* as I begin this chapter.[1] As few people living this side of the Greater Magellenic cloud are unaware, Chopra is a physician (trained in orthodox scientific medicine in his native India) whose American career has been a meteoric ascent to the top ranks of commercial guru-hood. This is a highly selective Olympus. Chopra has qualified by inspiring the public to buy millions of his self-help books and videotaped lectures and to attend his pricey seminars by the tens of thousands. How Chopra ranks in comparison, say, to *Celestine Prophecy* author James Redfield—which one is Bill Gates, so to speak, and which is merely Warren Buffett—isn't precisely clear to me. Yet there can be no doubt that Dr. Chopra is the proprietor and chief beneficiary of an extraordinarily successful operation.[2] The spiritual confection he offers his fellow Americans is a well-seasoned mélange. He carefully blends a dollop of orthodox medical wisdom, a heavy dose of supposedly traditional Indian spirituality and healing (whose authenticity must be taken on faith by most of his devotees), a strong infusion of New Age happy-talk, and a recurrent whiff of what sounds like cutting-edge science. Chopra is vastly fond of the word "quantum," and it gives a distinct coloration to his pitch.[3] In the context of his discourses, "quantum" is chiefly flaunted before audiences who would run in terror from any discussion of self-adjoint operators on a separable complex Hilbert space, though they might be lured back by the suggestion that Hilbert space is a domain of inexpressible spiritual bliss. Such terminology is intended to suggest that what is being offered in the name of health and fulfillment has the condign approval, if not of orthodox science, then at least of the boldest and most vaunting contemporary scientific sages. Chopra, after all, has been featured on the very same public television stations that gave us Carl Sagan's *Cosmos*. With that kind of imprimatur, who could accuse him of mere charlatanry?[4]

I am not particularly interested in the mix of charm, shrewdness, eloquence, and, for all we know, sincerity that has propelled Dr. Chopra to the top of his particular tree. What fascinates me, rather, is how the wide appeal of his message illuminates the tensions, ambiguities, and outright contradictions that permeate cultural attitudes toward medical science—and toward the philosophical and methodological axioms that have made "scientific medicine" an ongoing reality rather than a wishful vision. Contemporary medicine, with its armamentarium of basic biological knowledge, active laboratory science, statistical protocols, and emphasis on the links between research and clinical practice, commands the unambiguous loyalty of a substantial portion of the population. People like this certainly believe that contemporary professional medical practice is based on sound theoretical knowledge, and that it is on the whole highly effective. Moreover, they feel that in terms of both accurate insight and practical reliability, it so far exceeds any alternative claimant that its supremacy is not open to serious question. Antithetically, there is a far smaller, rather inhomogeneous, fraction that, for one reason or another, is wholly disdainful of modern medicine. For these people, the long climb of the healing arts into the realm of science has been an accelerating disaster, as most of them would declare science itself to be. But the largest part of the general population, I should guess, consists of people who are profoundly divided within themselves, to the point of being hopelessly inconsistent in their attitudes toward medicine.

These folk tend to trust the efficacy of modern medicine because they are aware that it works rather well. They are aware that orthodox physicians draw upon a well-tested body of knowledge that has the warrant of a spectacularly successful scientific tradition. At the same time, however, they are unnerved by the starkly reductive view of the human situation that infuses that self-same tradition, a view that makes itself clearly felt in the specifics of clinical practice. They are awed by the sheer splendor of the intellectual machinery that undergirds modern medicine, but abashed by their own helplessness to comprehend it. They are comforted that the terrors of superstition and necromancy have no place in the mindset of medical orthodoxy, yet terrified that the comforts of magical thinking have been thrown aside just as brusquely. They are reassured by the view that the human body is a complicated yet comprehensible mechanism which can be kept sound or restored to health through the systematic application of skills greatly resembling those of engineers and technicians. Yet they are appalled that such a viewpoint sees the moral resources that human beings typically try to draw upon at times of crisis—will, desire, faith, and hope—as irrelevant and, even worse, ridiculous. They are made easier in mind by a regime that minutely scrutinizes the parameters of their physical condition. Nonetheless, they are put off by a system of thought that allows no connection between the microcosm of the human body and the macrocosm of the vast physical—and spiritual—universe. They are impressed by the vast array of exotic substances and instruments employed to diagnose and treat them, yet repelled that these seem to exist in a realm so far apart from ordinary, natural experience. They are mollified by the vast quantity of medical information continuously pouring forth from the media, yet intimidated by their inability to make consistent sense of it. They are smug in the knowledge that, from an actuarial point of view, they are quite likely not to have to look death in the face before their seventies or eighties. Still, they are depressed at the prospect that literal decades of decline and infirmity await them (and their burdened children).[5] They are thankful for having been born into an age when medicine means more than hapless guesswork as likely to kill as to cure, but they are nostalgic, realistically or otherwise, for an age when illness and healing were the concern of a close-knit and supportive community. They are vastly grateful that such a huge and effective array of medical techniques is available, yet choked with outrage at how much the whole damn business costs.

A skillful reader of contemporary human anxieties like Chopra plays on these contradictions with stunning virtuosity. He offers the lingo (if not the substance) of modern science, but deftly relocates it to the realm of supposedly ancient wisdom. He offers advice that seems well grounded in the hard-headed, quotidian good sense of ordinary physicians, yet deftly interweaves exotic elements that suggest that numinous powers are also being conscripted. He blockades and nullifies the relentless monism upon which conventional medicine, like conventional science, is predicated. In its place, he offers a dramatically dualized vision of human nature, where a spiritual body, made of "energy" and "information," stands behind the gross impostures of mere flesh. He speaks with the authoritative confidence of objective truth, yet invites his disciples to believe that the strength of their subjective vision is the key to taking command of reality. He erects a facade of science to shelter nearly unlimited credulity. While satisfying the demands of his followers for a continuing stream of body-comforts, he cleverly insinuates that these are the mere tokens of processes deeply spiritual. He lets his admirers know that they will prosper well in this material world, yet offers them a refuge from the "superstition of materialism." He decries the malaise and decadence of a despiritualized culture, yet he is the very emblem of that culture. Like all purveyors of schemes for achieving health and vitality, he makes longevity a centerpiece of the benefits to be won by deferring to his wisdom. Yet, very cleverly, Chopra disarms the concomitant terrors of aging not simply by holding out the possibility of a quite unlikely continuation of health and vigor, but by inviting his flock to view old age not as decay but as an ethereal process wherein the waxing splendor of the spiritual body will more than compensate for the decline of the physical self. Most shrewdly of all, perhaps, he offers a convenient sliding price scale. One can opt for the full treatment, of course, with visits to clinics and retreats personally supervised by Chopra's deputies. But if that's too much for one's purse, then a few books and videotapes—an investment of dozens of dollars, rather than thousands—will serve perfectly well, if the spirit be willing.

Physicians of conventional views find little to applaud in the spectacle of

Chopra's success (although he has won a handful of them to his cause, for reasons that are grist for interesting, and possibly cynical, conjecture). He is the nightmare of orthodox medicine, brought vigorously to life with all the Barnumesque skill of a master pitchman. To credit his ideas is, ipso facto, to regard standard medicine as one-sided, incomplete and at least partially delusional. For whatever reason—ethics, perhaps, or mere prudence—Chopra doesn't press this point too hard. Sufferers who beseech him to cure serious physical maladies are insistently told that they must undergo conventional treatment as a complement to his more esoteric approach. What is most disturbing about the existence of a Chopra, then, is not necessarily the harm he inflicts on desperately ill people—there may be little, if any, of this sort of thing to worry about. Rather, it is the public distrust and demoralization, in respect to conventional medicine, for which his career stands as evidence and to which it certainly contributes.

Medicine is at an odd pass right now. By any sane historical standard, its powers are vast and rapidly growing. Its promise seems open-ended. Scientific medicine is far superior, in terms of efficacy, reliability, and safety, to any rival claimant or "alternative" practice. This is true in spite of the fact that medical orthodoxy has its own blind spots. It overrelies on some methodologies while neglecting others. It underemphasizes the psychological and personal. It is unable to intervene in the life and family situation of an individual in ways that go beyond immediate treatment of acute conditions. It strains the financial resources of patients to the breaking point. There is an enormous gap between the way it analyzes a patient's condition and the patient's own intuitive understanding. "Alternatives" play to precisely the anxieties engendered by these shortcomings, with results that are at best anodyne in the short term and at worst fatal.

The contrast between contemporary industrial societies in which scientific treatment is both the theoretical standard and the norm of conventional practice, and previous or current societies bound to prescientific notions of healing, is stark and unambiguous. Yet, at least for the past few decades, the prestige and authority of scientific medicine seem to have been waning, sometimes in subtle ways, sometimes in a manner terrifyingly direct. If there can be said to be such a thing as educated opinion, then one must note an overall slippage of the reputation of medical science within that spectrum of attitudes. Science writer Gina Kolata shrewdly notes the attitude that has grown up even within the most educated classes. Writing about the "miracle cures" supposedly obtained by patients with the courage to defy received medical wisdom, she notes:

> The problem, for me, is that too often these stories turn to dust when you examine them. The dying patient never really had the disease, or he had already had treatments that cured him, but didn't realize it and then credited

an irrational, expensive, and possibly dangerous treatment for his recovery. Or the patient died despite his treatment, but his friends and family decided that it had prolonged his life and would have saved him if only he had got it sooner.

A bigger problem is the moral of these stories: Doctors are not to be trusted when they tell you that your condition is incurable or that death is nigh. The onus is on you to go out and get yourself cured. If the doctors say that there is no cure, you can invent one, the way the parents did in the movie *Lorenzo's Oil*. And if scientists say your treatment doesn't work, you simply tell them that they're wrong. Scientific studies that test a hypothesis are worth nothing. All that matters is the story of the lone patient who believes he is helped, or cured.[6]

In recent years, vaccination of children against a long list of what were once dangerous, even deadly, endemic diseases has also come under increasing challenge. "With certainties fading, it is easy for people who feel medically vulnerable to build seductive hints and fragments into a coherent, if warped, belief system," notes one alarmed observer of the phenomenon.[7]

Quacks and nostrums we have always had with us. To be fair about it, although medical learning has been an integral part of the system of officially sanctioned Western knowledge since the founding of the medieval universities, for most of that long period there has been little to choose between medical practice predicated on approved doctrine and the various competing systems that have continually sprung up. Three centuries ago, "respectable" medicine and quackery were hardly distinguishable, a fact well attested in literature and in the stock comic figure of the physician that inhabits so many plays and operas. But what has evolved since, thanks to the firm alliance of medicine with both scientific practice and a scientific worldview, is a mutant entity of amazing strength and power which has so vastly outdistanced its historical precedents that to call both of them "medicine" is perhaps to mangle the language.

Asserting this, I doubtless run afoul of a new breed of historical relativists, for whom the story of modern medicine as a triumph of science and rigor over ignorance and prejudice reeks of "present mindedness," cultural arrogance, and the suspect practice of weighing all other systems of knowledge and practice in a balance just recently forged by the quirks of our own. It is, however, the quirks of historical relativists, rather than the posited idiosyncrasies of our culture, that worry me in this context.

In respect to medicine, as to scientific progress in general, I am an adamant and unrepentant Whig. What our physicians and medical researchers have achieved over the past century or two is objectively a huge triumph over disease, pain, and the ravages of age. It is anything but a shared cultural delusion. Moreover, terrifying as it may seem to a hothouse generation of supposed humanist intellectuals, the achievements of medical science *do* put its experts in a position to scrutinize, evaluate, and, quite often, condemn the practices of other cultures and other ages. The abashed modesty that faddish cultural relativism seeks to impose on us is not merely unwarranted on intellectual grounds. It is a species of grave moral delinquency. Our doctors not only "know different" about how the body works, and how its illnesses and injuries may be treated, they "know better." To deny this is to open the door to all kinds of avoidable suffering, in this society and elsewhere. The crotchetiness of those who insist on such denial is interesting—but only as a symptom and a minor instance of a more general estrangement from scientific medicine.

This estrangement certainly exists, and it is ominous as well as perplexing.[8] As I have noted, it always existed to some extent, even in the days when there was little true science to be estranged from. If, however, we narrow our focus to the last thirty or forty years in the industrialized societies of the modern world, we see that alienation has grown, deepened, and extended its tentacles into social and intellectual strata where it formerly would have found little welcome. A growing segment of the population—certainly, of the educated population—resorts to crank ideas about health and illness. Much of the impetus for these movements comes from a middle- and upper-class constituency.

This is the downside of what is generally regarded as a virtue: the self-confidence and moral independence of those schooled to believe in their own autonomy and intellectual competence. It can have the unfortunate, sometimes tragic, effect of inducing people to overvalue their ability to make judgments and decisions that require considerable technical training as well as a wide base of experience and knowledge. But deference to even the most well-warranted authority has become perilously dilute in a culture whose celebration of individualism is unconstrained. By contrast, the act of taking all decisions into one's own hands, of living according to one's own custom-built structure of theories and prejudices, is almost always seen as heroic.

Health insurance companies, for reasons which, one must suppose, combine gullibility and stinginess, are now willing to cover the services of all sorts of "alternative" practitioners.[9] Organized groups of nurses insist on their right to practice "therapeutic touch." Prestigious hospitals allow shamans into operating rooms so that they may join their self-proclaimed spiritual powers to the merely physical skills of surgeons. Most amazingly, the heart of the health-research establishment, the National Institutes of Health, has been conscripted (via the so-called Office of Alternative Medicine) into the fight to legitimate a range of practices that not only depart from but flagrantly contradict the assumptions of standard scientific practice.[10] As Leon Jaroff, a science journalist of wide experience, summarizes:

> Dr. Joseph Jacobs, OAM's first director, was friendly toward alternative medicine, but resigned under pressure after two years, complaining that [Sen. Tom] Harkin and his allies were demanding quick, scientifically unfounded valida-

tion of alternative therapies. Indeed, Harkin has had his way. Guided by an advisory council loaded with some of his favorite alternative medicine gurus, the OAM has repeatedly granted funds for patently ludicrous "investigations."[11]

The list goes on. Official panels bend over backwards to find some good in acupuncture.[12] The National Institute of Mental Health dispatches researchers to India to study the Ayurvedic theory of bodily humors.[13] Homeopathy, the "science" of pretending that nothing is something, flourishes. Its phantom elixirs can be found on the shelves of most neighborhood pharmacies. There has, in short, been a grave weakening in what used to be one of the primary default assumptions of our culture: that conventionally trained physicians, in general, tend to know what they're talking about and that claims from other classes of would-be healers, whether practitioners of folk medicine or proponents of unconventional new theories, ought to be regarded with considerable skepticism when they fly in the face of the consensus of the medical community.[14] Indeed, the medical community itself is more than a little shaky in defense of the science that supposedly underlies its own practices.[15] Commenting on the hostile reaction to an article of his,[16] Roy M. Poses, a specialist in analysis of clinical research, remarks: "Most distressing . . . was the influence of a certain kind of post-modern thinking on publications about qualitative methods in respected medical and health care journals. These contained arguments that there is no external reality, and scoffed at scientists' attempts to be objective as possible as futile and foolish."[17]

The Laodicean attitude of physicians in defense of their own scientific principles was embarrassingly highlighted by the recent publication in the *Journal of the American Medical Association* of what was essentially a fourth-grade science-fair project. Emily Rosa, the young heroine of the episode, demolished the central claims of "therapeutic touch" with a simple, elegant, and unambiguous experiment.[18] But more important than the (unsurprising) result itself is the poor light it casts on medical schools, nursing schools, and hospitals, which could easily have commissioned such studies themselves— and ought to have done so—before allowing advocates of the practice free run of their classrooms and clinics. But nonsense is not so easily exiled. One uneasy medical school professor, put on the spot by Ms. Rosa, acknowledged that there is no scientific evidence for the efficacy of therapeutic touch, but went on to declare that it could not be removed from the nursing curriculum because it was protected by academic freedom.[19] One is tempted to ask if "academic freedom" would also prevail if prospective nurses were being taught to bleed their patients after the fashion of eighteenth-century barber-surgeons![20]

The door has been flung wide open to virtually any would-be "healer," who has merely to claim the authority of an ancient or freshly contrived theory, while observing the few modest precautions imposed by an indulgent legal code, in order to set up as a purveyor of health and well-being. Given the spectrum of available nostrums, Chopra, for all his exoticism, seems a model of caution and modesty. Cranks and quacks of all descriptions are out there in force, and, unlike Chopra, most of them are unrestrained in their denunciation of standard medicine.

Forty or fifty years ago, there was much to complain of in the system of health care available to most Americans. Orthodox physicians largely monopolized what law and social custom recognized as legitimate medical practice.[21] In many parts of the country, local medical societies maintained veto power over who could and could not legitimately practice medicine in their bailiwicks. It was widely (and doubtless to some extent correctly) assumed by skeptics that such groups were not so much concerned about the quality of medical care as about the incomes of their members. It was far from uncommon for medical societies to blackball renegade physicians who had the temerity to charge insufficient fees. An old-boys network greatly influenced the system of medical education, keeping the number of new physicians below the point where overenthusiastic competition might begin to degrade the prosperity of individual doctors. The guild had strong hereditary elements as well, with sons (and, rarely, daughters) of physicians given the inside track as far as admission to medical school was concerned. The hellish conditions under which new M.D.s were obliged to work as interns, though rationalized as necessary to inure them to the moral and physical rigors of medical practice, constituted a kind of hazing ritual, bonding novices to the ethos of their clan.[22] Successful malpractice suits were exceedingly rare, largely because the reigning ethic condemned out of hand any physician who offered testimony against another. The American Medical Association, then as now the chief representative of the medical community, was widely regarded as a fortress, not only of selfishness, but of right-wing politics. In some progressive circles, it was as despised as the American Legion, if not more so. For decades, it opposed, consistently and effectively, any governmental attempt to widen access to medical care, and even private insurance schemes that defied the sanctity of the "fee for service" system.

Without nostalgia, it is nonetheless possible to see in this vanished age some real virtues which have since gone under. Most physicians, even in urban areas, were personally far closer to the communities they served than is now the case. The term "house call" had not yet dropped out of the vocabulary and, hard as this may be for those under fifty to believe, the ill were attended to in their own homes at their own bedsides. Whatever monopolistic medical societies may have done to inflate physicians' incomes, they at least had the desirable effect of moderating quackery, if not suppressing it entirely. Moreover, however mercenary the AMA and its allies might have seemed at the time, they have been far surpassed in that respect by the recent upsurge of "managed care" schemes, among other things.[23]

These days, the social position as well as the internal structure of "conventional medicine" varies enormously from what, in 1950, seemed the unalterable order of things. Most strikingly, the lone "private practitioner," serving a fixed community over the course of decades, seems a creature headed for extinction. Such solo enterprises have been supplanted by "professional associations" and "group practices," which, whatever their legal and economic rationale, have had the subjective effect of making medical services into an apparently fungible item, as easily to be had from one "provider" as from another. This tendency has been greatly enhanced by the proliferation of comprehensive health-coverage schemes. Though they vary greatly in specific design and in provisions for consumer choice, overall they encourage the propensity to see medical care as an impersonally produced commodity rather than a human interchange between people who know each other. Moreover, if medical treatment has been, in this sense, commodified, physicians themselves, much against their intention and very much to their horror, have been increasingly proletarianized.[24] Although there were glaring flaws in a system that delegated sovereign power to medical associations and the like, those associations were at least fully controlled by practicing physicians who had some direct concern, however skewed, with doctor-patient relationships. Much of their authority now resides de facto with for-profit health-care firms. The ethics of these combines and their sense of how the world should work come from the business community.[25]

Any attentive newspaper reader will be aware of how this has led to rationing of health services, based on ground rules that often override the medical judgment of individual physicians. Inevitably, this kind of profit-driven parsimony has kept physicians in line by offering them financial incentives to undertreat, and, even worse, by censoring and punishing them for putting patient interests above sharp-pencil economics. This forced conversion of doctors from entrepreneurs (of a very special type) to employees has generated great resentment and massive panic. At this moment, hospitals all over the country are scrambling frantically to merge with one another in order to obtain the economic heft needed to fight the dominance of the "managed care" companies, and to preserve some measure of autonomy for their medical staffs and associates. At this point, it is far from clear how the issue will be resolved. The health-insurance empires may become permanently dominant, or the aroused medical community may rally sufficiently to supplant them with its own pet institutions. Or perhaps a hybrid modus vivendi will emerge. In any case, the general public has been reduced to a largely passive position in determining the outcome. One is reminded of the adage that whether elephants fight or make love, the grass gets trampled.

Even beyond issues of raw economics, the privileges and immunities of physicians have effectively declined in other ways. By the 1970s, the aura of authority and the tacit codes of silence that had spared most M.D.s from the threat of malpractice suits had pretty much fallen by the wayside. A countersystem of legal specialists and medical "expert witnesses" grew up and often succeeded very handsomely in extracting large judgments from juries grown increasingly cynical about doctors and hospitals. Malpractice insurance rates skyrocketed, to the point that some physicians simply curtailed or abandoned their careers in frustration. More generally, "defensive medicine" became the watchword. One effect was to multiply greatly the diagnostic and preventive procedures routinely prescribed, notwithstanding the expense of the medical technology employed. As a consequence, medical costs soared. Even worse, in some sense, doctor-patient relationships began to take on a frankly adversarial cast. Even the kindliest doctors could not help but eye patients as potential sources of financial ruin, and even the most grateful patients took to furtively calculating their potential awards in the event of treatment gone awry.

Simultaneously, the ability of organized medicine to regulate and suppress its more or less heterodox competition went into decline. In part, this was due to a turn in judicial thinking that resulted in taking individual rights and procedural guarantees—particularly as they applied to "alternative" practitioners—much more seriously than had been the custom. The writ of medical societies in these matters lost much of its sovereign power. Likewise, the FDA, along with various state and local authorities, found it more and more difficult to procure and enforce judgments against even extreme forms of counterscientific "medicine." In part, this was due to the increasingly pugnacious and savvy litigiousness that had entered our culture. These days, any complainant against an alleged quack is likely to find the tables quickly turned. Quacks have lawyers. These lawyers are skilled not only in the defensive task of protecting the right of quackery to continue unmolested, but in converting complaints against such quackery, via libel and defamation suits, into further sources of income for quack and lawyer alike. Moreover, adherents of various pseudomedical cults have developed formidable lobbying skills, and often find it possible to recruit powerful politicians to help them keep orthodox fuddy-duddies off their backs.

It has become maddeningly difficult for the FDA to get the nostrums of "alternative medicine" off the market, even when these are positively dangerous as well as ineffectual. Diabolical legal loopholes—for instance, the device of labeling a substance as a simple "food supplement"—allow pseudophysicians to avoid scientific oversight of their favorite remedies. Because of the current craze for megadoses of vitamins and minerals, "Consumers are, in effect, volunteering for a vast, largely unregulated experiment with substances that may be helpful, harmful or simply ineffective," avers Jane Brody, the health columnist of the *New York Times*.[26] This poses an ironic contrast with the travails of standard drug companies, who constantly,

and with some justice, complain of the obstacle course of trial and evaluation which they must run to get official approval for new medicines.[27]

A further irony lies in the rhetorical tactics of many adherents of
antiscientific health fads. Often they profess to revere science, though it is
science in the heroically Galilean mode—the myth of a clear-sighted visionary defying fusty orthodoxy—that they have in mind. One of the quaint,
but grim, offshoots of the celebrity of Thomas Kuhn's *The Structure of Scientific Revolutions* is the proliferation of situations in which the notion of "paradigm shift" is offered up in defense of some half-baked claim, especially one
that frankly contradicts reliable scientific knowledge. It hardly matters that
the claimant, on such occasions, has no clear idea of what Kuhn intended
by the term.[28] The Galileo-Kuhn defense has become a standard ploy when
nonstandard medical practice is at issue. It remains quite effective in shield-
ing even the most egregious quacks.

One must also recognize that in subtle ways, science itself has undermined
scientific medicine, or at least the self-confidence and sense of authority of
those who hope to practice it. To an ever-increasing degree, the ordinary
physician these days finds himself beholden to scientific work whose methods and principles are not really within his competence to understand. Consider, by way of example, the CAT scan devices that have revolutionized
radiology by making it possible for doctors to view accurate cross sections
of the human body, or even three-dimensional displays synthesized from
these. It's a safe bet that only a handful of M.D.s has even a basic understanding of how a CAT scan works.[29] Even more humiliating, perhaps, medical ignoramuses like me, a lowly mathematician, *do* understand the principle
on which it works, but are frustrated when we try to explain.[30] This is one
instance where the science involved is particularly abstruse, but there are
thousands of others where a practicing clinician finds himself dependent
on research, ranging from lab work in molecular biology to the statistical
sophistications of epidemiology, whose details are impenetrable.

Just like the rest of us, the typical doctor is surrounded by devices which
must be regarded on faith as useful black boxes, since their theoretical underpinnings are unfathomable. In terms of actual power to diagnose and
comprehend, the modern practitioner is infinitely the superior of any physician of a century ago, so much so that the comparison is laughable. But in
subjective terms, contemporary doctors are far less the masters of their tools
than their nineteenth-century forebears, because the basis of a typical piece
of current medical technology is opaque to all but a handful of specialists.
At the same time, physicians often find themselves confronted by patients
who have made an intelligent attempt to absorb much of the current knowledge relevant to their conditions. Popularizations abound. They are, by definition, accessible and many of them are accurate. Moreover, even the primary
medical literature contains many articles that are within the competence of

well-educated people to grasp, at least in general terms. Confronted with a serious medical problem, many of them turn to that literature to bolster their judgment and enhance their morale.[31] Consequently, physicians frequently find themselves obliged to explain and justify their actions to patients armed with real knowledge.

In sum, contemporary medicine exists in a world where doctors are squeezed, as it were, between the imperatives of a technology that seems in many ways mystifying and the scrutiny of patients who resist being mystified. As one physician recently put it, "We are now in the age of scientific, hi-tech medicine. Machines do not make mistakes, spares can be bought for worn-out hearts and doctors have to contend with better informed patients, who cannot be fobbed off with evasions."[32] Ultimately, all of these changes in the matrix of knowledge and authority within which medicine is practiced must have some psychological effects on doctors, mostly in the direction of curtailing any tendency toward feelings of omniscience. Doubtless, much of this is healthy. On the other hand, the conviction of omniscience may occasionally be good thing in a physician. In any case, an overdose of uncertainty and tentativeness can certainly work harm. When considering the ability and even the inclination of scientific medicine to defend its (quite proper) authority, we must reckon that even subtle demoralization imposes inevitable costs.

It has to be recognized that the great majority of doctors practice scientific medicine without really being scientists. This is not to denigrate their intelligence or competence, but simply to point out that despite our automatic tendency to think of M.D.s as part of "the scientific community," they must function quite differently from research scientists if they are to do their job. Obviously, there is a cohort of trained physicians, often having Ph.D.s as well, who devote most, or all, of their time to scientific research. But these form a tiny minority in a profession made up principally of clinicians whose contact with ongoing research is intermittent or even nonexistent. The most obvious fact about clinicians is that their primary function is immediate and specific—to lessen suffering and to forestall death. The acquisition of knowledge, even immediately usable knowledge, is only a secondary commitment, if that. The speculative element of scientific thought is simply foreign to the duties of a healer who does not have the luxury of false starts and dead ends.[33]

The notion of professional acculturation is crucial here. If physicians are not formally trained in something we might wish to call the "scientific viewpoint," neither are apprentice scientists. At the level of degree programs, neither is tutored in the history or philosophy of science, or much exposed to abstract analysis of methodological principles. Neither profession leaves much time for this kind of reflection at an early stage of the typical career. The interesting thing about science, however, is the enormous osmotic pressure to absorb the essential criteria for scientific validity it puts on its practitioners, independent of whether they have a taste for philosophy. The epistemology of a working scientist, and, indeed, of the scientific community, is not principally a formal or explicit creed. Rather, it is a set of standards of practice and judgment that may well become internalized notwithstanding that it has never been explicitly stated. One may repeatedly appeal to these standards in discourse and debate without ever having been catechized in them. This is not to say that such standards are "social conventions," much less that they are arbitrary leaps of faith. They can be and frequently are articulated, both by philosophers and by scientists of philosophic bent. The point, however, is that over the generations, these standards have become embedded in the practice of science for the very good Darwinian reason that they are indispensable to the business of garnering accurate and exhaustive knowledge of the natural world. For the scientific community to ignore or discard them would be, in effect, to betray its entire purpose.

More to the point, when the individual scientist turns his back on them, he ceases, in the most crucial way, to be a scientist, since he is no longer capable of pleading for his ideas in acceptable logic, nor of effectively criticizing rival ideas. One's license to practice science cannot be revoked for the simple reason that there is no such thing in the first place. Anyone at any time can decide to call himself a "scientist," without penalty. But it is impossible to maintain anything that might reasonably be called a scientific career without accepting, at least in broad outline, the standards of evidence and inference that govern the field. This is what young scientists come to learn, usually without pain or embarrassment.

In medicine, the situation is different in a number of respects. One important point is that a physician is primarily a freelancer, who is kept in business (in many cases at least) by the loyalty of patients, rather than the approbation of peers. One may become an utter pariah in the medical community without the least penalty to one's legal status or one's income. These days (as those few physicians who try actively to fight quackery have found to their sorrow), it is nearly impossible to deprive a doctor of his license to practice for anything so problematical as mere scientific heterodoxy. To fly by the seat of one's pants, espousing or devising nostrums as one chooses, without regard to scientific validity, is an open option. The stock of patients who can be recruited by the force of personality or the lure of hope is essentially limitless. Fortunately, native intelligence and minimal ethics keep the vast majority of physicians from straying into this path. Nonetheless, a number of habits which we may call "subscientific," if not actually antiscientific, can still creep into medical practice. There is, for instance, the tendency to rely on anecdotal evidence and hunches, to remember selectively, to bias observations in favor of a pet theory. These are precisely the flaws that scientific methodology is designed to guard against, but they are inevitable when formal safeguards are absent. True, a working research scientist may also have a predilection for anecdotes, hunches, and pet theories. But, absent flagrant dishonesty, much of this is filtered out by the mechanisms that scientific procedure keeps in place. In the case of the ordinary clinician, however, there is little incentive to submit favored ideas to formal trial. Moreover, even a physician who is doing nothing amiss is often in the position of having to rely on protocols, treatments, and drugs that are beyond his understanding, whether because the science is beyond him or because the actual mechanisms are simply not understood very well. Medical usefulness is not precisely the same thing as rigorously established scientific validity, and for good or ill, most physicians are sensible of this fact.

It may be fruitful to compare physicians to engineers (and, for that matter, computer experts). Both are, in a similar sense, "subscientists," people who consider themselves as allies and colleagues of research scientists, who continually work with and apply a corpus of scientific knowledge, and who even modify or extend their repertoire as new science becomes available. Neither, however, routinely submits his actions and decisions to the formal procedures of scientific testing and peer scrutiny.[34] Both often depend on rules of thumb whose scientific warrant is unclear. The difference is that bad engineers will soon be checked by the results of their poor judgments through the inoperability, unreliability, or high cost of what they create. (Physical reality is a stern taskmaster!) By contrast, bad doctors—such is the way of the world—can often rely on the gullibility, suggestibility, fear, and desperation of their clients to get themselves off the hook. And, of course, the old adage still holds as ever: doctors get to bury their mistakes.

The point, finally, is that, strongly acculturated as physicians are to regard themselves as dependents and supporters of science, in the real world such an alliance has its limits. A typical doctor, even one who regularly participates in clinical research protocols, has restricted contact with systematic scientific thinking. Since so many research conclusions depend on essentially mathematical ideas—the principles of statistical and probabilistic inference—and since even the best-trained physicians tend to have only a modest mathematical education, physicians end up taking many of these conclusions on faith. They stand in the same position, relative to scientific specialists, as most laypersons, relative to doctors—appreciative, but somewhat mystified and perhaps a bit skeptical. Aspiring M.D.s come out of college with a degree in science and take a number of basic science courses in medical school. But physicians typically do not undergo the methodological acculturation that forges a practicing scientist. In the most urgent sense, medical doctors are concerned with effective application of a technique, and the only referee of their success is whether that technique betters the condition of a patient. Whatever knowledge or insight is gained in the process, valuable as it may turn out to be, is incidental.

It is not totally surprising, then, that a substantial (though not huge) number of practicing physicians wander from the orbit of scientific practice, whether in good faith or out of motives wholly mercenary. Very few "alternative" health theories, no matter how extravagant, fail to number some M.D.s among their supporters. Indeed, some of the ripest, most hallucinatory nonsense in this field comes out of the mouths of licensed physicians. Medicine is a field where "peer review" only constrains those willing to be constrained by it. Neither law nor pressure from patients requires a doctor to embrace only those theories that have decent scientific warrant. Medicine is thus an epistemological free-for-all to some extent and would be a much worse one but for the fact that, when all is said and done, the selection and training of medical students imposes the habit of taking science seriously in most cases.

Laypeople, for their part, have little inclination to examine the epistemology behind medical practice except under conditions that are not very conducive to good judgment. The average person is confronted with the question of deciding which theory, which diagnosis, which course of treatment might be most appropriate only when dire illness or injury threatens—distressing and disorienting circumstances, to say the least. Any doubts about the efficacy of orthodox medicine or the honesty of orthodox physicians are likely to be inflamed, rather than soothed, under such conditions. Likewise, any inclination to magical thinking and outright superstition that might be lurking in the emotional underbrush is likely to flare forth. Confronted with a prognosis that suggests that confinement, pain, and disfigurement are inevitable, a patient or relative is hardly receptive to a sober analysis of physiology and epidemiology, let alone to sermons on the philosophical grounds for accepting the superiority of science.

Given that the stock-in-trade of so many quacks and charlatans is their willingness to come up with "alternative" treatments that are less harrowing, in the short run at least, than orthodoxy demands (and are possibly less expensive), it pays to ask why so many people nonetheless remain faithful to conventional medicine, at least when life and health are at serious risk. The answer certainly does not involve anything like a deep respect for science or an understanding of its principles. Most people know science only superficially. When it comes to questions of methodology and verification, they know virtually nothing. If the rigorous use of statistics is somewhat off-putting to the average physician, it is utterly baffling and frustrating to the vast majority of laypersons. The only lesson they are likely to absorb from statistical presentations is that nothing is all that certain and that, far from having access to infallible guidelines, a physician, in many cases, is merely making an informed guess. Under such circumstances, the term "scientific medicine" resonates in people's minds mostly to suggest impersonality, technocratic aloofness, and an endless regime of uncomfortable and frightening procedures whose real worth is unclear. There are few more despairing phrases in the English language than "I'm going in for tests."

Still, most people stick by orthodox medicine when life is at risk or health is seriously on the line. This includes people who have flirted long and ardently with alternative modes of health care. Clearly, few who behave in this utterly conventional way could articulate convincing reasons. Rather, they are responding to a set of conventional evaluations and expectations that permeates the culture in a general sense.[35] These are precisely the myths that only conventional medicine is truly competent when the threat is dire, and that only authentic science has the authority to direct medical judgment at a fundamental level. In calling these postulates "myths," I am not in the least challenging or debunking them; they are *true* myths. I merely call attention to the way they function in determining the behavior, under certain circumstances, of the vast majority of people in this society. They are myths in that they invoke a shared, but largely unexamined, notion of how the world works. They represent the "common knowledge" to which people often resort when their individual judgments are not up to the task of answering vital questions. Even the most willfully perverse intellectuals, those who scorn the authority of Western culture in general and Western science in particular, yield to such socially sanctified doctrines when the fear of death is upon them.[36]

That such myths promulgate truths is indeed comforting. But it is terrifying to contemplate the fact that the predominance of such truths in the body of our cultural commonplaces is a contingent circumstance, dependent on a unique skein of historical chances. Certainly, it does not flow from the prevalence of reasoned discourse and dispassionate analysis in society as a whole. Consequently, the dominance of scientific medicine as our health "paradigm" of choice is fragile, and, indeed, already fraying. Of the physicians about to embark for India for the study of Ayurvedic theories, a reporter remarks, "Any medical tradition that is 2500 years old, they say, and has half a billion enthusiastic clients must be doing something right."[37] But to say so is to ignore the iron grasp of human folly and to misprize the best hope we have thus far evolved for escaping that folly. There is nothing in the record of the human species that suggests that wrongheadedness is self-correcting, or that the capacity for rationalization cannot sustain an illusion for 2500 years and more. Nor, sadly, is there anything in that record to suggest that scientific rationalism, in medicine or anything else, is self-sustaining and thus eternally embedded in the culture that developed it. This has little to do with the actual philosophical strength of the scientific outlook, which is enormous. Rather, it is to acknowledge that some law of intellectual entropy may be a given of the human condition. The present erosion of the prestige of scientific medicine, though still minimal and equivocal, may foreshadow a far more extensive decay.

Paradoxically enough, one of the villains may be the huge flood of medical news that ceaselessly pours forth from the media. As Marcia Angell, the editor of the *New England Journal of Medicine*, astutely points out in her book *Science on Trial*, it takes a considerable degree of sophistication to comprehend the medical research literature accurately. One must have a well-developed sense of how research really works. One must understand how hard it is to design, let alone carry through, a meaningful protocol, how vulnerable studies are to undetected biases and spurious correlations, and above all how vital it is to reserve judgment in the face of a single study, or even a pack of them. As she points out:

> Evaluating medical research is no easy matter. At the *New England Journal of Medicine*, we have seven full time physician-editors, six part time physician-specialists, three statistical consultants, and one consultant on molecular medicine to do it, and we call on thousands of outside peer reviewers who are experts in the subjects under study. Even with all this expertise, we are often not confident of the validity of a study, and we sometimes make mistakes in judgment that are not corrected until later.[38]

The lay public, with very little in the way of expertise to guide it, finds itself confronted with dozens of studies—or rather, with journalistic summaries of these, which are inevitably truncated and most often distorted into the bargain. Some of these are peer-reviewed studies, some merely claims put forth by ambitious researchers, but all of them have passed through a filter of journalistic judgment that selects for their ability to grab attention. Sober explanation is not usually part of the package. As Angell notes: "[T]he United States has a large scientific enterprise and its research findings are often widely publicized. But interest in scientific conclusions is not the same thing as interest in how they originated. The nature of evidence is simply not a front-burner item for most Americans."[39]

Even worse, from the point of view of public confidence, there are frequent cases where a succession of studies reaches the public's attention, each contradicting the previous one. Oat bran and fish oil have their day in the spotlight as dietary miracle workers, only to fade from view as further research moderates or abolishes excessive claims. Grave menaces appear, only to have their fangs pulled as statistics are amplified and clarified. To the expert, this is part of the normal course of scientific work. The apparent contradictions are evidence of little more than the effort and persistence needed to cut through the messiness and clutter of real life to reach sustainable conclusions. To the general public, however, these seemingly endless rounds of claim and counterclaim are thoroughly baffling and undercut respect for science as such, largely because science is falsely idealized as a direct high road to certainty. The result, unfortunately, is to heighten a tendency toward intellectual atavism as regards medical and health issues. With shaken confidence in the consistency and legitimacy of what they take to be scientific research, people revert to ancient and erroneous habits of thought—*post hoc ergo propter hoc* reasoning, reliance on anecdotal evidence, false analogy, and simple rumor chasing.

The tendency toward neglect of scientific standards only grows stronger when the issue appears as a tale of trusting innocence betrayed by technocratic guile and establishment arrogance. The long-standing and still-unsettled breast implant controversy illustrates this perfectly. Here, the underlying story was that tens of thousands of women, having placed their faith in assurances from doctors and manufacturers that silicone is an essentially harmless substance, underwent cosmetic breast enhancement with silicone implants. Only a few years later, so the story goes, they were beset by mysterious ailments presumably provoked by an allergic reaction to the stuff. A number of mythic archetypes were evoked by this narrative, primarily the callousness of greedy megacorporations willing to poison the multitudes for the sake of mere profit. This was reinforced by a concurrent feminist morality tale: women, desperate to conform to the impossible ideal body type demanded by patriarchal fantasy, had submitted themselves to a dangerous procedure whose evident risks the medical establishment chose to ignore or minimize.

Beneath this, another, more ancient myth is operative. People are seen to be suffering; suffering comes of malice; the agent of suffering must be found out and forced to make amends! When lawsuits based on claimed injury from silicone implants have gone to a jury, the result has always been favorable to the plaintiff, sometimes to the extent of huge monetary awards. Angell's book is principally concerned with the silicone controversy. She makes clear that the claims of the women purportedly harmed by implants have been thoroughly, even irretrievably, refuted so far as the scientific community is concerned. But the real lesson is how far the public is willing to let sympathy for the ill and reflexive assumptions about establishment wrongdoing overcome the great weight of scientific evidence. "The breast-implant controversy shows the proclivity of the public—unable to tolerate uncertainty, unwilling to make minimal efforts to evaluate scientific stories, and reinforced by the media—to embrace uncritically the medical scare story of the day."[40]

It hardly needs to be pointed out that, in instances like this, the initial prejudices of jurors are easily able to control the outcome because the typical juror is not only ignorant of the particular issues of chemistry and biology relevant to the claims at issue, but equally at sea when it comes to appreciating the importance of statistical and probabilistic inference. Of course, plaintiffs' attorneys, as a matter of course, make every effort to exclude scientifically literate veniremen from jury panels. But, given the utter scarcity of the scientifically literate, and, even more, the statistically literate,[41] their task is hardly a difficult one.

Medical and health issues constitute the arena in which awareness and appreciation of science and its strategies are most urgently needed to insure personal and familial well-being. Ignorance concerning evolution, or the significance of quantum theory, or the likelihood that aliens make interstellar voyages in order to kidnap us, is a childish deficiency and perhaps even a blameworthy one, but hardly a direct threat to those so afflicted. Susceptibility to quackery and other medical delusions is a far graver matter. Yet the very emotions put in train by serious health crises are potent antagonists to the frame of mind necessary to approach things from a scientific point of view, especially if such an attitude is not habitual and is not based on close familiarity with science. We would like to think that human beings are genuinely educable, that sloppy or biased habits of thought are fully corrigible, that with a little goodwill and hard work any interested layman can equip himself to follow and participate in all discussions that concern his well-being. This, however, represents a leap of faith which sorts ill with the evidence available to date. Pain and danger, even when merely contemplated rather than directly sensed, seem to call forth aspects of our mental proclivities that work at cross-purposes to logical and systematic thought. Doubtless, there are good evolutionary reasons for what seems, in the context of modern society, a gross dysfunction. This does little, however, to disarm the danger. That danger is even greater because, throughout history, such instincts, if we may call them that, are not merely individual tics, but deeply embedded cultural suppositions.

The alliance of science and medicine has succeeded, to a modest extent, in neutralizing these assumptions, in abolishing them from official discourse, in training a stratum of the population to treat them with suspicion and contempt. But a strong undertow is always present, tending to erode this achievement which is, after all, hardly more than a few centuries old, and by no means universally prevalent in a culturally diverse world. It is not part of conventional medical training to make a doctor into an advocate for the empiricist philosophy underlying contemporary medicine or a missionary on behalf of the epistemological superiority of modern science. Yet, as we have come increasingly to see, insofar as such ideas have failed to take hold, society is in darkness and, worse, in danger. Amid all the contention and confusion about medical insurance, managed care, and loss of professional autonomy, physicians, as individuals or through their organizations, have hardly had the leisure to ponder such issues. If they are able to find time to address public attitudes toward medicine, then their first priority, surely, is a crash program to repair the distrust sown by the cold-bloodedness of modern medical practice. Its compartmentalization and its tendency to reduce treatment to a sequence of impersonal mechanistic procedures, rather than an engagement with the human situation of a patient in all its aspects, have played hell with the trust between physician and patient. Alienation at this level must surely be allayed before the medical establishment, if we must call it that, can begin to deal with such abstruse issues as the public understanding of methodology and its philosophical underpinnings. But inevitably these seemingly recondite matters will make themselves felt in the way in which society deals with sickness, injury, and mental affliction. The funny thing about philosophical stances is that people have them whether they think they do or not. In the case of attitudes toward medicine, the philosophy that reigns in the minds of most people, notwithstanding muddiness and inconsistency, is in many ways antagonistic to the scientific way of approaching things. The danger to the legitimate hegemony of orthodox medicine may come in the form of a slow leak, a gradual erosion of the societal premise that physicians conventionally trained and credentialed are entitled to a special measure of deference. It might even build to the point of portending a serious overall collapse of "medicine" into methodological anarchy, hard as that may be for us to imagine. It is anyone's guess whether scientific medicine will be able to respond adequately to whatever threat emerges.

TEN

Law

A good part of the authority of science rests on the assertion that science is "public" in its character. Theories and findings are openly published and debated, and are liable to critique, as well as confirmation, by any competent investigator. The openness of the scientific literature is one of the most sacred tenets of the profession and has been defended with great success against bureaucracies, governments, and corporations which, for one reason or another, have tried to impose secrecy or to forbid publication. Yet, paradoxically, the public nature of science is a rather private affair. Few people outside of the profession have ever seen a scientist actively at work. College students, of course, often meet working scientists as teachers in their science courses, but teaching, especially in the case of nonspecialist courses, is far removed from the practice of scientific research. Occasionally, a television documentary features scientists at work—more often than not naturalists in picturesque settings—but these shows have a limited appeal and, by the nature of the medium, tend to put the emphasis on the drama and human interest that can be found in the situation. The dry discourse of science is carefully rationed.

The one real exception to this situation, the single venue where the lay public gets to scrutinize scientists as they display their expertise, where they watch as scientists expound at length on their methods and reasoning and are forced to defend their conclusions against skepticism, is the courtroom. The use of scientific experts in cases of all sorts has grown apace over the past few decades. Civil matters, as well as criminal, increasingly bring scientists onto the witness stand. This tendency is amplified, moreover, as courts are increasingly called upon to adjudicate matters that arise directly from the workings of science and technology. The strong and growing role of science in all aspects of our economic and political life is inevitably mirrored in the number of lawsuits growing out of professional scientific activity, including medicine and engineering.

It would be cheering to report that the performance of contemporary science in this arena has helped convince the public of its integrity and intellectual splendor. Unfortunately, that would be pretty much to reverse the truth of the matter. The typical layman could be little blamed if the spectacle of science in the courtroom, as we have seen it in recent years, has convinced him that scientists are dogmatic, pigheaded, nitpicking, arrogant, closed-minded, myopic, deaf to common sense, envious, spiteful, and—worst of all—mercenary and hopelessly corrupt. In a number of notorious recent cases, the public has looked on with growing cynicism as the opposing legal teams have bludgeoned each other with competing, antagonistic "experts"—on DNA evidence, on "recovered memory" evidence, on infantile brain trauma, on autoimmune disease, and much else. Expert scientific opinion has come to seem as much a commodity, crafted to the order of those who can afford it, as the eloquence of the lawyers who put the experts through their paces. The props that accompany these performances—charts, photos, videotapes, and so forth—seem to testify not so much to the authority of the science on exhibit as to the theatricality of the entire show.

All this has been amplified, sociologically, by the recent practice of putting television cameras in the courtroom. Every few months, it seems, one court case or another seizes the attention of the public so strongly that the televised proceedings become fiercely competitive with the other delights of daytime TV. Complainants, defendants, lawyers, jurists, witnesses, and even jurors are promoted to instant celebrity as the sweet incense of prospective book-deal cash enlivens the performances of all concerned. Assorted characters who, a year previously, were nothing more than workaday courtroom hacks find themselves better known, better paid, and more widely courted for talk-show appearances than sitcom stars. Most of the live courtroom broadcasts, as well as the now-numerous programs that greedily follow cases current or pending, are hosted by attorneys and rely on other attorneys for expert commentary. It is hardly surprising, then, that legal skills and lawerly prestidigitation get heavy stress. In practice, this means that when an expert witness appears, the main emphasis of the media is not on the reliability of what he has to say, as judged by independent experts with appropriate scientific or technical credentials. Rather, the focus is on the virtuosity of the opposing lawyers, their deftness in making the testimony dramatic and compelling or, alternatively, belittling it and minimizing its effects. The most intense praise goes to the cross-examining advocate who, like a graceful torero, succeeds in tormenting and confusing the scientific heavyweight whom the opposition has put on the stand, befuddling him and inflicting upon his evidence wounds numerous and deep enough so that it will bleed to death before the jury's eyes. Sometimes, logic and evidence are the weapons that do the trick. All too often, however, it is a matter of rhetoric, showmanship, bamboozlement, and blue smoke. If truth is well served by such proceedings, this is a matter of mere accident. It hardly inheres in the process itself.

One can hardly visit this territory without making some reference to the

O. J. Simpson murder case, which hurled itself in the public's face just as the media machinery that was to make it the most notorious legal matter in the nation's recent history came into its own.[1] I shall not revisit this noisome travesty, except to note that despite the endless displays of criminalistics and forensic technology, especially the hours of claim and counterclaim concerning DNA traces,[2] very few of the hundreds of millions of people who followed the trial on television came away from the experience having learned much about the science involved.

Even more distressing (from my parochial mathematician's point of view) was the fact that the basic principles of statistical inference underlying the interpretation of DNA evidence were opaque to all concerned except the witnesses themselves.[3] The lawyers (even the supposed experts), the judge, the dozens of commentators (attorneys and journalists alike), and certainly the woozy public—all seemed utterly ignorant as to what statistical independence (for instance) might mean,[4] and what it might have to do with the probability estimates associated with this or that sample. None of the ponderous hundreds of hours of interpretation deigned to offer a quick primer on the elementary mathematics involved.

Needless to say, all the other scientific issues encountered the same combination of neglect and evasion. No one in the media seemed to care about the basic science. Nowhere in the special TV coverage did we ever see a panel of scientific experts explaining what scientific issues were in dispute or expounding the scientific consensus (or lack thereof) on those issues. Rather, the emphasis was all on the legal legerdemain employed to burnish or disparage the scientific testimony. The host of commenting attorneys seemed to believe, at least in public, that there was no scientific truth apart from what the rhetorical and obfuscatory powers of legal craftsmanship could create.

Attitudes like this persist far beyond the Simpson affair. Forensic science continues to emerge in other well-publicized cases, but when it does, the attitude of lawyers, loyal as they are to the code of their guild, seems to be that it is a lawyer's plaything. There is nothing wrong, from this standpoint, with putting nonsense on the stand in the guise of expertise if the resulting muddle helps one's client. By the same token, the only proper thing to do with damaging scientific testimony is to go after it with stiletto and meat ax, no matter how impeccable it might be, as viewed by competent scientists. Science—or pretend-science—is like any other witness. It is to be flattered or crucified as circumstances dictate. The advocate is under no ethical obligation to defer to the consensus, no matter how firm, of the scientific community.

A philosophical excursion into the role of law might be in order here. There are, I think, three principles that are in perpetual contention within our legal system: truth, justice, and order. It is not so much that these potent ideals are necessarily mutually inconsistent in the abstract. Rather, the problem is an institutional one, arising from the ways in which our society tends to operationalize these notions and to translate their claims into concrete policy. Our societal consensus idiosyncratically distorts these ideas and frequently formulates real-life questions so as to put them at odds with one another. The courts play host to especially intense conflict since they are charged in fiercely unconditional terms with the duty of serving all three ends. This results in irreconcilable tensions that make our judicial system awkward, inconsistent, and often, to all appearances, outrageously foolish.

Nominally, and in the rhetoric of attorneys and jurists, our court proceedings have the primary duty of "seeking truth." In practice, however, no sooner does one hear a lawyer utter these words than the suspicion arises that truth is in for a rough day. The quotidian practice of our courts makes it clear that it is often not only permissible but mandatory for our courts to turn away from the pursuit of truth, and even to exclude demonstrable and highly relevant truth from their considerations.[5] Our criminal trials, for instance, are constricted by a web of exclusionary principles that frequently disbar direct evidence, not out of any reasonable concern for its accuracy, but on purely procedural grounds. Here, the implicit rationale is justice, for society at large if not for the individuals immediately involved. The principle being vindicated is that certain practices, usually on the part of police or prosecutors, are inconsistent with the standards of a just society, as stated or implied by the Constitution, or at least by its recent judicial interpreters. The reigning theory is that accepting evidence acquired by tainted methods, even on a case-by-case basis, would so far erode our civic life in the long run that the only remedy is to turn a blind eye in every instance to the truths to which such evidence attests. Thus, justice, in the abstract, trumps the right of truth to determine the outcome of particular cases. (A cynic might be tempted to speculate that the Byzantine complexity of our exclusionary rules is sustained not only by a modern-day concern for civil liberties, but equally through the influence of the defense bar, whose arsenal of tricks is thereby greatly extended.)

Truth can also take a beating in tort cases, again with the rationale that justice is better served if truth is compromised. Hundreds of juries have given substantial awards to plaintiffs in the full knowledge that the weight of evidence and law favored the defense. Here the logic has been that the suffering of the victim, whatever the real cause, is manifestly unjust. Therefore—given the deep pockets of the defendant or his insurance company—why not use the legal process to repair the injustice, even if truth, in the immediate sense, is scorned? Juries frequently produce "compromise" verdicts as a way of squirming out of the hard choice between "guilty" and "innocent." They are a way of saying, "We think the defendant did it, but we're not sure."

Thus they are absurd, on any literal reading of the jury's duty with respect to the facts. Yet attorneys, judges, and the public easily accept that justice is thereby done.

However one views these practices from the point of view of justice, their existence makes clear that neither courts nor juries consistently adhere to what might be called a scientific epistemology. As Norman Augustine puts it, "Most juries seem to give about as much weight to the opinions of astrologers as to astronomers."[6] Accuracy of verdicts, in terms of the objective truth of the matter in hand, often has to take a back seat to other considerations. There is something medieval about the assumptions embodied in these practices. They reflect a world in which systematic pursuit of truth, if it existed at all, had yet to acquire the methodological and philosophical insights that grew up concurrently with the rise of the natural sciences. They echo a frame of mind that looks to omen and to ritual to arrive at acceptable "truth." The mentality of prescientific societies survives here.

Even justice and truth can be put out of countenance by the practical requirement that our court system must impose a measure of order on the turmoil of contemporary life. Order can't be defined in any absolute sense. It refers both to the objective fact of a society which responds obediently to the decisions of its supposedly legitimate authorities and the subjective belief of citizens that the machinery for keeping civil peace is not wholly ineffectual. It emphasizes efficiency and closure, often at the expense of accuracy and justice. It is difficult for our system to rectify even flagrant abuses of fact once they have been sanctified as the outcome of trial proceedings. Appeals against criminal convictions are enormously hard to win, even with the help of overwhelming exculpatory evidence. It is far easier to sway appellate courts through allegations of procedural error than through citation of material fact. Acquittals, according to our sacred axioms, are utterly beyond appeal, and thus forever immune to contravening truth. Given the adversarial nature of our justice system, the subversion of truth is a positive duty that advocates of one—or both—parties to a conflict owe their client. This applies even to prosecutors who, theoretically, are sworn to serve strict and impartial justice, but who, given the realities of politics, often take the lead in drawing a heavy veil over the facts. Under such circumstances, justice is as much a victim as truth. The requirements of order impose an unstated but powerful doctrine of expediency on our court systems. This is attested to by the now-ineradicable plea bargaining that dominates our criminal proceedings. This practice demonstrates that proper procedural justice, case by case, is far beyond our social means.

A New York reporter's tribute to the late Judge Harold Rothwax, one of the rare jurists who let his impatience with the absurdities of the system spill over into public view, recalls Rothwax's weary cynicism: "Trials had stopped being searches for truth, he argued, and had instead disintegrated into gamesmanship between rival lawyers throwing dust at every turn in the eyes of judges and jurors alike. . . . He had decided that trial lawyers . . . abuse the system to select the dumbest jurors possible, people who are gullible and easily manipulated."[7] We can't build enough prisons, nor fund enough courts, nor provide enough jurors for the necessary trials to subject most criminal acts to the retribution that law and fact demand. The unspoken consensus is that it is best to deliver exacting and exemplary justice only in cases where the alleged acts are unusually vicious, or where notoriety demands that the actual behavior of the justice system be consistent with its nominal obligations. To demand much more of the system, to insist on genuinely proportionate punishments in many more cases, would be to risk paralysis and collapse, and thus a massive degradation of public order overall. The tort system shares many of the weaknesses of the criminal. Here, too, the volume of cases flooding the courts creates countervailing pressures for informal out-of-court resolution. Sometimes, the result reasonably approximates equity, but often truth is buried or justice is otherwise short-circuited. The kind of case where a company pays damages to a presumably injured individual, but insists on silence and suppression of the evidence as part of the deal, is an instance of this.

It is not my purpose, nor am I particularly qualified, to offer a vision for the general reform of the justice system in this country.[8] It is nonetheless clear to me that the practices, as well as the principles, of that system are insulated from criticism by a mantle of sanctity (which exists in both "liberal" and "conservative" versions), and that in reevaluating these practices, we might do well to emulate the resolute empiricism of scientific work. In particular, we must learn to accept the painful distinction between "is" and "ought" that is the wellspring of the epistemological genius of the modern era.

Science and scientific expertise have become more and more important to our legal system in recent years, a role that is unlikely to diminish in the foreseeable future. Thus, the poor fit between scientific epistemology and the justice system's often awkward, sometimes bizarre version of the "search for truth" will become increasingly significant and increasingly deleterious. Truth, justice, and order all stand hostage to the inability of our legal machinery to take proper account of what science knows and how it works. Countless errors flow from the inability of jurists, jurors, and attorneys to make good sense of the scientific evidence that comes before them. The neglect of scientific fact, as well as the habit of allowing unsupported speculation to disguise itself as science, has poisoned the outcome of numerous cases, sometimes with severe injury to the most basic notions of justice. Worst of all, the inability of courts to establish and police meaningful standards for scientific testimony threatens our ability to make sober and reasonable public choices about the deployment of technology. The untamed capriciousness of the legal system has the potential at any time to undermine even the best-informed, most thoroughly disinterested decisions in this area.

Instances of these mischances abound. One needn't reach for the O. J. Simpson case to unearth examples of deaf or purblind juries. An ancient and undispelled prejudice still predisposes many people to accept "eyewitness" testimony over even the most scrupulously authentic forensic science. This certainly was a factor in the reluctance of Texas governor George Bush Jr. to free convicted "rapists" when DNA evidence subsequently cleared them completely.[9] The victim's adamant certitude about her "identification" of the accused carried great emotional weight, and the mere scientific certainty of innocence was hard pressed to overcome it.[10]

On the other hand, pseudoscience has had a field day in other criminal cases. Courts and juries have been extremely gullible about "expert" testimony in connection with the supposed recovered-memory syndrome. Defendants, particularly in cases where sexual abuse of children has been alleged, have been pilloried by the heads-I-win, tails-you-lose logic of the "psychologists" put on the stand by prosecutors. If toddlers confirm a story of abuse, they must be believed, no matter how many obvious absurdities they allege, no matter how much coercive coaching has gone into shaping their testimony. On the other hand, if a supposed victim denies having been abused, this must be interpreted as hysterical suppression of the memory, and hence as concrete proof that the accused is guilty. Judicial indulgence of such haywire reasoning, justified by the naive equation of trumped-up "expert" credentials with scientific competence, has often led to disaster. Many innocent people have languished behind bars for years because of the naive enthusiasm of prosecutors and the irresponsibility of trial judges in putting this "spectral evidence" before juries.[11]

Such wide-eyed credulity is not limited to rubes and yokels. A few years ago essayist Janet Malcolm published a notorious book intended to raise doubts about the justness of a famous murder verdict.[12] Malcolm insisted, among other things, that the concrete scientific evidence developed in forensic laboratories was virtually worthless and ought to have been disdained by the jury. On the other hand, she claimed that a Freudian analyst's testimony judging the accused temperamentally incapable of the crime not only ought to have been admitted, but should have carried enormous exculpatory weight. People like Malcolm, a leading intellectual on the New York scene, do not usually come to mind when we speak of unsophisticated thinking and rampant scientific ignorance. But in this instance, the label seems to fit. As we shall see, there are other intellectuals whose hostility to the courtroom authority of science is comparably strong—and for these, the derogation of that authority is not a side issue, but the central theme of their life's work.

The willingness of courts to disregard accurate science and to indulge the crank version is not just a matter of an occasional miscarriage of justice. Large issues of public policy have turned upon it. Marcia Angell's thorough and compelling account of the silicone breast implant controversy reveals how much money pseudoscience, conjoined with outright greed, was able to extort from the legal process.[13] As Angell convincingly demonstrates, and as subsequent evidence continues to confirm, the supposed epidemic of immunological disorders afflicting women who underwent cosmetic implant surgery was a figment foisted on the legal system. The chief culprits were rapacious attorneys, assisted by a few physicians who were either deluded or dishonest. There is virtually no hope now that legitimate epidemiology will at last have a strong voice in the final disposition of the issue.[14] So much has already been conceded to the plaintiffs that the debacle is irreversible. Women who have suffered from illnesses (but not on account of silicone) or merely from the fear of illness (but not justifiably) will walk away with substantial awards. The attorneys who have organized them into a powerful political force will become even wealthier than they already are.

The possibility of similar legal grotesquerie looms in connection with the current hysteria over "electromagnetic radiation" and its supposedly dire effects on human health.[15] Most of the fuss, as I noted in an earlier chapter, has arisen from claims that the very low frequency EMR generated by power lines, transformers, and household wiring can induce cancer.[16] Physicians and other scientists not only responded rapidly, but with a remarkable lack of dogmatism (given that the theory seemed quite unlikely a priori). Numerous epidemiological and laboratory studies of low-frequency radiation were initiated, carried through, published, carefully scrutinized and compared. At this point, the data seem conclusive: low-frequency EMR has no pathogenic effects on adults or on children.[17] For once, the scientists seem to have achieved a head start over the tort lawyers. Nonetheless, given the litigiousness endemic to contemporary culture, it would be the height of premature optimism to think that reasonable scientific certainty will remain immune to lawyerly ingenuity. Remember that a substantial lobby already exists to maintain the alarm over low-frequency radiation and related allegations concerning cellular phone relay antennas and the like.[18] Jurors—and jurists—are prone to endless quirks. Human sympathy for suffering is widespread, scientific discernment is not. There is thus some hope, but nothing like a guarantee, that science will prevail should lawsuits alleging great harm from low-frequency radiation make it into court.

How, then, should courts and the legal system act to make their processes consistent, or at least not wildly inconsistent, with the best scientific judgment? It is possible to subsume this question within the larger one of wholesale reform of the justice system, both criminal and civil. The role of the jury, for instance, has been increasingly questioned by some scholars. Might it not be a good idea to curtail the absolute authority of juries as "triers of fact" if not to abolish them altogether? This will seem sacrilegious to most Americans, who think of the jury as having been divinely decreed along with the Decalogue. It will be hardly less offensive to most trial lawyers, who have specially shaped their skills to manipulate juries. But, historically speaking, our version of a jury system is the product of a long series of historical contingencies, only some of which involved rational debate and design. We are left with a system that, far from appearing natural and enviable to most of our fellow-democracies, seems peculiar and idiosyncratic. Most countries dispense with civilian juries or use them only in conjunction with assessors who are professional jurists. Certainly, many of our most recent courtroom lunacies would have been avoided under the kind of system that prevails outside the Anglo-Saxon tradition. It is reasonable to assume that scientific evidence, in particular, would be more soberly and accurately weighed by a system that was not so committed to giving lay juries the final word.

It is, however, a pipe dream to aim for serious modification of the jury system per se, whether or not it is a good idea in principle. In practical terms, we are committed to something very like the current jury model for the foreseeable future, if only because it is built into the provisions of the Constitution itself. Rules of evidence and other aspects of court procedure seem to be more malleable. It is through these channels that we might hope for the kinds of reforms that might make our courts and, more generally, the quasi-judicial proceedings of regulatory agencies better able to make intelligent use of scientific findings. In practical terms, the most we might hope for in the short run is a shift in the underlying assumptions that judges bring to cases in which science, forensic or otherwise, plays an important role. This might take the form of more rigorous scrutiny of supposed "expert" testimony (and of the supposed experts who are to proffer it) before such theories are put in front of juries. Judges might take it upon themselves to be far less passive, ruling out lines of argument that offend the common sense of the intelligent layman. This could generally be justified by the principles put forth by the Supreme Court's *Daubert* decision, which endowed judges with the power and responsibility to scrutinize proffered "expert" evidence for scientific plausibility before allowing it to be heard by juries. Although it contained ambiguities, *Daubert* was widely viewed as a promising beginning to the campaign to keep "junk science" out of the courtroom. Recent rulings suggest that a strong interpretation of the *Daubert* principle is likely to survive the inevitable legal challenges from disgruntled litigants whose pet just-so stories are barred.[19] Pretty clearly, *Daubert* invites judges to acquire a modest expertise in the canons of scientific judgment, and to apply that knowledge to questions of admissibility. If judges respond by becoming less easygoing in the face of improbable theories, appellate courts will probably see things their way.

Such attitudinal changes, however, answer only part of the challenge. A daydreaming scientist might envision a more systematic, firmly institutionalized methodology. It should not be left to an industrious judge to take on the ad hoc burden of becoming familiar with whatever branch of science becomes relevant to a case. Such learning will usually be no more than impressionistic, and will often turn out to be inadequate or misleading. In a much more reasonable world than this one, judges would be able to obtain scientific information and advice from forums instituted for that specific purpose. Associate Supreme Court Justice Stephen G. Breyer has suggested as much in his concurrence to a recent decision that stiffened the *Daubert* principle.[20] Even more, he has begun to recruit the support of the scientific community for such proposals, declaring that, "we must search for law that recognizes scientific validity."[21] My personal preference would be for an arrangement where experts, even legitimate ones, would simply not appear before juries for direct testimony, let alone cross-examination. Attorneys for both sides would be obliged, in advance of trial, to offer their scientific claims before qualified tribunals, allowing their experts to testify and be challenged there. Such tribunals would be made up both of experienced jurists and qualified scientists (working pro bono, perhaps). They would have the power to recruit further personnel as circumstances required. They would hear out the hypotheses mooted by the contending sides, and invite comment and criticism from the respective opponents. Even more, they would have the authority to do their own questioning and to call their own experts. Having scrutinized the presumptive evidence and listened to the various lines of argument from proponents and skeptics, they would then formulate advisory opinions for the guidance of the trial judge.

This would not necessarily be a definitive vindication of one side or the other. In many cases, one assumes, even the most exacting scientific evaluations will leave ambiguities and unresolved questions. It would be the duty of the tribunal not only to stipulate what will have been proved to a reasonable standard of scientific certainty, but also to sketch out the range and degree of reasonable uncertainty. If competing hypotheses were both viable, it would take note of that fact too. The trial judge would then be obliged to prepare and deliver an advisory to the jury at an appropriate point in the trial itself. This instrument might well be modified by such omissions and elaborations as the judge thought proper, but it would strictly adhere to the tribunal's conclusions on technical matters.

Clearly, this scheme challenges the notion of the jury as "trier of fact" in a significant way, but only insofar as it dispenses with the palpable fiction that a random selection of laymen can appropriately evaluate highly technical matters. To the degree that it removes a bewildering array of ancillary matters from their purview, it frees them to direct attention to the sorts of questions that jurors, in the eyes of defenders of the current system, are best suited to evaluate. It certainly defers—unapologetically—to the idea that mainstream science ought to be accorded special epistemological privileges. But to do so is no more than to concede what the past four hundred years of human history has made ineluctably obvious. In some theoretical sense, it might give more social "power" to scientists as a class. But in practical terms, it is much more likely merely to burden them with tedious and (from the point of view of researchers) largely uninteresting work. The hardest questions to deal with would probably arise in connection with the softest sciences—psychology and the like. This is an area in which some of the most notorious swearing contests have arisen, and in which the most outlandish ideas have been advanced without a blush. I suspect that the net effect would be to minimize the role of psychological expertise as such, while pointing up the vast uncertainties that exist in any theory purporting to analyze the human mind. This should not be a source of great regret, all in all. True, it will bar certain ingenious quasi-insanity defenses in criminal cases. On the other hand, it will bring an end to such outlandish spectacles as the notorious Texas "psychiatric" expert who moves from one first-degree murder trial to the next, mindlessly proclaiming that every convicted killer is a hopeless recidivist who must be promptly murdered by the state.

In proposing such reforms, I have, I admit, an ulterior motive going beyond the desire for just and accurate verdicts. The recurring spectacle of "dueling experts" is extremely harmful to the reputation of scientists. It presents them in the worst possible light, as mercenaries or fools who, prompted by greed or sentiment, will offer up convenient opinions on demand. Viewed thus, science itself seems an arbitrary and capricious game, a pretext for a war of egos and illogic. In particular, when psychiatrists and psychotherapists are on the stand, acknowledged officially as scientific "experts," it can seem that science is not only the most arbitrary, but also the most palpably ridiculous way to try to come to grips with reality. The situation is hardly better when cranks, quacks, and pseudoscientists are credentialed by complacent judges. Either they seem credible, to the detriment of honest science, or their evident eccentricity taints the reputations of real scientists. It may be mere selfishness or vanity for a scientist to give such weight to the public perception of science, but, in my opinion, respect for, and even deference toward, science—orthodox, "establishment" science, at that—is an incomparable social asset. It stands, if I may be melodramatic, between our civilization and the propensity to commit costly and dangerous errors of all sorts. It is at best a fragile bulwark. But that is precisely the point. Further erosion may have deeply damaging consequences, quite apart from making our justice system unreliable.

Of course, this scheme of mine—indeed, any reform, no matter how mild, tending in the same general direction—is certain to horrify and outrage most of the bar (and therefore a good part of the bench as well). It forbids the appearance of the trained-seal expert witnesses in whom trial lawyers take particular delight. Yet I am persuaded that it does not unduly diminish the role of the jury, or turn it into a mere puppet of judicial functionaries. Indeed, there is precedent for the procedure I have outlined: the process by which a judge prepares and delivers a charge to the jury before it retires to consider its verdict. Here, too, "experts"—the attorneys themselves—argue technical points (of law) out of the presence of the jury, and the presiding expert, the judge, accepts or rejects these arguments and reaches the final conclusions that will guide deliberations. The process I suggest simply allows qualified experts to canvas the scientific points in contention and to devise a comparable set of guidelines.

This idea, even in modified form, is unlikely to be implemented in the near future. The institutional inertia of our court system is far too formidable an obstacle. The notorious obstinacy of attorneys in defending their perceived self-interest is an even greater impediment. But perhaps it is not too soon for jurists, scientists, and scholars to begin a sober examination of the merits of such proposals. It is, at any rate, encouraging to note that Justice Breyer has been advocating reforms incorporating similar notions.[22]

It is simultaneously encouraging and depressing to know that for more than twenty-five years similar ideas have circulated in connection with suggestions for reforming regulatory as well as trial proceedings. The proposed institution, first devised by Arthur Kantrowitz, is often called a "science court."[23] Unfortunately, aside from a few brief experiments, it has not been much studied in a realistic context, much less institutionalized. The justification offered by its advocates pretty much seconds what I have said previously. As Peter Huber, an implacable opponent of phony science on the witness stand, puts it: "The real debate is to what extent scientists will be required to validate their claims with other scientists before peddling them in court. Requiring scientists to write down their claims and expose them to scientific peers for possible rejection, before offering them in court, is not suppression. It is a vindication of the scientific process."[24]

Unfortunately, the science-court idea has evoked as much opposition as support in the narrow circle of public-policy scholars who study such ideas.[25] One objection, which seems to echo postmodernist scorn for objectivity, is that proponents of science courts posit the ability of a tribunal to separate "facts" from "values" in a way that cannot be achieved by any merely human institution.[26] "Social constructivism" is the theme that prevails among the critics. Sheila Jasanoff, one of the leaders of this school of thought, avers:

> An important insight emerging from the social studies of science in recent years is that scientific claims are to a large extent "socially constructed." This argument holds, in brief, that claims in science do not simply mirror nature but are subject to numerous social influences. These include, most obviously, the theoretical and methodological constraints imposed by prevailing scientific

paradigms in a given discipline or historical period. More controversially, how-ever, scientific claims also seem to incorporate factors unrelated to the pre-sumed cognitive concerns of science, such as the institutional and political interests of scientists and their organizations.[27]

For these thinkers, "ought" is always hopelessly intertwined with "is." Any move to deny this (especially on the part of a scientist) must be regarded as a disguised power play, an attempt to endow a limited perspective serving the interests of one faction with the prestige of "objective" reality. This point of view has been quite successful in infiltrating the humanities and social sciences in recent years. It has even penetrated the fastness of our law schools.[28] Clearly these are not views to which most scientists will readily, or even reluctantly, assent. One must grant something to the cynics, of course. Scientists put on their trousers one leg at a time, and are no more immune to greed, ambition, egotism, or spite than any other clan. Notwith-standing this truism, one must reject the implication that science is a murky transcription of individual and group interests, rather than an effective pro-cess for obtaining and codifying knowledge about the material world. Sci-ence is indeed "social," but one of the paradoxical consequences of that sociality is the way it constrains scientists to strive—with frequent success—toward the kind of objectivity the social constructivists are desperate to deny. The "method" of "science" is really another name for the ability—of indi-viduals and of the culture as a whole—to make precisely the kind of fact-value distinctions that are declared impossible by the facile cynicism of the constructivist stance.

The insight Jasanoff claims for "social studies of science" tends, under scru-tiny, to melt away into self-serving dogma.[29] A typical example may be found in a recent Jasanoff tirade on AIDS: "That drug development nevertheless was ruled for so many years by the 'baby-talk story' [i.e., HIV is the cause of AIDS] may reflect the dominant political and cognitive status of virology in the political economy of the US pharmaceutical industry."[30] Note how many bizarre theories and simple factual mistakes are woven into one sentence: it is "baby talk" to say that HIV causes AIDS, despite a mountain of evidence accumulated over the course of fifteen years. Virology has dominant *politi-cal* status in this country. (What a peculiar notion—virology on a par with balanced budgets and prayer in public schools!) Theories of viral etiology of diseases play into the hands of drug companies! (In fact, those companies historically have had little interest in developing antiviral agents because of the difficulty, expense, and doubtful profitability of doing so.) Even worse, Jasanoff absurdly ignores the fact that scientists' dogged pursuit of antiviral agents has paid off spectacularly in the emergence of protease inhibitors specifically designed to counter the HIV retrovirus—a treatment that shows enormous promise. Errors of this kind are flatly inexcusable—the sentence in question was written in mid-1997. But Queen-of-Hearts logic is fully acceptable to social constructivists whose ferocious dislike of science over-rides all other considerations. Most recent opposition to tightening standards for purported "scientific" evidence into the courtroom is rooted in this doubt-ful intellectual base.

Social constructivist laments notwithstanding, there is no reason to be-lieve that it will be particularly difficult to maintain the appropriate degree of objectivity in science courts (or under the more comprehensive system I outlined previously). This certainly does not require ideal neutrality. It merely means making the system objective enough to function "for all practical pur-poses"—in other words, like science itself. As one close student of the prob-lem observes:

> The first lesson is that, contrary to the fears of critics, we never had a serious problem in separating questions of scientific fact from questions of policy pref-erence. It puzzles me that this complaint is still raised today, in this journal, since I thought we had compiled enough demonstrations to put it to rest. As a practical matter, the isolation of scientific questions is easy, despite philo-sophical arguments to the contrary. I claim this objection is a red herring, the gut response of those who deny that science is more objective than any other road to knowledge.[31]

Of the scholars cited above, two—Sheila Jasanoff and Peter Huber—are par-ticularly prominent in the quarrels concerning the use of scientific evidence in courts and regulatory proceedings. Both have written recent important books on the subject.[32] As the passages cited previously make clear, Huber is emphatically determined to keep fake science and its proponents from hav-ing a significant effect on the outcomes of legal proceedings. Jasanoff, true to her constructivist ideology, insists on blurring the distinction between well-confirmed science and shaky speculation.[33] For her, "Historically, so-ciologically, and politically, the proposal that courts should increase their reliance on a value-neutral mainstream science is . . . extremely problem-atic."[34] For Huber, by contrast, "Science does, indeed, offer less than abso-lute certainty. But it has proved to be vastly more accurate, reliable, stable, coherent, and even-handed than the alternatives."[35] There is every reason, therefore, to let the canons of scientific acceptability play a large role in de-termining what courts are to receive as evidence. Huber's book gives telling evidence of the damage done by allowing offhand speculation—some of it previously refuted in detail—to dress itself up as scientific expert opinion and thus to control the outcome of lawsuits. In a clearheaded world, his evaluation of the situation would not only make grimly entertaining read-ing—which it assuredly does—but would lead to new ground rules for judi-cial practice as well.

Unfortunately, clearheadedness will not have easy going. A wing of the scholarly community—Jasanoff is an important representative—insists on magnifying minor philosophical quibbles and relatively minute imperfections in scientific practice into pretexts for what Huber calls the "let it all in" approach. This unseemly spectacle dampens any hope for swift and appropriate reform in the way the law deals with science. If highly esteemed scholars at the most prestigious academic institutions insist on finding wiggle room for scientific fakery in hearings and trials, what are the chances of convincing the reluctant journeymen of bar and bench to banish it?[36]

Perhaps the difference between the views of Huber and Jasanoff is most tellingly illustrated by their respective attitudes toward the pseudodiscipline of "clinical ecology." This is a body of poorly confirmed or specifically refuted medical doctrine which purports to trace the etiology of many diseases to a variety of environmental insults. Supposedly, these trigger pathological reactions in the immune system. As Huber wryly notes, however, "The clinical ecologists can connect anything to anything."[37] Tort lawyers find such a flexible doctrine extremely useful, for reasons all too apparent. The resulting symbiosis between these attorneys and clinical ecologist "expert witnesses" has obvious consequences: "[I]f the clinical ecologist does not routinely deliver $49 million verdicts, he can quite often provide a fair shot at one."[38] Unfortunately for those who became skilled at this game, the courts are beginning to take a dim view of a field which so conveniently provides diagnoses of immunological disease in cases where orthodox immunologists find nothing at all.[39] For Huber, this might provide a small, tentative victory for common sense.

Not so for Jasanoff. For her, the exclusion of clinical ecologists as certified experts is merely one episode in an ongoing war between different social formations—conventional medical science and clinical ecology—where truth cannot be ascribed to either faction, because it is never more than mere "truth," that is, the temporary state of a shifting consensus negotiated within a subculture: "This decision [to exclude the evidence of clinical ecologists][40] may be applauded from the standpoint of efficiency, but it again illustrates a propensity on the part of appellate courts to adopt and refract back into society the scientific community's own perception of its own cognitive and social credibility. The law in this way plays a powerful role in upholding the legitimacy of science in society."[41]

Jasanoff's indignation is palpable on the page. The idea that there should be a general social recognition of the epistemological legitimacy of science ignites a resentment whose ferocity is clear, even though its sources are obscure. For Jasanoff, "science" is merely the consensus of one limited, if powerful, group. Therefore, when issues involving science come up in courts or in public policy decisions, "official" science should be only one among the multiplicity of voices to be heard, even on questions that seem narrowly scientific. On this view, proper decision making—in the courts, the regulatory agencies, and elsewhere—consists in brewing a new consensus from a larger, more diverse spectrum of groups than is encompassed by the narrow world of professional science. Just as science has "negotiated" its version of "truth" by social struggle within its own world, so the legal and governmental decision process must negotiate a broader truth among a variety of factions, of which science is merely one element: "[A]llowing orthodox scientific practice systematically to dominate over other types of meaningful [*sic*] knowledge production may not be the best way to bring closure to such controversies. [The silicone implant scare is the case in point here.] Instead, we need to search for mechanisms that strike a better balance between scientific and subjective [*sic*] knowledge in toxic tort litigation."[42] It is not only scientists who find this view perverse. Its claims, however, are buttressed by appeals to the supposed empirical findings of the sociology of science— desperately slender stuff, in my view—by the epistemological theories of Thomas Kuhn (properly understood or otherwise) and by a variety of postmodern clichés which lend it the superficial semblance of philosophical sophistication.[43]

Jasanoff is passionately fond of the word "deconstruction," which she has made the leading element of her lexicon. In her hands, the term is far more earthbound than in the dizzy verbal confections of literary critics, although her use of it may be a ploy to get the English Department on her side in the campus cultural wars. When Jasanoff "deconstructs" the behavior of scientists and their organizations, she merely performs what most people would call a skeptical or frankly hostile examination of selfish motives and concealed interests. It would be at least as accurate, and considerably less pretentious, simply to call this "debunking" or perhaps "demystification." However, eschewing the neologism might interfere with Jasanoff's own efforts at mystification, which clearly rely on an aura of au courant knowingness. How far this is from authentic sophistication will be obvious to readers who note her use of the phrase, "the logical positivists' view that science simply mirrors nature."[44] This assessment of what logical positivism is about reveals a deep ignorance of the matter (which seems to be endemic to the postmodernist community, even at its fringes). All by itself, it irrevocably dooms any pretensions to philosophical maturity. Yet such shallowness does not prevent Jasanoff from expatiating on "*Daubert's* fine disregard for a philosophically coherent decision rule." Her disdain for *Daubert*, however, arises not from its incoherence, but from its very coherent and very sensible view that mainstream science ought generally to be accorded serious respect. This is what Jasanoff finds insupportable.

Nonetheless, Jasanoff has been the beneficiary of a waxing reputation, even among scientists.[45] This is partly because of her claims, via slightly *outré* terminology, to have tapped new sources of insight, but mostly because of the emphasis she puts on the process and the psychological atmosphere of dispute resolution. True to her constructivist roots, she regards "knowledge" as an artifact of consensus. Therefore, she urges that the consensus through which the law resolves disputes involving scientific claims ought to be as inclusive as possible. Even more, it must be *amicable*, marked by mutual respect and empathy on all sides for the deeply held beliefs of every faction. Jasanoff's aim, in short, is to make the process of conflict resolution as painless as possible, with each of the vying parties convinced that he has won something from the process. The promise is that nastiness and injured feelings are to be minimized, if not eliminated, and that social solidarity, over a wide spectrum of individual and group perspectives, will ensue. Under her notion of conflict resolution, settlement of a dispute should suffuse the disputants in the lambent glow of reciprocal esteem. This, I submit, is the core of Jasanoff's appeal. She proffers a vision of blunted resentment and social peace. Even scientists can be seduced by such a pleasant prospect, especially if it releases them from their accustomed role as spoilsports.

Consider, however, that the purpose of the game ought to be not genial social interaction of the participants, but *truth*. This is a criterion that insists on having its say despite the unremitting efforts of postmodern thinkers to banish it altogether. Agreement cannot create truth; at best, it can acknowledge it. The most amicable conflict resolution in the world is still hostage to the reality it cannot abolish. However mellow it may make the contestants feel, mutual regard is a poor substitute for accuracy. Morris Shamos, a supporter of science courts and an astute critic of the naive populist ethic in respect to scientific questions, makes an important point, disquieting as it may be to the celebrants of demotic wisdom:

> Among those who hold that public debate can eventually resolve such issues, one senses an underlying belief that direct confrontation of "experts" by the uninformed public is the best *means* of reaching an answer to a policy question involving science or technology, even if it may not be the best *answer*. That is, even if the public does not understand the technology involved, the opportunity to air its views and to challenge the experts provides comfort that its commonsense judgments are correct. In other words, consensus, even though it may be arrived at incorrectly, would appear, under this theory, to be more important than truth. But while a given use of technology may properly be decided by popular vote, it is folly to believe that reasonable scientific conclusions can also be reached by this means.[46]

To put it concretely, the outcome of deliberations over global warming, say, ought to depend maximally on reliable knowledge about the prospects for global warming. Such knowledge may be excruciatingly difficult to get hold of, but it is inane to pretend that accuracy is secondary, or chimerical, or a mere corollary of the widespread societal satisfaction that ensues from a well-conducted negotiation. Jasanoff is fond of legal mechanisms because "one of the greatest strengths of legal proceedings is precisely the ability to produce localized, context-specific epistemological and normative understandings that are not subordinated to inappropriately universal claims and standards."[47] In other words, truth takes a back seat to the accommodation of the local community's conceptualization of truth. Under such a formula, consensus may well evolve, but the consensus thus produced is likely, in many cases, to curdle sooner or later into recrimination for the direst of reasons: it will be superseded by the grim reign of fact. To reject a priori the notion of fact and the possibility of objectively achieving at least a good approximation thereof is ultimately to reject the deepest social value of law and government. It is to favor a fleeting notion of fairness that is contingent upon treating all factions as equally competent, epistemologically speaking. This may be regarded as democracy à la mode in certain academic circles, but it is anything but wise. Reality is a jealous, not to say savage, god, and will be revenged on those who deal so blithely with objective fact.

Beneath the surface of Jasanoff's irenic prose there fulminates a deep and genuine disdain, perhaps even mounting to loathing, for science. This is illustrated by an aspect of Jasanoff's thought which is not, so far as I know, in print, but which is still publicly attested to. (Here, I hope the reader will admit anecdotal evidence to which I was a direct witness.) Not long ago, Jasanoff delivered a lecture at the annual meeting of the American Association for the Advancement of Science, the purpose of which was to defend the constructivist viewpoint as an appropriate ground for determining governmental science policy.[48] As a corollary, it naturally follows that the constructivist school of academic science studies is the proper source of the experts who are to mediate between science and other social institutions—an inference that cannot have escaped the speaker's notice. (Here, I must lay my own cards on the table. Jasanoff's address was punctuated by repeated disparaging references to *Higher Superstition*, a work skeptical of science studies which I coauthored with Paul R. Gross.) However, the most interesting moment came when she responded to a question from the audience which asked how her concept of public science policy would play out in the case of the long-festering dispute over the teaching of evolution and creationism in public schools.[49] The answer was florid and lengthy, and piously invoked Jasanoff's dictum that the issue is one of "locally constructed scientific agreement versus more universalizing notions of science."[50] Ultimately, however, it was unequivocal. Jasanoff believes that in communities where fundamentalist sentiment prevails, local school boards infused with that sentiment ought to be able to mandate teaching of creationist doctrine in the schools they control. This, it turns out, is the real implication of her esteem for "localized, context-specific epistemological and normative understandings that are not subordinated to inappropriately universal claims and standards."[51] Even in a clear-cut case where the consensus of scientists has been emphatic and stable for decades, scientific opinion is "inappropriately universal" and can be set aside for the convenience of a narrow sect bent on using the public school system as an instrument of indoctrination!

At this point, I can only resort (or descend) to rhetorical questions: How can it be that one of our most highly lauded authorities on the role of science in public policy is enmired in such an eccentric philosophy? How could so many of the perfectly reasonable people who have praised Jasanoff have failed to recognize the deep-set contempt for science—indeed, for rationality itself—that infests her theories and her recommendations? It may seem that I am placing inordinate weight on a single, transient incident, but I have reported it accurately, and to me it is diagnostic and damning. The same sentiments recur, sometimes emphatically, sometimes in judicious rhetorical disguise, throughout Jasanoff's oeuvre.[52] There is also painful irony here; Jasanoff appeared at the AAAS meeting not as an outsider and critic, but as a member of the organization's governing board. One can only shudder at her singular notion of the "advancement" of science.

Of course, there is more to the story than the idiosyncrasies of one thinker. It is not because her view of science and its epistemological status is so original that Jasanoff is important. That view, appalling as it may seem, is standard stuff in large sectors of the academy. But Jasanoff has been uniquely clever in creating a place for that view in the councils of law and government, and uniquely successful in disguising a bizarre doctrine as a commonsense route to civil peace and equity. Thus she deserves special scrutiny. There are many other scholars who share her views and whose ambitions are beginning to turn in the same direction. It is not too much to say that their ascendancy in the world of science policy would eventually prove an unequivocal disaster.

In the end (arrogant as it may appear for a scientist to say so), there is only one sound approach to rectifying the troubled relationship between law and science. This consists of adjusting legal doctrine and practice to accommodate the unique success of science—the very orthodox, mainstream version of science whose perceived smugness and self-certainty is the real wellspring of so much "philosophical" challenge. There is no charitable symmetry in this view. Science is simply incomparably better at addressing the kinds of questions it was developed to address than any actual or potential rival. For our legal machinery not to recognize this is ultimately to disdain truth, deform justice, and compromise the stability of our social order. This may seem overly histrionic, but consider the consequences we have already suffered as a direct result of our legal system's tenderheartedness toward junk science. Major corporations have been destroyed,[53] billions of dollars have been wasted through needless caution over harmless technologies, and other billions have been showered on dubious claimants on the basis of even more dubious evidence.[54] Cynicism about the prostitution of both law and science has reached enormous proportions. Quacks and charlatans have flourished in an atmosphere where the legal system functions as their active accomplice. In short, the time has come for the paladins of bench and bar to show a bit of unaccustomed humility in the presence of scientific competence. True, this will do nothing to augment scientists' decidedly meager store of humility! But we are nonetheless obliged, by logic and history, to accept the science we now have, as done by the scientists who, with all their human flaws, currently live and work among us.

ELEVEN

Journalism

It would be silly to accuse journalism of systematic hostility to science. On the contrary, the news media tend to dote on science and even on scientists. Science, technology, and especially medicine make good copy. Among those who follow the news, there are many people quite eager to hear about scientific discoveries, whether in the form of new and curious dinosaur species or new and curious galaxies. Newspapers and television news programs are happy to oblige them. Nonetheless, the relation between science and journalism remains equivocal, and scientists are often dissatisfied by news accounts, even those that flatter them. Journalists tend to like science, but they like it on their own terms. It is a mine of stories that liven up a newspaper or a broadcast, create a change of pace for readers or viewers, and play well with the large constituency of "science buffs." But the culture, as well as the frankly commercial purpose, of newspaper publishing creates a filter that selects stories—and emphasizes elements within stories—that do not always accord with scientific judgment or the reality of scientific work. As one knowledgeable science journalist puts it: "Science writers may suggest stories that should be covered, but it is the editors who decide what will be covered, and how. Editors are looking for stories that are new, interesting, hot, scandalous—and also stories they and readers can understand. Editors and readers like conflict and personalities. Moderation is inherently boring."[1]

In other words, as a topic of journalistic interest, science is a source of high-class entertainment. The purpose of running science-oriented stories is frequently to startle and amaze, sometimes merely to entertain, although dire warnings of one kind or another are also popular. This is not to say that such coverage is necessarily exaggerated or overly alarmist—often it is accurate and valuable. There is no doubt, however, that the way in which stories are worded and shaped reflects journalistic values rather than the canons of accurate scientific exposition. Distortion can and does ensue. One should not underestimate the restrictions engendered by the demand for "stories readers can understand." This is a very severe boundary condition on the accuracy and range of science coverage. It frequently creates "readability" at the expense of actual logical coherence. It often actively precludes the establishment of an adequate context for real understanding—a very different thing from the "understanding" which editors aim for. Science journalism in daily newspapers—and, even more so, in the broadcast media—is, ipso facto, superficial. "With a few notable exceptions," observe Jim Hartz and Rick Chappell in their recent report on the problem, "most American newspapers and television news operations basically ignore the accomplishments and failures of science, overlook the nation's investment in science and take for granted the tangible benefits science provides."[2]

To say so is not necessarily to cry out for reform, although that is the clear aim of Hartz and Chappell. The superficiality is embedded in the very nature of the beast. A cynic might be forgiven for concluding that the situation is beyond immediate reform. But we—especially the scientists among us—have to be aware of the consequences. The worst of these is a public consciousness of science issues that is both vague and skewed, that is plagued by a propensity to neglect important issues while overemphasizing unimportant ones, and that overrates its ability to comprehend matters for which it lacks the basic conceptual tools.

If the prevalence of journalistic over scientific values generally distorts science coverage, some of these distortions can be rather innocuous. The popularity of everything pertaining to dinosaurs is one such instance. To dig up a notably large, fierce, or bizarre dinosaur is to win instant celebrity, quite apart from the scientific importance of the find. A doting public (that emphatically includes my own guilty self) will eagerly devour story after story about the beast. If it represents a new species, its newly minted name will be immediately annexed to the lexicon of third-graders throughout the nation. Plainly, as a culture we can't get enough of these animals, for reasons that intrigue cultural analysts as well as devotees of Jungian archetypes.[3] However, viewed soberly, from the killjoy vantage point of the disinterested scientist, the emphasis on dinosaur lore is greatly misplaced. With all respect to vertebrate paleontologists, who work just about as hard as any scientists, the subject is a marginal one, even for biologists and evolutionary theorists. The study of dinosaurs tells us little about anything aside from dinosaurs. Even within the realm of paleontology, the most fundamental questions have little to do with dinosaurs. The fossil record of tiny, inconspicuous creatures discloses far more about the history of life and the pattern of its development than the petrified bones of fifty-ton monsters. But such matters almost never come within the purview of newspaper readers; diatoms simply do not enthrall us, particularly when presented in terms of something as saturated with mathematics as population biology.

The point is not to deplore the emphasis given to dinosaurs (by museums and fossil hunters, as well as journalists). True, it helps to distort the picture of what science is up to, so far as the casual observer is concerned. On the other hand, there are probably thousands of scientists who have no current professional interest in archaic reptiles, but who first began to contemplate scientific careers because they were fascinated by prehistoric monsters as kids. I draw attention to our dinosaurian preoccupations merely to illustrate what governs the priorities of science coverage in the news media. It has little to do with the values of the research community.

To take another example, this time a negative one, consider the "search for extraterrestrial intelligence" (SETI to its fans). Given the public's obsession with alien visitors and the like, the attention given to SETI by science journalists is decidedly minimal. It pays to ask why. In my opinion, the answer lies in the frustrating nature of the search—maddeningly unproductive since it was begun thirty-odd years ago—but also in the nature of the stories that can be gleaned from the SETI community. Experts in this area don't go out of their way to provide good copy for journalists. They don't speculate very much on alien psychology or anatomy or, indeed, on any of those fanciful aspects of imaginary "aliens" that make them a staple of science fiction. Rather, SETI researchers are the ultimate techies, forever tinkering to improve the sensitivity of their listening devices, the speed of their computer hardware, and the efficiency of their search algorithms. This is reasonably engaging stuff for those able to follow it, a category which comprises physicists, astronomers, mathematicians, electrical engineers, and computer scientists. But it provides rather dull reading for everyone else.[4] Consequently, SETI makes poor press fodder. There is nothing much to see in the observatory, no breakthroughs to report (thus far!) beyond false alarms. SETI is speculative, but its speculation is expressed within the constraints of near-monastic self-discipline. Of course, if the search were ever to succeed, it would create a journalistic firestorm unparalleled in history, but that doesn't make the ongoing search process any less dull to journalistic eyes. Consequently, SETI is almost invisible to the public that flocks to *Independence Day*, *Starship Troopers*, or even *Contact*.[5] In fact, government funding for the project died nearly unmourned, despite the fact that the amount of money involved was almost absurdly small.[6] I might add that, in like manner, interest in the once-famous Mars meteorite—initially reputed to contain traces of microscopic Martian life forms—also faded rapidly once the discussion of the pros and cons turned technical. Laymen want to be amazed when they read about such things, not bogged down in a minute examination of chemistry and electron microscopy.

These examples sustain a more general point: the science press is, by and large, interested in *results* rather than process. Specifically, it is interested in results that journalistic word painting can gloss with a glamorous aura, whether they involve insights into the social behavior of primates or into the quantum-mechanical behavior of electrons. Consequently, the reality of doing research, with its quotidian frustrations and *longeurs*, its inevitable interruptions, distractions, blind alleys, and irrelevancies, tends to be obscured or hidden altogether. Hartz and Chappell set a high, indeed an intimidating, standard for competent science journalism:

> To be effective, a science journalist should be comfortable with a variety of scientific disciplines, as well as with engineering. His or her education should include experience in the laboratory, where the ambience of the scientific research process can be felt and learned. The curriculum should combine training in science and engineering with courses in mathematics, including probability and statistics, balanced by coursework in communications, writing, history, economics and political science.[7]

The number of people meeting this standard is, however, depressingly small. One obvious consequence is that "everyday" science, science as a typical middle-class career, is invisible in the media. In one way, this omission is quite flattering to science, which thereby comes to be portrayed as an unremitting stream of breakthrough successes and deep revelations achieved by a few rare and unaccountable talents. The bitter aspects—failure, triviality, mere redundancy—are excluded from science's public image. The scientist who attains modest success, or no real success at all, is one of this culture's invisible people. But there is a downside to this cosmetic presentation. On the basis of most popular accounts, science is perceived as a realm detached from the snarls and confusions of ordinary life. There seems to be an unearthly, frictionless quality to this highly edited version of a scientific career. This, to be sure, is mystification, elevating the everyday behavior of working scientists into a sort of solemn ritual, as impenetrable to outsiders as it is efficacious. Mystification of this kind creates admiration, certainly, but it also creates resentment.

By the same token, scientists, since they are usually portrayed at their moments of triumph, rarely come across to the public as frustrated, morose, or nasty. Media accounts tend, in fact, to sanitize them, to present them as rather saintly worshipers of nature, faithful souls whose devotions are ultimately rewarded by privileged glimpses of the truth. The cantankerousness that is as much a part of a scientific career, on average, as of any other, is airbrushed away. So, too, the anxieties, envies, and resentments that are as intrinsic to research labs as to corporations, political parties, or church hierarchies. Seldom is a scientific finding depicted as having been driven by a fierce desire to beat out rivals, if not actually to humiliate them, though such motives are as effective a spur to scientific achievement as to success in any other human calling.

Likewise, it is a rare scientist who is publicly quoted to the effect that a major discovery was motivated by a desire for a bigger salary, a larger office, and greater deference from his colleagues. Motives that are perfectly acceptable in lawyers, businessmen, and athletes are not supposed to influence the scientist, who is assumed to be unworldly when it comes to money and power. In all, it would be fair to say that most science writers—the lazy ones, certainly, but even some of the not-so-lazy—perpetuate a model of the scientist as single-minded, somewhat obsessive about ideas (though not about people), emotionally detached, rather adolescent, and largely insulated from the hard truths of the world.

Religion is something of a wild card here. To the extent that it comes into the picture, it poses a bit of a problem for science reporters. If we are to believe the surveys, an absolute majority of scientists have no belief in supernatural religion, while most of the remainder are religious in only a vague and approximate fashion. This estranges them, on average, from the official theistic consensus that is continuously and fervently restated in American public discourse, and puts them at a truly enormous distance from the religious literalism that is still so depressingly commonplace in contemporary life. Consequently, journalists finesse the issue as much as they can, taking special note if a scientist is more or less conventionally religious, but ignoring a gross social defect like outright atheism.

What emerges is not an unkind portrait. Indeed, it is flattering. But it also puts a barrier between the mass of people who accept it—who demand it—and the scientists thus portrayed. Scientists are seen as preternaturally intelligent, as extremely valuable to the society, and as paragons of many other virtues as well. But there is a pervasive sense that they ought to be kept on the reservation, so to speak. Their constant immersion in formalistic thinking and their unique insulation from the messiness of life, as well as their uncanny ability to take seriously things that are hopelessly abstract to the average citizen, enfeebles their judgment in practical matters, or so it is widely thought. Part of this legend is embodied in the stereotypical distinction between the woolly-minded scientist and the hardheaded engineer. The former is envisioned as preoccupied by the quest for the "Why?" of things at the heart of nature, the former by the "How?" of things in the heart of society. The totemic figure of Einstein (as distinct from the real man) is the icon that testifies to the stereotype.[8] In our culture, he is the adult version of Santa Claus, an elf with a twinkle in his eye, uncannily wise in some matters, hopelessly naive in others. I would even go so far as to say that those other patriarchs of atomic weapons, J. Robert Oppenheimer and Edward Teller, symbolize, in their own way, what it is that people distrust in the scientific personality. The tenderhearted Oppenheimer, hero of the left, disconcerts the average citizen through his apparent guileless pacifism. But his archenemy, the tough-minded Teller, hero of the right, is equally distrusted, I should guess, for his heedless pugnacity. Here, again, I speak of the images that shape public legend (still potent after almost half a century), not biographical accuracy. The point is that in either case, the idealized "scientist" is perceived as being too tied up in abstract, theoretical schemes to have a good sense of the realities of life.

Journalists are not wholly responsible for this iconography, nor are they slavishly constrained by it. However, most journalism is a near-term operation, built around rules of thumb and shortcuts. Reporters are not psychotherapists nor even novelists (at least, not good novelists as a rule). They create brief sketches, not intimate, revealing portraits. In depicting businessmen, athletes, and judges, as well as scientists, they tend to fall back on the common stock-imagery of our society and the background assumptions that undergird it. Why not? It gets the job done, usually without excessive distortion of the truth in any given instance. Cumulatively, however, the continuing resort to this practice creates a definite bias in our cultural perception of science and those who engage in it.

One of the most unfortunate aspects of the unrealistically positive image of the scientist is that, when it crumbles, it crumbles with a vengeance. Lately, this has been all too evident in the journalistic fad for exposés of "scientific fraud." For more than a decade, science reporters have constituted a rather unanimous claque in support of the efforts of, for instance, the Office of Scientific Integrity, whose missionary zeal to sniff out any slightest hint of fakery, duplicity, or plagiarism has resulted in a number of highly publicized misconduct cases, most notably the charges lodged against Nobel laureate David Baltimore and his associates.[9] The recent vindication of Baltimore and, even more important, his collaborator Imelda Imanishi Kari has taken much of the wind from the OSI's sails, but the episode was instructive in that it brought to the surface an unexpected streak of viciousness among science journalists.[10] The heart of the matter was the "revelation," if one can call it that, that scientific research is not always, indeed, not very often, a pristine model of textbook protocol. Data are not always recorded in ideal format at the time they are gathered, records are sometimes misplaced and must be reconstructed from fragmentary notes, and cleaned-up notebooks are sometimes backdated for the sake of form. None of this is desirable, but neither is it fraudulent nor sufficient to impeach the validity of arduous research. It is the inevitable result of the slog and grind of real life in a laboratory where there is much more work to do than hours available to do it. Nonetheless, trivial misdemeanors like this have sufficed to evoke the jackal in a fair number of science writers. Having often exalted science as the model of absolute rigor, they turn upon real-life scientists who fail, in relatively trivial ways, to live up to this standard, and denounce, as near-criminal, behavior which in any other walk of life would be regarded as quite consistent with probity.

The psychological mechanisms behind this turnabout are not hard to divine. The idealized image of science which journalists and popularizers have set in place has been much admired, but within the hymns of praise one can almost always detect a counterpoint of resentment. This ubiquitous, if muted, note of protest is the inevitable rebuke to an outlook that sets apart and exalts a class of savants who have knowledge, presumably denied to the rest of the culture, that endows them with profound and inscrutable powers. If you are a scientist, it is acknowledged that the spirits you "call from the vasty deep" really *do* "come when you do call for them." Such is the image and, often enough if far from inevitably, such is the truth of the matter. It is an awesome reputation to have, and the admiration it invites is rarely unmixed with the suspicion and hard feelings that envy engenders.

Thus, when the public, or even the publicists who have crafted the public's image, perceive a lapse from the inviolate purity of the scientific ideal, vindictiveness emerges. The logic is that if a given scientist cannot live up to the Olympian standards that supposedly authorize his sweeping claims to knowledge, then the most cynical, black-hearted fraud must be involved. Moreover, the indignation evoked by acts of supposed scientific dishonesty rarely remains focused on the individual perpetrator. It diffuses outward to taint the reputations of the colleagues, supervisors, and subordinates of the accused, with the implication that cronyism is rife. The standing of science itself is often called into question, especially if the misbehavior involves a scientist with a substantial reputation.

There is a close analogy between accounts of scientific misbehavior, as the press tends to promulgate them, and tales of errant clergymen, whether adulterous Evangelical preachers or pedophiliac Catholic priests. In either case, once the veil of sanctity begins to fray, it is not long before it is entirely ripped away by violent hands. The sustaining myth of the unique virtues associated with the calling is rather quickly swapped for an alternative myth, which looks on the known malefactors as a mere token of the thousands of others who simply have not yet been exposed. What had been viewed as a bastion of rectitude is now reviled as a stew of corruption and hypocrisy. It is self-evident that this kind of debunking is valuable to a wide spectrum of journalists. The press, as self-appointed warden of openness and probity, sets the tone for public discussion and puts into place the lineaments of the emerging countermyth. Naturally, reporters also get much of the credit for exposing the bad guys. It might seem somewhat facetious to compare the tabloid newshounds who sniff out the sexual dirty linen of erring preachers (or presidents) with the highly educated mandarins who write for the science sections of our "serious" papers and newsmagazines. But the prosecutorial zest and the delight in catching important personages with their literal or figurative pants down is much the same in either case.

Some of our most reputable and knowledgeable science writers have devoted considerable effort to the question of fraud and deceit among scientists. William Broad and Nicholas Wade, two experienced science journalists long associated with the *New York Times*, may be said to have started the trend as we have known it in recent years.[11] Since then, the idea that science is ridden with fraud has gained considerable currency—there is even an Internet academic discussion group, "Sci-Fraud," dedicated to the notion. In point of fact, fraudulent claims are extraordinarily rare in science.[12] Compared to most professions, the field is remarkably pristine. This doesn't amount to a claim for the moral superiority of scientists as individuals. I should assume that their tax returns, for instance, are no more accurate, on average, than those of taxpayers of any other calling. It merely points to the fact that enlightened self-interest usually nullifies whatever tendency a researcher might have to cheat outright. If a discovery is important enough to earn significant credit for the discoverer, then sooner or later replication will be attempted in some form, and retribution will likely follow if fakery is detected. Nonetheless, many science reporters are convinced that cheating is endemic to research science and that every laboratory hides its guilty secret.

An interesting variant of this attitude appears in the work of John Horgan, a senior editor of *Scientific American* and a science journalist with a prolific output and wide experience. Horgan is unconcerned with fraud in the quotidian sense, but asserts instead that contemporary science is permeated by delusion. He insists that science—on the scale of "grand ideas" at least—has come to an end and that many of our most celebrated scientific thinkers, particularly those who work in areas like cosmology, foundations of physics, and evolutionary theory, are, as it were, perpetrating a grave fraud upon themselves. They are engaged, claims Horgan, in a kind of speculation that has outrun any possibility of empirical confirmation. Therefore, what they do is not real science but "ironic" science, a fit counterpart for the irony that prevails in the humanistic scholarship that has emerged under the spell of postmodernism.[13] In his recent book, there are points where the scorn he heaps on scientists whom conventional wisdom regards as brilliant easily outdoes the worst imprecations of those who are merely concerned with deceit and fakery.[14] Describing an intense discussion among cosmologists, he scathingly declares, "There was something both grand and ludicrous about grown men . . . bickering over such issues."[15] To his way of thinking, such speculation is as futile as that of medieval scholastics. It is detached from any real possibility of confirmation, and thus it is an arid private mythology rather than a sober description of the knowable universe.

Horgan's theme is interesting, but ultimately decidedly wrongheaded. The most astute of his critics have been as scornful of his thesis as he is of what he takes to be the epistemological hubris of contemporary science.[16] However, his work may be of unintended significance in that his sneering, contrivedly skeptical attitude toward contemporary science is a token of a broader vein of restiveness that afflicts quite a few members of his profession. Certainly, there have been similar eruptions among other experienced and, in the usual sense, sophisticated science writers. I might mention, by way of further example (and with the caveat that I was personally deeply involved in the incident), the recent resignation of the book review editor at the prestigious journal *Science*, precipitated by her favoritism toward the school of postmodern thought that stresses the delusory nature of scientific "truth" (always, of course nesting that term in the ironic window dressing of quotation marks).[17]

Like Horgan's book, this discloses, if not unalloyed hostility, then at least a profound ambivalence toward science itself on the part of some of the writers and intellectuals who create the accounts of science that most influence the general literate public. A portion of this surprising cynicism probably stems from the fact that these writers are in an intermediate position between the scientific community and the world of academic "humanism," from which many of them emerged and where science has been in bad odor lately.[18] But similar sentiments undoubtedly affect many science journalists who have no use for current academic fads. It's safe to say that resentment, if it exists in the first place, will create its own rationales, whether highfalutin or in the form of mere street wisdom.

Whatever its inadequacies, the comparison of scientific fraud with clerical sinfulness also reveals that the public's capacity to be obsessed with wrongdoing, virulent as it may be from time to time, is finite and fitful. Despite the spate of Tartuffery that dominated headlines and enthralled the media only a few years ago, the clerical profession seems still to be thriving in all the modes that brought forth our most noted miscreants. Jimmy Swaggart still holds forth as influentially as ever, and even Jim Bakker seems destined for a modest comeback. (This is to be deplored, at least by anyone who shares my dyspeptic view of religion as such.)

Likewise, science has not been indelibly disgraced by the publicity given to its own scandals, though there are doubtless residual scars on its public esteem. Both religion and science answer to perceived cultural needs, though those needs are vastly different. Whatever one thinks of the idea that religion (at least formal religion) might ultimately be dispensable, it is clear that society desperately needs science, engineering, and medicine for the day-to-day functioning of our social apparatus. The problem, for most would-be critics, is that once their antiscientific jeremiads have been launched, they still find themselves inescapably dependent on the very science, sometimes the very scientists, they have criticized.

The Baltimore case affords a clear example. Much of the press delighted in the tale—an apparent morality play with well-defined heroes and villains—of a stiff-necked, overweening scientific satrap brought crashing down by an underling who remained loyal to the methodology of rigorous science and to the code of personal honesty that it demands. The apparent denouement—Baltimore's forced resignation as president of the prestigious Rockefeller University—rounded off the story nicely, with justice triumphant and the mighty brought low. Critics of high-powered science exulted.

A number of the most important people in American science condemned the search for truth in the Baltimore case. They wanted to ignore the question of fraud and let the so-called replication process sort out whether the central claim of the *Cell* paper was true or not. Believing that science is inherently self-correcting, they were convinced that the truth would ultimately win out. It is an approach that pretends to seek the truth, but in fact rewards lies.[19]

Unfortunately for those who prefer their narratives to end on a note of unambiguous closure, the truth was more convoluted and Baltimore emerged at last as the hero of the sequel. This was yet another morality play, but this time one in which a man loyal to the truth as he knows it finally outfaces the scheming bureaucrats and politicians plotting his destruction. Here again, an appropriately symbolic final scene, with the vindicated Baltimore newly appointed as president of the even more prestigious Caltech, rings down the curtain with a flourish.[20] The problem is that the journalists who took such delight in the first episode must now find a way to reconcile the public—and themselves—to the subsequent chapter. Baltimore, once a reprobate, has to be reconstituted as a great scientist, an important policy expert, and a national intellectual treasure. He is, to be sure, all of these, but the scars inflicted in his days as a semiofficial outcast will not fade soon, nor will the confusion of a public that was whipsawed by the press's instinct for wrapping its narratives in sanctimony.

If we turn our attention to issues, rather than individuals, the same general pattern is easy to find. A few years ago, a crusading environmentally minded press congratulated itself on having brought to light a previously unsuspected danger, the supposedly carcinogenic effects of low-frequency electromagnetic fields. The story was framed, to a considerable extent, by a prefabricated narrative articulating a myth that resonates in contemporary culture: humanity, in its greed for the comforts of technology, had (once again) disrupted the natural balance of things (this time with power lines and transformers) and the bill had come due through the emergence of yet another plague, in the form, most poignantly, of juvenile leukemia. As is usual with stories of this kind, the alarmist elements were given the heaviest stress, while doubts and objections were underplayed or ignored. Needless to say, detailed analyses of the methodological pitfalls that lie in wait for overhasty epidemiological generalizations were absent from popular accounts. This is easy to understand, given a corps of journalists who, on average, are no more conversant with statistical subtleties than the public at large. But lack of mathematical sophistication is not the only besetting sin of the science press. Far more significant is the yearning, common to most journalists, for a simple, strongly inflected theme, preferably one in which sin, retribution, and possible redemption are markedly present in appropriate costume. The implication—that the people who have brought the danger to light are Galilean in their courage, their insight, and their enemies—is also very welcome. Such was the case with the power-line scare, especially as presented by its chief proponent, science writer Paul Brodeur.[21] Once again, however, the story initially promulgated turns out to have been savaged, if not utterly routed, by a thorough examination of the facts. The science press, which once prided itself on raising the alarm, must now put out the fire it helped to light.[22]

The mechanisms involved in these examples and many others like them are fairly obvious, but it seems that there is little that can be done to disarm them in a society where the mass media flourish vigorously and do what they must to catch and hold the public's attention. There is, moreover, a strongly moralistic element to the ethos of professional reporters, which influences other kinds of science writers as well. The ideal reportorial coup is the exposure of misbehavior on the part of the powerful, especially when it can be combined with the disclosure of grave threats to public safety. This is an archetype that continues to direct the activities of science journalists and to provide the template for many of their stories. It is rarely modified by the countervailing impulses that might arise from firsthand familiarity with scientific research or training in appropriate methodology. The recent record is therefore a very mixed one. Some exposés are triumphs of journalism in the fearless service of the public interest, but others are tall tales promulgated by a profession all too beguiled by negative stereotypes of science. In the latter instances, the initial accusations of fraud or duplicity rebound upon their makers.

Notwithstanding the undeniable damage wrought by baseless charges of scientific fraud or false alarms about imagined threats from technology, the curious thing is that even more damage has probably been done, in the long run, by the media's naive enthusiasm for science. Consider the long, uneven history of the American space program, which never would have thrived had it not been for the unremitting enthusiasm of the mass media. Yet that very enthusiasm—presumably shared by much of the public—is responsible for distorting that effort and turning much of it into an enormous waste of money, time, skill, and energy, so far as scientific priorities are concerned.

The main culprit has been the obsession with manned space flight, driven by the media's thirst for engaging and dramatic images, and by bureaucratic calculations of the benefits to be derived from slaking that thirst. It is safe to say that virtually no scientists, in any field, believe that space exploration has lived up to its promise for augmenting scientific knowledge. The reason is that the needs of the manned space program have been given priority over all other considerations. There is virtually no scientific goal accomplished through the manned effort that couldn't have been done much sooner and far more cheaply by unmanned craft. Similarly, the technology set in place by the space program—communications satellites, weather satellites, earth-survey satellites, even military intelligence devices—could easily have developed in the complete absence of astronauts and manned space-craft. The main effect of manned flight has been to beggar the purely scientific and technological efforts, filching the enormous amounts of money necessary to ensure the comfort, safety, and survivability of pilots and passengers. Indeed, if one examines the rationales for crewed vehicles, one finds that a rather loony circularity reigns. The chief purpose of putting people into space is to make it even easier and safer to put people in space in the future! Comments one appropriately cynical science journalist, "It is as if an inquiring fifth-century reporter asked Saint Simeon Stylites why he spent decades sitting atop a pillar and was told that Saint Simeon wanted to 'learn more' about pillar-sitting."[23]

This tissue of illusion is sustained by the fact that politicians and bureaucrats are convinced, rightly or wrongly, that the only way to keep the public interested in the space program and willing to foot the bill for it is to provide them with a continuing stream of pictures of jumpsuited astronauts cavorting in space a few hundred miles above the Earth's surface. This, it is assumed, provides space exploration with a dimension of human interest, without which the NASA empire would be hard-pressed to win funding. Perhaps this is a wrong assumption, but in any case it is one fostered and perpetuated by the media, which have never failed to find the image of a shuttle taking off or landing worth at least a few minutes of network airtime. Of course, it is easy to pass on the blame to the prejudice and incomprehension assumed to be endemic in the public mind. If the average layperson is more impressed by space-suited acrobatics than by even the most far-reaching feat of the Hubble telescope, broadcasters and reporters can hardly be faulted for taking these tastes into account, inappropriate as they may seem to most scientific professionals. Such tastes dictate what the public will watch, read, and buy.

This rationale lets the press off the hook far too easily, however. Much of the widespread ignorance or confusion to which journalistic coverage presumably panders is, in fact, the product of distortions and misrepresentations that the media have incessantly repeated. Every aspect of the manned space program is presented to the public as important scientific progress when much of it is a mere display of engineering tricks having no purpose outside themselves. Astronauts are invariably touted as creative scientists, even when they are basically military pilots with some background in nuts-and-bolts engineering.[24] The actual value of experiments and observations carried out by astronauts is typically overrated. So far as I know, none of these has come close to matching the research achievements of unmanned vehicles. Moreover, even if it could be shown that having human researchers in space has yielded important results unobtainable in other ways, this would hardly justify the extravagant expense (in time and money) of the Space Shuttle itself. In fact, one of the early effects of the priority given the shuttle was the premature abandonment of the Skylab space station. The Shuttle is no more needed for low-orbit manned missions than an eighty–foot yacht is needed to go bass fishing. The basic motivation for developing it in the first place was merely to give pilots something interesting to do while the news cameras looked on. A far cheaper system could have sustained whatever manned space flights might have been necessary to a judicious research program. All of this has been largely concealed by a science press corps interested, like NASA itself, in glamorizing the entire business.

The damage to scientific research resulting from NASA's grotesque overemphasis on manned space flight hasn't been confined to space-based research in astronomy and earth science. Science in general has felt the pinch. Nothing makes this clearer than the relative fates of the superconducting supercollider, on the one hand, and the proposed multinational space station, on the other. The supercollider, a project expensive enough by conventional reckoning but a mere fraction as costly as the space station, was killed by Congress in the early 1990s. The space station survived. Not all scientists were convinced that the supercollider, a particle accelerator designed to explore the structure of matter at an unprecedentedly profound level, could justify its great cost.[25] But a clear majority supported it and, certainly, no one denied that it would have made possible scientific work of the highest importance. On the other hand, there was virtual unanimity concerning the space station. It is an outright boondoggle from the point of view of pure science and, moreover, has very little to contribute to practical technology. Like the Shuttle (which is now primarily justified as an element of the space station program), it will bring profit to the contractors who develop it while providing NASA with further opportunities to exhibit astronauts disporting themselves on the nation's TV screens. Other than that, it is an enormous white elephant. The disparate fates of the supercollider and the space station provide a potent symbol of what has gone wrong with the nation's science policy. Even if the space station never gets built—its ultimate fate is still far from certain—an ocean of cash will have been poured down a rathole in the attempt to get it off the ground.[26] On the other hand, the contributions that the supercollider might have made to basic physics will be deferred for years.

The media undoubtedly played a role in precipitating this ongoing debacle. There was little effort to give the general public a sober sense of the relative value of the two megaprojects. Indeed, to the extent that science journalists fostered a point of view, it was one that emphasized the weirdness and incomprehensibility of basic particle physics, whose public image came to resemble that of alchemy and the Black Arts. Simultaneously, the press stressed the all-American derring-do of the astronauts who were to assemble the space station. This contrast merely replicated the all-too-familiar iconography of strapping jocks versus pencil-necked geeks. It was, in effect, anti-science tarted up to look like pro-science, with a bread-and-circuses version of science displacing the real thing. Scientists will have to endure the consequences for decades.

For different reasons, the performance of the science press has been just as unsatisfactory in another important area—the endless conflict between honest science and the claims of cranks and pseudoscientists. This is not merely a matter of flattering the self-esteem of scientists or reassuring them of their favored place in the scheme of things. In certain contexts—medical quackery for instance—it is literally true that lives are at stake. The press, however, seems increasingly incapable of making the distinction between authentic science, science fakery, and raw superstition. It is not only a question of sneering at science by admittedly escapist TV shows like *The X-Files.* Ostensible documentaries "document" the dissection of the supposed aliens that crashed at Roswell, New Mexico, or unblushingly instruct us that dinosaurs were contemporaneous with modern humans.[27]

There is worse. On one occasion, the respected ABC news-in-depth program *Nightline* devoted itself to touting the virtues of yet another perpetual-motion machine. There was no hint given that if the gadget performed as advertised, it would violate the law of conservation of energy and thereby overthrow the best-confirmed, most reliable science of the past three hundred years.[28] Nonsense combined with piety gets an even better press. A recent story in *U.S. News & World Report* celebrated a claim by a fundamentalist geologist that his computer model of convective flow in the Earth's mantle "proved" the validity of the Noachian flood myth.[29] In its eagerness to exploit the novel, startling, and (to believers) comforting qualities of the tale, the writer, one C. Burr, failed to note (among other things) how far the computer's "conclusions" depended upon the gross cooking of input data.[30] "Garbage in, garbage out" is the colloquial name for this kind of swindle— but not according to Mr. Burr and his editors. It is amply clear that so far as most of the media are concerned, credulity pays, skepticism does not. "Anything goes" seems to have become the epistemological watchword in journalistic precincts. As one astute and disillusioned science writer puts it, "We have moved from the concept of equality of individuals to equality of ideas and beliefs. . . . These days it is politically incorrect to call something a dumb idea."[31]

In this, as in many other things, Tocqueville seems to have been an inspired prophet of the dynamics of egalitarian democracy. He knew quite well that the emerging democratic ethos that he anatomized in nineteenth-century America harbored the postulate that since any one person is as good as any other, any one person's idea must be as valid, or at least as entitled to as much respect, as any other's. This rather questionable formulation of the fairness principle has become naturalized in the methodology of many working journalists. Arguments from authority are regarded with instant scorn and suspicion—a natural development, perhaps, in a corps of journalists nurtured on the duplicities of the Vietnam War, Watergate, Oliver North (not to mention Bill Clinton—and Kenneth Starr), but almost always a fatal mistake when dealing with scientific issues. The implicit logic is that since many who have held political authority have proved to be hypocrites and scoundrels, it follows that any claim to authority must be self-serving and probably false. This is hopeless gullibility posing as deep sophistication. It is also extraordinarily lazy, a way of letting myth do the work of thought. It avoids the difficult task of making important distinctions and the unpleasant necessity of alerting much of the public to its own shortcomings of judgment. Here is how Wallace Sampson, an inveterate crusader against all kinds of health quackery, sums up the typical journalistic formula for dealing with disputes in their area:

> The press amplifies and exacerbates the problem with its attempts at balance. Its ethic is ostensibly to present objective and balanced articles. But in reality, the technique for reporting medical pseudoscience is to find a proponent or satisfied patient, quote that source for two columns, and then "balance" that encomium by quoting a skeptical physician or scientist for one or two paragraphs. Of course, the piece concludes with a rebuttal of the skeptics by the original proponent. This formula is ubiquitous; it is widely demanded by editors and apparently taught in journalism schools.[32]

The methodology that Sampson scornfully describes is not limited to disputes over health claims, though it is probably there, more than anywhere else, that real damage is done. The same formulaic version of the claim-counterclaim ritual is observed in stories about UFOs or past-life regression, always with the views of the unconventional claimant given pride of place and the lion's share of type or airtime. For many writers, the mere brief citation of critics' views, without any extensive analysis of plausibility or consistency on either side, suffices to meet the canons of fair coverage. A number of factors, joined to the reflexive hostility to authority already cited, create this unwholesome pattern. From the point of view of writers as well as editors, a story that reports on an unconventional claim, only to shoot it down promptly, is an inevitable dud.[33] The readers who dote on these stories want to have their credulous propensities flattered, not derided. They want their hopes for miracles stoked, not doused. Those who delight in debunking comprise a specialty market, served by such periodicals as *Skeptic* and *Skeptical Inquirer*. They do not form a very desirable target group, from a commercial perspective. Far more important are the readers who faithfully consult the horoscope column, or the viewers who respond positively to Psychic Hotline ads. Even among the publications and broadcast media that specifically appeal to the most educated portion of the public, it is realized that these upscale groups nurture their own upscale version of credulousness, often with New Age resonances.[34] It was PBS, after all, that gave Deepak Chopra a broadcast outlet and that gave the earnest Bill Moyers leave to enlighten college-educated suburbanites about *Healing and the Mind*.[35]

In addition to case-hardened suspicion of authority and the cynical calculus reader and viewer statistics, a few other factors are responsible for the dismal performance of most of the media when reporting on the conflicts between science and unorthodoxy. What one might call the "Galileo effect" comes into play. No writer or journalist wants to be numbered among the mockers if a reviled maverick indeed turns out to be the herald of a revolution in knowledge. The fact that authentic Galileos are a good deal rarer than honest politicians, and that most real revolutionaries have emerged from the ranks of a scientific orthodoxy that quickly accommodated itself to their insights, does little to disarm the false caution that leads every crackpot claim to be treated as a potential breakthrough.[36] Another factor is that "controversies" with the potential to engage or entertain readers are often assigned to general-interest reporters with a presumed popular touch rather than specialists in science reporting. Even among the latter, however, weaknesses often appear. Many have educational backgrounds, both graduate and undergraduate, in the humanities rather than the sciences they write about. Nowadays, this usually means that they have at least a casual acquaintance with the postmodern epistemological skepticism that has made a comfortable home for itself among academic humanists. This doctrine is notoriously reluctant to concede that science and scientists deserve credit for reliable knowledge. But this is probably only a minor factor, though in some instances it has played a role.[37] More important by far has been the paucity of science writers with firsthand experience in scientific research, or even with a good background in history and philosophy of science.[38] This almost always entails shortfall in direct understanding of day-to-day research methodology and makes it difficult for the writer to grasp the reasons for the extraordinary epistemological success of contemporary science.

The result is a foreshortened perspective in which the differences between genuine science and pseudoscience appear much less drastic than they really are. Of course, this deficiency might be made up by directing people with advanced scientific training, and even some professional research experience, into journalistic careers. But, for the average scientist, young or old, who has put so much heart and soul into meeting the exacting requirements of a scientific, medical, or engineering career, the temptation to retool as a science reporter cannot be that strong. Moreover, while the successful scientist has no reason to consider a change of vocation, the unsuccessful scientist, who might welcome such a change, may also be embittered against the science he is supposed to cover as a journalist or popularizer. This is arguably what has happened in a few cases at least. All these factors have weakened the ability of science journalism to give accurate and informed accounts of the squabbles between science and pseudoscience, between medicine and quackery, that continue to erupt on virtually an everyday basis. Science has had too hard a time of it in the press, while fakery (or honest crankishness) has had something of a free ride. Science reporters have let us down, by and large. In the name of clear-mindedness, objectivity, and courage, they have engaged in behavior that reeks of muddle and bias, and that is indistinguishable in its effects from pusillanimity.

Science also plays a peculiarly understated part in the world of people who take ideas seriously—intellectuals, for want of a better name. Folks who take pride in being closely acquainted with contemporary developments in politics, literature, and art, who are well-informed about history and social theory, who bear all the earmarks of extensive education and wide travel, are nonetheless often only vaguely conversant with contemporary science and its principal themes. The province of scientific ideas is a partially sequestered precinct of the larger world of ideas, one that remains largely unvisited by nonresidents. Of course, specific events in science, technology, and medicine often spark strong interest among the broad public, which certainly includes the narrow constituency I am describing. As we have seen, questions like cloning or global warming or deforestation of the Amazon Basin elicit responses from a wide spectrum of thinkers and writers, representing every degree of scientific expertise (or lack thereof). But nonscientists usually come alive to scientific issues only when these are seen to have great public import. Occasionally, scientific ideas will catch the imagination of nonscientific intellectuals when they are used as emblems of currently fashionable philosophical positions. The recent surge of interest in "chaos theory," much of it naive and dreadfully ill-informed, is an example.[39] In general, nonscientists lose interest when the cultural resonances don't seem quite so compelling, or when detailed knowledge, rather than nodding familiarity with paraphrases, is required.

Familiarity with science seems like a grace note rather than an essential element to an intellectual's packet of credentials. It is a special touch that some members of the clan can bring to the conversation, but not a staple element. Scientific motifs may enter the discourse, but they must not overstay their welcome, nor make importunate demands. The attitude is perhaps best encapsulated by the title of Jacques Barzun's book *Science, the Glorious Entertainment*. Although a generation has passed since this was written, and the interval has seen a number of violent revolutions in intellectual fashion, his attitude—sometimes disguised, sometimes proudly displayed—is still widespread. Indeed, one point of the cult of postmodernism—an extravagant and transient set of conceits, no doubt, but one that reflects, in funhouse fashion, many more widely felt concerns—is that science is a mere "narrative," whose ostensible "truth" becomes a mere phantom outside the charmed circle of practitioners. This kind of folly is swallowed whole only within its own charmed circle, but, on an emotional level, it echoes what many nonscientists might wish to believe, did it not collide so violently with common sense. Aside from the more antic elements of the academy, the estrangement of intellectuals from science is not often stated in severe terms. Rather, it is the constant, if almost inaudible, harmonic of an attitude that resents the marginalization of most intellectual pursuits, other than science, in the world of money, politics, and power. This gulf has slowly developed over the centuries as science became more and more indispensable to the functioning of contemporary society, and more and more intricate and mathematical as well.

Contrast the current situation with the seventeenth century, when the divergence of scientific intellectuals from the more general community of scholars and thinkers had just begun. During that period mathematics was as much a gentlemanly accomplishment as the ability to read and write Latin. Science and mathematics were part of the general conversation. Recall that at its inception, the Royal Society contained such figures—highly "unscientific," as we would now think them—as the poet John Dryden, and the antiquarian John Aubrey. The state of affairs may be symbolized by a celebrated publication of the time, the journal *Acta Eruditorum*, which, for a generation, informed its readers about all sorts of important developments in European intellectual life. Its purview included geography, ethnology, jurisprudence, medicine, philosophy, theology, philology—and, of course, science and mathematics. It represented the compass of the subjects—they would hardly have been called "disciplines" at that point in history—that even modest erudition required. Its existence is testament to a lively and curious society in which all avenues to knowledge were assumed to be open to all discerning men (and, indeed, to some women).

Contrast that situation with today's intellectual life, which, for all its cosmopolitan pretensions, remains rather cloistered. Obviously, no publication now current plays the role of *Acta Eruditorum*. I have already discussed the flaws and virtues of the science coverage found in major newspapers and newsmagazines. The specialized science-for-the-general-reader publications—*Scientific American, Discover, The Sciences, American Scientist*, and so forth—do a respectable job, but obviously have little appeal to anyone not obsessed by science and technology.[40] These magazines constitute a pleasant, but isolated, intellectual ghetto. In any case, the "general reader" they appeal to often turns out to be a professional scientist.

The interesting thing is that the farther one goes up the scale of "intellectuality," the sparser and more fitful coverage of science becomes. Journals of opinion—the *New Republic, Nation, National Review*, and the like, give considerable coverage to general cultural issues but very little to science, unless it becomes commingled with current politics. Indeed, none of them even has a science editor, although volunteers could easily be found. The same holds of most of the "little" magazines—*Hudson Review, Partisan Review, Granta, Wilson Quarterly*, and so forth—wide ranging as they are in most other respects. Major bookchat publications—*New York Review of Books, Times Literary Supplement, London Review of Books*—cover science at best fitfully, more or less as a sideshow, and certainly not in proportion to the importance it holds in the contemporary worldview. It seems that a certain "gentlemanly" code prevails here. It is acceptable to know something about science, if one is not too ostentatious about it. It is also acceptable to know not very much science. One may declare one's perfect ignorance about the relation between DNA and proteins, or the fundamental theorem of algebra, or buckyballs, or the Einstein-Podolsky-Rosen *Gedankenexperiment,* without compromising one's status as a card-carrying intellectual. Even magazines like *Lingua Franca* and *Academe*, though supposedly addressed to the professoriate as a whole, generally give short shrift to science professors and are even more disdainful toward professors of engineering, who remain decidedly *infra dig* among the right-thinking.

Although the rather haphazard coverage of ongoing science by "high end" periodicals points to a number of interesting things—the generally poor science education of humanists and social scientists being one of them—even more interesting is the studious avoidance of an ancient but still potent philosophical theme: the implications of science, both as a body of knowledge and a subculture, for our understanding of humanity's place in the scheme of things. Rarely is this issue addressed head-on these days, though up to a few decades ago it was an obsessive theme in Western intellectual life.

A recent piece in *Salmagundi* by literature professor Marilynne Robinson is a rare exception,[41] which only underscores the general absence of this kind of discussion. Professor Robinson's piece is a fascinating, if faintly repellent, throwback. It flays "Darwinism" as the instrument that has desacralized our notions of the world and withered the moral authority of the Judeo-Christian biblical tradition. (Robinson is no garden-variety reactionary however; for her, "Darwinism" is the authority behind the endless campaign of the wealthy against the poor.) In 1897, such a piece, even with its political slant, would have been a standard item in a public discourse replete with questions of science versus belief and faith versus fact.[42] Nowadays, however, such dichotomies seem to be hidden under a veil of reticence. As a general rule, the media outlets of this country are terribly uncomfortable with such deep issues, especially if they challenge the shallow pieties that are the common coin of public discourse. There is obviously a very real disjunction between the materialist view of the world, which is central to the practice of science (if not the personal convictions of all scientists) and the religious view, whether sectarian and sharply defined or vague and generic. When is the last time that this disquieting question, which endlessly roiled the public discourse of the nineteenth century, was seriously addressed in a forum available to the broad public?[43] As with so many other contentious issues, the mass media seem to deal with this one best in jocular, just-kidding form through cartoon shows like *The Simpsons*.[44] "Serious" venues avoid it.

This might be understandable for television or for mass-market magazines, which have to take into consideration the fervent piety of the general public. But how might we account for a similar absence in the elite journals whose readership consists of decidedly secular academics and intellectuals, who, for the most part, will be happy to tell you that they have no use for churches and ritual, or even for religion as such? The answer, I think, lies in the fact that undiluted materialism, particularly when its reductionist premises are made clear, retains its power to dishearten even the resolutely secular. The argument against materialism is no longer heard, in other words, because it has essentially been won by the materialists, at least in the sense that there is a consensus that there is no effective way of assaulting it or even blurring its implications. A good part of its immunity derives from the sweeping success of physical and biological science over the last century. The only arguments against materialism are, paradoxically, those that come from the heart of the most materialistic of sciences, that is, those that appeal to the conundra that haunt physics under our present (very incomplete) understanding of quantum mechanics and cosmology. "Quantum mysticism," as we may call it, is a device that has frequently been used, even by some prominent physicists, to ward off the intolerable psychic consequences of unbending materialism.[45] The so-called anthropic principle in cosmology has recently come to play a similar role.[46] But these are poor shields at best, both because they lack the reassuring ontology and teleology of religion, and because they are rather desperate, ad hoc lunges for an escape hatch rather than confidently reasoned theories.[47] There is a wistful defiance in this quasi-theistic position: "What we are suggesting here is that those of us who do not wish to conclude, as Stephen W. Hawking apparently does in *A Brief History of Time*, that the scientific world-view is such that a belief in the active presence of God or Being in the cosmos is rather effectively disallowed now have good reasons for arguing and, more importantly, 'believing' that this is simply not the case."[48] But the conclusions are saturated with the desperate desire to believe. The resolutely materialistic positions of Hawking and Steven Weinberg are far more characteristic of scientists' beliefs and don't require such a procrustean struggle with the evidence.[49]

Secular intellectuals may be as free as they think they are of any desire to reintroduce theism into the basic assumptions of their discourse. But the fear of seeing emptiness where they look for ontological and teleological reassurance still haunts their dreams. Without any conscious, let alone malicious, intent on the part of scientists to do so, the inexorable rigors of the scientific worldview remorselessly undermine the ground on which cherished foundations of artistic and moral judgments try to stand. This kind of subversion is frequently threatened, of course, by philosophers and critics of the postmodernist persuasion. But it is a threat they cannot and really do not wish to carry out. Scratch a postmodernist and you will find a sentimentalist solidly welded to sturdy moral, political, and esthetic principles. The celebrated irony of the postmodernists increasingly dissolves away the nearer one gets to these cherished, and very foundational, tenets. The corrosive nihilism that decorates the surface of postmodernist manifestos is, in fact, a shell game, a subtle ritual against having to confront frankly the many huge leaps of faith implicit in the political nostrums simultaneously propounded. The scorn of science frequently uttered in the name of "anti-foundationalism" is a disguise for deep fear of the true anti-foundationalism of which science is the relentless agent.

Against their will and desire, therefore, postmodernists are actually leagued not only with traditionalists and the untraditionally religious, but equally with the vast majority of secular, liberal intellectuals. All across this broad spectrum there is a longing, disguised or open, for a restored sense of the world as linked, through its essential mechanisms, to human feeling, judgment, and hope. It is this that the scientific view of things simply disallows. Science, again without any conscious intention on the part of scientists, is an unrelenting gatekeeper against these assumptions, no matter how they are smuggled into the conversation. The dire implications of science may be contested, but the chances that any challenger might prevail seem, at this point, vanishingly slight. Consequently, humanists and others who might at some level want to elude these implications are disinclined to challenge them explicitly. Most scientists, for their part, even those quite comfortable with adamant materialism, are not particularly eager to rub anyone else's face in it—what would be the point of doing so? Thus the debate between the materialism of science and the widespread yearning for transcendence, between monism and dualism, has been resolved by a grudging peace where one side has pretty much triumphed on the merits, while the other side keeps the implications of that triumph decently hidden. This, in an ironic reversal, is the "Victorianism" of our generation of intellectuals. The Victorians may have shied away from human lubricity, but were all too eager to pursue the issue of how the ascendancy of science might or might not erode faith. Our own aggressive mongers of words and ideas, by contrast, are a heedlessly lickerish lot, but the bleak corollaries of materialism stab at their sense of decency, and hence must be shrouded although they can never be banished. For this reason, careful evaluation of the consequences not only of scientific ideas but, even more, of the scientific mode of scrutinizing the world, is a very infrequent visitor to the pages of most "serious" journals. Consequently, it is highly unlikely that a new avatar of the spirit of *Acta Eruditorum* will be reborn any time soon.

TWELVE

Plutocracy

Money is, of course, the working fluid of science. Even if scientists were as unworldly and monastic as popular imagery suggests—and, assuredly, they are not—funds have to flow in a rather steady, ungrudging stream, on one pretext or another, for research in any area to proceed. The need for money varies widely among scientific fields, depending upon the need for apparatus and overhead. In pure mathematics and theoretical physics, the outlays are almost absurdly small, compared to the quantity and quality of knowledge generated.[1] On the other hand, in high-energy physics or astronomy, or in anything that relies on satellites or space probes, the costs can be huge, that is, a measurable percentage of the national budget. In molecular biology, which many people would now regard as the most exciting aspect of science, both intellectually and in terms of practical promise, the costs of any individual project may be rather modest compared to the megaprojects of particle physics, but the number of such projects, together with the number of researchers involved, creates an enormous, ever-growing thirst for financial support. Notwithstanding these disparities, there is scarcely any area of scientific research that does not require a high degree of grantsmanship from its participants. At the same time, there are areas where a shrewd researcher with entrepreneurial instincts can parlay skills and results into a substantial fortune by becoming a businessman as well as—or instead of—a working scientist. This possibility has come to play an increasingly large role in the personal calculations and ambitions of scientists over the past few decades. The sense that science is a lifelong vocation set apart from the marketplace, pursued for its own sake with no financial expectations beyond modest middle-class comfort, has correspondingly eroded during that period. This has raised some rather justified fears for the disinterestedness of research and even for its intellectual integrity. It evokes the disconcerting possibility that profitability, rather than intrinsic intellectual interest or usefulness to the community as a whole, will shape the contours of research priorities.[2] If money is the lifeblood of science, it possesses considerable corrosive power as well.

The scenario I have just laid out reeks a bit of nostalgia, perhaps. It sug-

gests that there was, not very long ago, a golden age when scientists constituted a quasi-monastic order of truth seekers who, having renounced worldly
ambition, spent their lives shuffling between lab and lecture room with no
thought for personal wealth. It suggests a world where knowledge for its own
sake, curiosity-driven science, was the unchallenged ethos, and where the
notion that scientific knowledge might be proprietary, that it might be a
valuable asset to be held secretly for as long as possible, would have offended
the conscience of all right-thinking scientists. This cannot be sustained in
its naive form, at least not when we speak of the last sixty years or so. For
one thing, it is all too obvious that the long engagement of science with
military research has accustomed scientists to think of knowledge as a commodity, one with enormous value outside the scientific community in some
instances and therefore subject to hoarding and monopolistic exploitation.
Moreover, corporate science is hardly new. It has been a concomitant of in-
dustrialism since the late nineteenth century. The General Electric Corporation, for instance, owes much of its preeminence to a research effort led by
the great Charles P. Steinmetz (himself a dedicated and principled socialist). Hordes of chemists contributed to the ascendancy of the German firm
I. G. Farben. The skills of oil geologists have been at a premium for a century, as are the skills of computer experts today.[3] Even the supposedly unworldly Einstein transformed himself at times into a very practical inventor,
with every intention of reaping the profits from his ingenuity.[4]

Nonetheless, throughout this period scientists have constantly made a distinction between work done for the advancement of general knowledge and
that done for the specific purposes of business or government. In part, this
follows the rather snobbish distinction between "pure" and "applied" science, but this can be misleading, since much of what is usually designated
"applied science" is essentially curiosity driven and openly published, and
is not intended for the immediate benefit of any specific corporation or government organization. A special sanctity has always been attached to the
notion of basic research. This is because such work is dictated by the broad
logic of the discipline and the search for fundamental understanding, rather
than by the imperative for quick, useful solutions to narrow practical questions. Even beyond this, however, basic research has an undeniable appeal
to scientists as a realm which respects their autonomy and their sovereign
power to direct their own work as they see fit. If I may have recourse to Marxian categories, basic research—curiosity-driven research—can be described
as "unalienated labor," the fruits of which—psychic, more than financial—
accrue entirely to the researcher. By contrast, even the most brilliant work,
when done "to order" for a corporate or governmental sponsor, carries, however faintly, the taint of subordination to a purpose other than that of the
scientist who carries it out.

For decades, the ethos of science as done in this country and most of the industrialized world has demanded, and received, a special refuge for basic research and science for its own sake. Universities, sometimes augmented by special research institutes, constitute the heart of this system. Before World War II, much of this effort was privately funded through wealthy donors and philanthropic foundations. The war, however, brought in the era of "big science," characterized by large-scale projects employing hundreds of scientists and technicians. It also ushered in a philosophical shift in government policy, which stressed the new and vital role of direct government funding to nurture and sustain basic research. The National Science Foundation was the flagship of this effort, making direct grants to thousands of scientists in every discipline, although in medical research and much of basic biological science the National Institutes of Health took the lead. Their role was augmented by the contributions of several other government agencies, including, of course, the Department of Defense.

A word might be said here about the diverse quality of Pentagon support of science. In the standard left-liberal view of things, which is the received version within the university community, the military is the font of all wickedness, and every move it made between 1945 and 1990 is instantly assumed to have been linked to one Cold War machination or another. Certainly, most military-supported research had a direct military end in view, in the form of weaponry, intelligence-gathering capability, or encryption and code-breaking technique. Nonetheless, in the 1950s and '60s the Pentagon developed a modest tradition of supporting pure research with no immediate prospect of applicability in or apart from a military context. Many pure mathematicians, for instance, benefited from this largesse, without tainting their work with the smallest hint of applicability.[5] It might well be argued that DoD motives in doing this were hardly so beneficent as they appear on the surface. Perhaps the idea was to buy the goodwill of a "reserve army" of scientific talent, against the day when it might be needed for immediately useful research. Possibly, the aim was simply to humanize the Pentagon's grim image among an influential group of citizens. My own view is that, in some cases at least, there was a genuine desire on the part of both uniformed and civilian bureaucrats to contribute to the development of American science as such, without any hidden agenda in mind.

Whatever the true motives of the military, Pentagon money was only part of a support system that enabled American science to flourish on an unprecedented scale in the decades following the war. The centerpiece of scientific research was undoubtedly high-energy physics, a complex and expensive enterprise requiring elaborate machines served by armies of experimental physicists, urged on by the explosive development of theoretical understanding at the subatomic level. Again, conventional left-wing mythology often gives a sinister interpretation to all this, assuming that "high energy" equates to weapons technology. In this it is supported by the rationales offered to the politicians who were importuned to vote funds for these projects. Scientists and their bureaucratic allies incessantly affirmed that particle physics was a vital element of "national security." This was, however, a delusion; in point of technology directly useful to the military (among others), solid-state physics was infinitely more valuable than the esoteric investigations of the high-energy community, a fact that eventually surfaced in public debate and played a role in the downfall of the superconducting supercollider. For several decades, however, particle physics was the flagship of a vast and generously supported national research effort that vaulted American science into world leadership across the board.

Most scientists who were active in the 1950s and '60s look back on this period with special fondness as an era when the horizon open to an ambitious young scientist appeared boundless. It seemed that jobs in university science departments or in research establishments would always be there for the asking, and that funding from one source or another would always be available for worthwhile research projects. Since that time, there has been a gradual, sometimes hardly visible, but nonetheless remorseless decline in this ebullience, as job prospects have shrunk and funding has become more and more stingy. The reasons for this shrinkage in support for science are many and varied. Disparate and apparently unrelated forces have converged to diminish the ability of American science to conscript the resources it needs to continue the great postwar expansion. As was noted in the previous chapter, the financial demands of the space program, especially the Space Shuttle, though glossed with declarations of sterling scientific significance, turned it into an enormous drain on funds that might otherwise have been available for far more productive research. Moreover, according to some observers, the use of increasingly elaborate technology has driven the cost of all kinds of scientific projects into the stratosphere, thus limiting the number that can be supported: "[R]esearch is becoming more and more technologically sophisticated and more and more dependent on elaborate, and expensive apparatus. This apparatus has a very high rate of obsolescence, and has to be written off over just a few years. Labour costs, also, are becoming less stable, with larger numbers of research and technical staff employed on short contracts to carry out specialized tasks."[6]

It is clear, however, that neither the increase in the pool of research scientists nor the growing "per capita" cost of supporting these scientists can fully account for the diminution in financial support. Though the 1970s and '80s are sometimes reckoned as a period of relatively poor performance for the American economy, the aggregate result of the past quarter century is that both total and per capita economic output have grown considerably in this country. Consequently, the United States is now the dominant economic power in the world to an extent not seen since the 1960s. In short, financial resources certainly exist to match easily the level of science support prevalent during that decade.

Many people advance the argument that support of science was primarily a weapon of the Cold War. Supposedly, the demise of the Soviet Union and its allied Communist governments, the rapprochement of the West with a Chinese regime that now seems quite estranged from the ideal of working-class revolution, and the overall recession of Marxist movements throughout the world allowed the ruling elite of this country to stop pampering scientists as it once did. This is far from convincing. One might have expected, overall, that the relaxation of the international atmosphere and the consequent reduction in military research would have freed up at least a modest amount of funding for pure science and for nonmilitary technological development. This worthy outcome appears to have been frustrated, in the short run at least, by a widespread political and cultural uncertainty about the role of science, a phenomenon in which the liberal left and the conservative right have proved unwitting collaborators.

On the left, science has been under suspicion since the 1960s because of the entanglement of science with the military, particularly in the context of the Vietnam War and the nuclear arms race. Added to this, the upsurge of an urgent environmental movement, usually with liberal support, also added to the myth of science as an institution inimical to essential human values (although the severity of the critique is by no means uniform among "Greens," many of whom actually are scientists). Further, the liberal community, especially within its university base, nurtured a brand of academic radicalism that pugnaciously challenged orthodox science as part of the wider spectrum of institutions venerated by Western culture. Science, from this perspective, emerged as the impermissibly hegemonic embodiment of a legacy of male dominance, racism, and imperial rule over the non-white peoples of the world. This mood was defined by aphorisms equating "the universality of science" with "cognitive colonialism."[7] It drew much of its doctrinal support from the poststructuralist dogmatics that had gained ascendancy among French intellectuals in the "post '68" atmosphere.[8] It was responsible, as well, for the growth of the quasi-discipline of "science studies," particularly that branch characterized by a radically social-constructivist theory of knowledge (or, perhaps, "knowledge"). Whatever the causes, however, a discernible current within liberal ideology had turned against science for the first time since the Enlightenment.[9]

At the same time, clouds were gathering in the form of the increasingly dogmatic right-wing economic theories that came to dominate conservative and Republican politics during the Reagan era. It became an article of faith that any federal expenditure—any governmental expenditure, for that matter—for anything other than national security and police functions is confiscatory and utterly reprehensible. As has always been noted, the Reagan administration largely avoided putting this idea into practice, except in symbolic ways, but the theory took firm hold of the rank and file.[10] Thus, in the 1990s, conservatives in Washington and in state governments became much more draconian in their determination to dismember "big government." In contrast to liberal ideology, conservative doctrine in this country seems very strongly favorable to science—but with a catch. "Science" in this context usually means science that can be swiftly exploited as marketable technology or as military hardware. "Curiosity-driven" science, by contrast, draws marked skepticism. The unqualified worship of the market has the effect of further cooling the ardor of the right for purely theoretical knowledge. The tendency is to think that if research cannot sustain itself through its commercial possibilities, perhaps it ought not be done.

The liberal and conservative versions of skepticism toward science made common cause in the fight over the superconducting supercollider. The concern of the left—a not altogether dishonorable concern, shared by some scientists—was that the project siphoned off moneys that might otherwise be used for vital human services to the needy. Factions on the right, on the other hand, argued against the expense as such.[11] Moreover, support for the supercollider was drastically weakened on every front when politicians belatedly realized how extremely theoretical high energy physics is, and how far distant it is from foreseeable practical applications. The right was chagrined to learn, at long last, that virtually no military technology had or ever would emerge from this research. The liberals were almost equally put out that no social benefits would ensue. Commentators on both sides denounced particle physics as a costly way to answer purely philosophical questions about the ultimate—or, worse still, less-than-ultimate—nature of matter. In consequence, the supercollider was canceled in an especially emphatic way. The noses of particle physicists were rubbed in their own arrogance!

This decision was in part a symbol of wider developments, in part a harbinger of new ones. Research funding shrunk across the board during the 1990s, especially when the conservatives took control of Congress following the 1994 elections. The deification of the marketplace and of private funding as the only truly legitimate source of research money remains a strong theme among conservatives, even those who claim to be science enthusiasts. They are perversely echoed by at least some critics on the left, who also oppose government funding of science, but for antithetical reasons. Herewith, Richard Sclove, a prominent crusader for a science "accountable" to the values treasured by the left:

> But assume for the sake of argument that government funded science does markedly accelerate commercial innovation. Is the obvious conclusion that we should increase public funding of science *as it is presently organized*? Only if you believe that commercial innovation is tightly correlated with the overall

public good. Of course, we all know that commercial innovation delivers many useful products. . . . But it also contributes to plenty of social results we don't want (cardboard-flavored tomatoes, toxic wastes, global climate change, unneeded military weapons systems, job insecurity, and everyday stress and speedup) while failing to deliver other results we do want (a just and environmentally sustainable economy, vibrant communities, healthy families, adequate leisure time, humane medical care, deep insight into social problems, etc.).[12]

It is hard to read a passage like this without rueful reflection on how far this petulant, indiscriminant, almost adolescent, brand of utopianism has estranged the American left from any prospect of significant political influence. But then, one is disconcerted to note that even some scientists have joined the right-wing bandwagon, arguing for the ineffable wisdom of the market and the ineluctable evil of everything funded by the state. Terence Kealey, a British biochemist, is particularly absolutist in his doctrine and florid in his rhetoric:

The decision to fund science through the State rather than voluntarily is, moreover, morally flawed, since the State forces all people to support an activity to which some individuals are opposed. There are people such as religious activists, or believers in animal rights . . . who find the activities of scientists offensive, and yet who are compelled by the State to support it through their taxes. That is an immoral imposition on individual will by the collective. Furthermore, *dirigisme* destroys individual morality, which thrives when it is exercised under *laissez-faire*, but which atrophies with disuse when the State takes over the responsibility for moral decisions.[13]

And what is to take the place of government funding? "Science will attract generous patrons—if taxation spares them. Almost everyone in a wealthy country believes that pure science should be funded. . . . The historical evidence shows that the empowerment of wealthy men by money breeds a sense of responsibility, which inspires them to endow research and universities."[14] Ignoring the obvious retort that there are also people who object to being taxed for schools, police forces, armies, and streetlights, we may note that recent experience among American scientists suggests that pure science has thrived when government funding has been open handed, while private funding of such work has shrunk in tandem with recent government cutbacks. But then, from the point of view of market idolatry, says Kealey, pure science might not be that important anyway: "By pouring money into academic science, the French followed the wrong model. But *laissez-faire* Britain, freed from centralized planning, could follow the market, which correctly directed investment into technology, not science."[15]

This flies in the face of long experience strongly indicating that fertile and inventive technology often depends on serendipity rather than on development projects undertaken with the aim of short-term profitability. A wide, growing base of basic knowledge seems to provide the surest warrant for technological advance. Critics of government support of science often point to the failure of touted governmental R&D programs here and abroad to produce promised breakthroughs, whether in the "cancer wars" or in the development of "fifth generation" computers. (Oddly, the right mutes such criticism when addressing the development of purely military technology.) What is overlooked is the fact that stunning technological advance can arise from the opportunistic exploitation of basic research developments originally undertaken with no such end in mind. The examples I know best come from pure mathematics and foundational physics. The basic idea behind CAT scans is embodied in a theorem of N. Radon, proved eighty years ago as part of a purely abstract investigation in the branch of mathematics called functional analysis. A more recent example is the emergence of the "open key" encryption systems for communications security. These emerged from work by mathematicians in the esoteric area of number theory. New systems of data compression, representing vast economic efficiency in all sorts of telecommunications and data retrieval systems, have grown out of the mathematical investigations leading to "wavelet" theory. Above all, the virtually limitless bounty provided by solid-state physics grew directly from the development of quantum mechanics, an achievement spurred originally by the most esoteric of motivations. Examples like this—exploiting theoretical results from chemistry and biology as well as mathematics and physics—could be multiplied endlessly.[16]

It must be noted that, as of this writing, the immediate prospects for government funding of scientific work, including basic research, have brightened, but the issue is still in doubt. Congressional and administration moves to increase appropriations for traditional funding agencies have been countered by the continuing budget-cutting fervor of some fiscal conservatives and it is not clear what accommodation will finally be reached.[17] Some elements of even the conservative leadership are probably well aware of the economic leverage provided by a healthy and active research community. But long years of ritual denunciation of federal spending make it difficult to act forthrightly on this insight. It is quite possible that research science will continue to be undernourished, a sacrifice to right-wing dogma.

There is another aspect of the entente between financial interests and right-wing politics which is clearly injurious to science. I trust that few, other than Republican spin doctors, will be greatly offended by the observation that the core ideology of the Republican party is essentially plutocratic, that the central aim of the party is to preserve and advance the interests of a rather small fraction of wealthy Americans. (The Democratic party, on the other hand, can hardly be said to be antiplutocratic, but it does have a tradition of trying to broker a social contract that gives working people and the poor at least a modicum of protection from the vicissitudes of the unconstrained market.) Although this is a truism to anyone who keeps an eye on the workings of the political process, especially in areas like taxation and regulatory policy, it is not the kind of thing useful for Republicans to proclaim outright at election time. It hardly catalyzes the kind of coalition necessary for electoral success. The strategy of the Republican party is to remedy this defect by, first of all, playing on the race issue (in its varied disguises) as much as possible in order to portray itself as the party of white people. Further, it has entered alliances with a number of single-issue groups: the gun lobby, the so-called Right-to-Life movement, and the theologically and socially conservative religious forces represented by the Christian Coalition. Increasingly, the appeal to conservative Christians has been cemented by the endorsement of the creationist position in one guise or another. Conservative intellectual journals which, only a few years ago, fastidiously steered clear of the issue, are now providing a platform for purveyors of creationist fantasies.[18] How much additional damage to the prospects for responsible science education will be done by this cynical opportunism is anyone's guess.

Coincident with decreasing funding for scientific research, and, arguably, partially caused by it, there has been a major change in the relationship between researchers and private enterprise over the past couple of decades. More and more scientists have come to regard it as natural that their work should dovetail with profit-making enterprises, either their own companies or corporate sponsors and benefactors. Where once universities tended to keep their precincts free from the taint of commercial connection, they now actively court it. Similarly, the propensity of university officials to reckon the academic reputations of their science faculty purely in terms of scholarly publication has given way, in many respects, to outright demands for research that can be linked with the interests of corporations and lead to remunerative financial arrangements.[19] Deans now routinely canvass professors to see if something commercially exploitable might be in the pipeline and offer to broker agreements with business firms. Corporate-sponsored research funds more and more labs. Venture capitalists have become a standard source of funding as well.

By the same token, many scientists have begun to think of themselves as having dual careers. To have founded a commercially successful enterprise exploiting one's work is just as honorable, and considerably more profitable, than having one's own endowed chair. Researchers who, twenty years ago, would have spent their working lives ensconced in academia are now being sought out by aggressive private firms that flaunt the lure of substantial wealth along with the opportunity to do cutting-edge science.[20] The link between science and money, never entirely absent, is, at this point obvious to all, and unapologetically welcomed by many.

The question, then, is how far this frank embrace of commercial values threatens to distort or inhibit scientific work. There are a number of obvious dangers. First of all, the increased stress on work that will have short-range applicability, results that can be quickly turned to profitable purposes, obviously decreases attention given to basic problems of largely theoretical interest whose solution cannot be directly exploited in the marketplace. Even more, it threatens to lure budding scientists away from fields where the business links are scant or nonexistent to other areas where money abounds. This is not precisely a new problem. There has always been a tug-of-war between the subjective satisfactions of purely academic work and the high salaries of corporate science. But the monetary disparities have grown greater, even as the ethic which accorded special prestige to work undertaken purely "in the interest of science" has weakened. We are still in a transitional era when the changes obviously underway have not completely gelled, so it is impossible to predict the precise consequences for basic, curiosity-driven science. It would be foolish, however, to rule out the possibility that these will be drastic.

A further problem that occurs with more and more frequency is the alteration of the traditional relation between academic institutions and research faculty. The publish-or-perish ethos, by which most science departments in major universities are governed, has always drawn criticism, particularly from the perspective of undergraduate education. In its defense, however, one could always point to the resulting productivity of the scientists thus motivated. At least in the United States, the higher-education system, taken as a whole, is the historical mainstay of "pure" science, science done with no thought for practical applicability. Furthermore, it has provided a substantial part of the infrastructure of applied research as well. The historical justification of this system has always involved a white lie—the assertion that a faculty engaged in cutting-edge scientific research is necessary if undergraduate teaching is to be first-rate. This is certainly a gross exaggeration and, in many instances, palpably false. I frankly confess that 90 percent of the students I teach—mostly freshmen and sophomores taking "service" courses from the mathematics department—could be taught just as well by people having only a fraction as much training in mathematics as I do. Moreover, the advanced undergraduate and foundational graduate courses I teach from time to time, while necessitating a considerably deeper understanding of mathematics, certainly don't require a "research active" instructor. I assert that the situation is similar in science and engineering across the board. It is only at the level of advanced doctoral work that a student needs to be brought into contact with the frontier of current research. If supporting the educational needs of students—undergraduates, in particular—were as important to their "missions" as most institutions claim, the "research culture" would receive far less emphasis. Fortunately for the health of science, university administrations, as well as faculties, have quietly but firmly adhered to another ethic, one that emphasizes the utter necessity of ongoing inquiry, supported by enough time for research and publication (in the humanities as well as science). This is viewed as an obligation of our civilization as a whole to its own best traditions. However, in a culture as saturated with the urgencies of commerce as ours, it is surpassingly difficult to create the social and economic space necessary to sustain such activity. Even the philanthropic foundations which once contributed to its support are now largely governed by other considerations, more along the lines of social reform and direct action on behalf of the dispossessed. *Faute de mieux*, the task has fallen to the American research university. The traditional, unspoken bargain is that professors are given light teaching loads and paid far more than their teaching is worth in return for a heavy commitment to research and publication. Promotion and tenure are largely predicated on research success. Status and prestige are firmly linked to celebrity in the research world.

It can be argued with some justification that in many humanities fields the amount of research that really needs to be done is considerably less than would justify all the professors with "research" careers. It can be argued with even more justification that in many humanities fields, this ethos has produced not valuable scholarship, but an elaborate parody thereof, where triviality, novelty, theatrics, and political grandstanding have largely expropriated the perquisites that belong to honest research.[21] Chauvinistic as it may seem, I would reject any attempt to apply the same argument to the sciences. Here the moral, as well as the practical, exigencies of our society require more, rather than less, work in science. This means, in practical terms, that universities must continue to embrace their traditional role, providing incomes and titles to thousands of people who spend the greatest part of their time on research. Unfortunately, however, that bargain is beginning to fray at the edges and may well unravel within a decade. More and more universities are coming to look on scientific research not as something to which they owe financial as well as moral support, but rather, as a source of financial support to them.

Increasingly, the hiring and retention of faculty has come to be conditioned on grantsmanship, the kind of "soft money" support that not only covers the salary of the primary researcher, but those of subordinates and graduate students as well. Moreover, the institution itself is often in line for a percentage of the grant, on the basis of "direct" or "indirect" costs. Coming at a time when direct government and foundation support has been grudging, this impels universities to invite corporations to step into the role of sponsor of "sponsored research." This is something that corporations are reluctant to do, obviously, except on a quid pro quo basis, where the research sponsored must be of immediate tangible benefit to the firm that puts up the cash for it. Moreover, quondam scholarly institutions have grown eager to enter into direct working partnerships with businesses, agreements which convert university labs into the development arms of companies. Faculty members are now urged to let administrations know about patentable ideas they may have come up with, with a view to a cozy profit-sharing arrangement. This was certainly one of the factors that led to the cold-fusion debacle at the University of Utah a few years ago. But even where the results are considerably less risible and considerably more remunerative, something deleterious has happened to the culture of knowledge and science. The idea that intelligent curiosity is itself worth nurturing and rewarding within the academic cloister has been degraded, a decline that continues day by day.

Inevitably, the entire research culture is being skewed from one in which the internal logic and intellectual imperatives of a field are paramount to one which mimics the short-run value system of industrial research. "Pure" science, especially when it requires expensive physical resources, is increasingly seen as a luxury item. Administrators these days are far less inclined to view such work as an ornament to their institutions and far more inclined to view it as a burden. Unless it can pay its own way through grants and contracts, research comes to seem like a hungry white elephant.

Moreover, pressures from another direction are eroding the traditional status of the research-oriented faculty. Costs are soaring, and students who can afford to pay their own way in full have become relatively scarce. Most state-supported schools have felt the chill of increasingly stingy legislatures on their necks, a result, to a great extent, of the ideological success of the right in promoting the idea that taxes are the ultimate evil and that all government expenditure is frivolous. Consequently, at a time of unparalleled prosperity, higher education paradoxically finds itself in an increasingly desperate financial squeeze. The result is that paying students are now prized clients for whom institutions, except those at the highest level of prestige and financial autonomy, compete with increasing fervor—and cynicism. More and more, students are thought of as "customers" whose preferences and whims must be accommodated. At best, this might be viewed as an imperative for improvement in the quality of classroom instruction, and thus a boost to the status of gifted teachers who are not particularly productive researchers—something many thoughtful critics of higher education have long sought. More often, however, competition for students endangers educational standards more than it enhances professorial classroom performance. The idea that the university represents a long-standing tradition which has the duty, as well as the right, to impose disciplined intellectual standards on raw young people comes to seem quaint and irrelevant. Courses of study—indeed, individual courses—are increasingly shaped to accommodate preexisting adolescent tastes and pleasures. This accounts, even beyond the standard effects of academic trend chasing, for the success of such innovations as "cultural studies." It is convenient to have an ideological excuse for plying students with courses on sitcoms, horror flicks, and rock albums. The idea of the instructor as authority gives way, insensibly, to the notion of a professor as a glorified peer counselor, whose classroom is to be used for the sharing of emotional confidences. Needless to say, grade inflation—an ancient and intractable problem—continues apace.

Of all the major areas of study, science is the hardest to convert into this "user friendly" mode. No doubt, reforms in the format and methodology of elementary courses are possible. Presumably, they can make these experiences more valuable, as well as more agreeable, to the typical student. Some of these useful reforms may even involve modifying the impersonality of traditional science classes, and encouraging more direct personal engagement between instructor and student in freshman calculus or freshman chemistry. But the process has limits, and these are soon reached, especially for science and engineering majors. One of the things a student must master in a technical field is the ability to absorb and integrate information from formal presentations, in which time for student-teacher interchange is limited, and from which divergations into the subjective, emotional end of life are wholly excluded. To put it plainly, students must learn, in the time-honored way, to take lecture notes and to put time into decoding and assimilating them, which usually involves studying formally written textbooks and doing taxing exercises. There is no royal road circumventing this process, and the best efforts of eager-to-please administrators will not persuade responsible scientists to convert their undergraduate courses into pleasure jaunts. But these administrators, especially at private institutions, are in a bind because of the enrollment problem. The likely outcome is simply that serious science will become a more and more marginalized part of the curriculum as it answers less and less to the recruiting needs of the school. Even worse, cost-conscious officials will be drawn to the solution of altering faculty demographics, so that tenured, research-oriented scientists will decline in numbers, to be supplanted by adjunct faculty or academic gypsies who, far from being confronted with the classic "publish or perish" imperative, will face utter indifference to their research aspirations.

Of course, this phenomenon is hardly limited to science. But again, disciplinary chauvinism compels me to believe that the effects are far more disastrous when the victim is research science rather than, say, literary criticism. As this change of priorities bites deeper and deeper into the assumptions of academic culture, science graduate programs—expensive in terms of resources and difficult to fund—will also come to seem like dispensable luxuries. Why pay an arm and a leg for graduate programs in mathematics or astronomy, when the institution can show black ink by providing master's degrees in business or marketing?[22] I stress here that I am talking about a process that has been going on for more than a decade, but which now threatens to become truly virulent.

There are countervailing tendencies now at work or at least latent. The country simply cannot allow its sources of scientific innovation and scientific talent to dry up. Business leaders, even when they are reactionary outright, generally recognize this fact. Consequently, there is a very good chance that the rulers of our economy will step in to rescue most of the present system of academic research and training. Corporations will probably use their influence and even their own money to keep the most vital parts of the university science community from withering away. But there is, most likely, a cost to be paid. Increasingly, priorities and directions, even inside university departments, will be aligned with the needs and desires of the plutocracy. Jobs and titles will be salvaged, financial well-being will probably be protected, but the autonomy of research science will be eroded. The quid pro quo ethic will be starkly obvious to all. This doesn't mean that "pure" research will stop, but it is likely to be rationed. Fewer careers will run solely on that track.

The deification of market values has further untoward consequences for science. Traditionally, most of science, especially the research results coming out of universities, was not considered an economic good. It was published openly—indeed, that was the only way for a scientist to get due credit—and universally available, gratis, to other scientists throughout the world. There were exceptions, but these were few and special. The idea that scientific knowledge might be proprietary, the trade secret of an inner circle primed to take commercial advantage of it, was obnoxious to the university community. Even in the case of research deemed to be of military or national security value, the ethic of scientific openness often prevented sequestration.[23] Again, this view of things is leaching out of university life as the ethic of commercial exploitation seeps in. With more and more researchers acting as virtual freelancers, as much the employees of the firm that provides their contract money as of the university that provides their academic base, market values threaten to become enshrined in the research community. As a recent study points out, "more than half of the university scientists who received gifts from drug or biotechnology companies admitted that the donors expected to exert influence over their work, including review of academic papers before publication and patent rights for commercial discoveries."[24] Scientists still pay lip service to open publication, but for certain types of research this is in the process of fading to mere ritual genuflection. With labs functioning as fiefdoms of corporations and university administrators eager to reap royalties, and with university scientists equally eager to parlay their talents into entrepreneurial success, swift and open publication of work starts to look like a perversely profligate and incautious practice. Why give away a salable process for free? Why tip off competitors to a valuable insight?

Even more problematical is the fate of scientifically interesting but commercially unexploitable lines of research. As pressure mounts to set aside scientific imperatives in favor of market considerations, fewer scientists can afford to be "curiosity driven." Commercial entanglement with university laboratories threatens even worse ethical disasters. In a famous case not long ago, a school brokered a contract between a medical researcher and a pharmaceutical firm. The researcher was to test the company's product against nonproprietary generic analogues, in the expectation that the patented product would prove superior. But her work showed otherwise. In reaction, the company, with the full backing of the school, forbade publication of the results, until the case came to public light and the institution was forced to relent and change its policy.[25] Yet the school's embarrassed backtracking was merely a reprieve. Such arrangements are bound, eventually, to seem less and less out of line as more schools become de facto partners of corporations and as more scientists take on the ethical values, as well as the concrete interests, of the commercial world.[26]

Corporations seem equally willing to try to stifle researchers whom they have not subsidized. The case of Dr. Deborah Swackhamer is quite instructive.[27] Swackhamer is a University of Minnesota environmental chemist whose work has cast suspicion on the paper industry as a source of toxaphene pollution in Lake Superior. Initially, her work came under strong attack from a number of scientists on the payrolls of large paper companies. Subsequently, things became much nastier when she was the victim of a massive raid on her research materials, notes, and financial records. This was carried out by a corporate law firm (on behalf of an anonymous client) brandishing the broad authority of the Freedom of Information Act. Swackhamer was subject to these demands because her research support comes from the Environmental Protection Agency.[28] Obviously, this seizure, which intruded on Swackhamer's personal life as well as her professional freedom, was meant to harass and intimidate, and also, obviously, to distract her from her work. It was clearly an act of blackmail. In this case, the attempt to intimidate a researcher may have blown up in the faces of the would-be extortionists because of the wide publicity generated. Nonetheless, it constitutes an ominous precedent that will be carefully noted by thousands of other scientists whose work might run afoul of wealthy interests.

The image of scientist-as-mercenary is a durable fixture of populist demonology. The idea that it is commonplace for researchers to craft their opinions to serve the interests of their paymasters is routinely rolled out in debates over matters ranging from low-frequency radiation to the danger of silicone implants.[29] Obviously, in the case just cited, some of Dr. Swackhamer's enemies fall under that suspicion. Given the history of medical "researchers" who have colluded with the tobacco industry to disparage the idea that cigarette smoking causes lung cancer, it would hardly be fair to dismiss this image as the mere product of paranoia.[30] Scientists, after all, are not immune to corruption ex officio. But on the whole, there has been more fantasy than truth to such accusations. However, the more closely research becomes integrated with the world of investors and speculators, the more reasonable it will seem to regard science as potentially, even probably, tainted by the pecuniary interests and corporate commitments of scientists. The reputation of scientists for rendering judgments that are at least fair-minded, if not inevitably correct, has rested on the perception that science, as an institution, occupies a social domain where the narrow calculations of bankers and business people are largely scorned in deference to a higher ideal of truth and disinterestedness. Nonetheless, as collusion between science and rampant capitalism becomes more blatant, this image is likely to fray. And more than an image might be at stake for, indeed, the monastic picture of science, mythical as it may be, contains a germ of truth. There is in fact a scientific ethic, and disinterestedness is a vital component of that ethic. Imperfect beings as they doubtless are, over the years scientists have nevertheless done a reasonably good job of living up to that ethic. But financial temptation can be a powerful solvent of principle. How fragile that ethic is, how vulnerable to the corrosive force of common greed, has yet to be established. The results may well prove disheartening.

In America, as in much of the world, plutocratic values, marking out wealth and its concomitant power as the only measures of success, have become predominant and nearly unchallengeable. The faint distaste with which popular sentiment traditionally views extreme wealth has faded almost completely as our organs of publicity ceaselessly ram home the message that the market is the ultimate arbiter of all value. Science has been partially insulated from this tendency, historically, and it is hard not to believe that such insulation is valuable. Perhaps it is sentimental to insist that there must be a core of unselfishness to science, that scientists must retain a sense that mundane affairs are relatively inconsequential if they are to be psychically attuned to the toughest demands of the calling.

Perhaps it is mere prejudice to believe that a single-minded desire to become wealthy inevitably coarsens the delicate mechanism upon which the deepest scientific insights depend. Nonetheless, I suspect that something like this is the case, at least in the long run. But the barrier, such as it is, between the traditional ethos of science and the unfettered market worship that infuses our present cultural mood seems to be eroding. Increasing numbers of scientists have frankly embraced the rather giddy *enrichez vous* mentality that drives fund managers and venture capitalists. For them, science is more and more an instrument for reaching the monetary big leagues, less and less a vocation whose values transcend the obsessions of a particular era.

Scientists who scorn the gravy train are, for the moment, more the norm than the exception. But the signs are ominous, especially when one considers the possibility that the every-man-for-himself ethic which right-wing politicians have foisted upon us may become ineradicably engraved in our culture. This will certainly wither the public and governmental openhandedness on which disinterested science, curiosity-driven science, science as a cultural achievement rather than as an adjunct to the marketplace, has always depended. It might well establish entrepreneurial science as the increasingly cynical norm and thereby turn science into the kind of monstrosity its most paranoid critics have always warned against. The best hope for blocking such a catastrophic development is probably not within reach of the scientific community merely acting on its own. It depends on the emergence of a moral and political counterforce that would at least be capable of contesting plutocratic values and opening up ideological space for a more generous notion of public good. How that might happen remains, for the moment, a vexing mystery.

THIRTEEN

Democracy

Is it possible to love democracy and yet remain vigilantly suspicious of the demotic? This position—one might call it with some justice "Tocquevillian"—obviously involves chronic psychic discomfort if one is determined to maintain it honestly. It is a continual balancing act, performed to the resounding clash of conflicting ideals and insights. In our public rhetoric, democracy is the ultimate trump card. It is the touchstone of all political virtue and the arbiter of all values. Consequently, it is the idol in the public square, which no politician with a taste for survival will ever be seen to deface.

Nonetheless, there cannot be many thoughtful people who have never found an occasion to question the absolute virtue of the democratic ethos or to wonder whether the interests of the general community might be best served, in some instances, by institutions that are frankly undemocratic. The most militant advocate of unsullied participatory democracy occasionally slips into elitism, if only unconsciously. Yet in our society, the most fervid elitist has no choice but to accept at least some form of majoritarianism as a political given, immune to direct challenge. So, while our politicians, and even our most sophisticated theorists, sound a universal paean to democracy when they speak publicly, in private there are always doubts. For some people, these misgivings are transient and incidental, put aside as quickly as possible, mere fleeting shadows on a noble ideal. For other thinkers, more cynical about democracy if not hostile to it outright, they are deep and scornfully felt, even if cautiously uttered. The middle position is a tricky one. It envisions democracy as a system of political life that is destined to produce serious, even dangerous, errors of judgment from time to time, but also as an ideal that is indispensable to any ethically palatable view of society.

To celebrate democracy uncritically, on the assumption that for every issue of public interest its mechanisms will generate the best solution, or at least avoid disaster, is to make a leap of faith that in one form or another invokes the mystical *telos* of the popular will. It is to assume that, in point of intelligence and intellectual responsibility, the mass of humanity is fully equal to any self-constituted elite, and that, in terms of ultimate wisdom, it is far superior, since it incorporates interests, perspectives, and sentiments that any narrow oligarchy inevitably excludes. On this view, the only thing that frustrates ideal justice is the existing disparity in the distribution of power. Once this is abolished, say the democratic optimists, all people, even the most humble, will instantly display the moral and reflective faculties necessary to meaningful participation in the social order.

The resolute skeptic, on the other hand, views intellectual independence as a rare virtue, given only to a desperately small minority at any point in history. From this perspective, civic responsibility is indivisible from carefully nurtured disinterestedness, which, since it runs against the grain of human nature, is in chronically short supply. Unrestricted democracy, therefore, is truly the war of all against all, a melee in which the unrelenting cacophony of contending interests obliterates the possibility that policy can be steered along a consistently sensible course. In real life, the pessimist insists, democracy inevitably coarsens into populism, and is chronically susceptible to the temptation to abandon reasoned discourse for demagoguery.

Yet this seemingly hardheaded position involves its own leaps of faith. In its fastidious distrust of the stampeding herd, it is forced to assume that whatever currently stands in the way of the egalitarian distribution of actual power embodies wisdom as well as prudence. Anti-democrats are thus inclined to legitimate the existing power elite, however it is recruited and maintained. Fear of the disenfranchised mass is transmuted into enthusiasm for the ruling caste. Skepticism about unbridled democracy is therefore liable to curdle into adulation of whatever oligarchy currently occupies the top of the tree.

Revulsion against democracy as a political ideal is certainly real, even in this country where democracy has been the official civic religion for two centuries. It quietly infuses the ideology of many public intellectuals, including politicians, and troubles the reflections of many private citizens. But in the public forum it is inevitably encoded as rhetoric that begins by paying lip service to the democratic ideal. Even more confusingly, it is frequently alloyed with its apparent opposite, a steadfast celebration of equality. To take the most obvious example, modern American conservatism, as embodied by the Republican party (and more than a few nominal Democrats), is unconditionally pledged to the interests of a plutocratic elite. Yet it also celebrates, with apparent—and often genuine—sincerity, the ideal that the population ought to constitute a yeomanry of tough-minded, self-reliant individualists who defer to no one, particularly not an intrusive government.

At this point, there are very few conservatives who really want to see the monied class ossify itself into a classical hereditary elite. Indeed, though it is by no means despised in practice, inherited wealth, for many of the conservative faithful, still carries a faint whiff of effeteness. The self-made man remains the hero for most of the American right. Furthermore, conservative rhetoric incessantly dwells on populist themes—the institutionalization of "Christian" cultural values, unrestricted access to firearms, an exclusively punitive approach to criminal justice, the degeneracy of contemporary art, and so forth. In this lexicon, "elitist" is a term of opprobrium, not praise. It is usually rolled out to denounce theories and theorists of the left, especially the "cultural" left. In part this is a ploy, a device necessary to conscript an electoral majority around what is essentially a program studiously crafted to serve the interests of the extremely wealthy. But there is no denying, at this point, that incessant repetition of these themes as well as a strategy of using them to recruit a corps of fervent activists, including office holders, has fused a traditionally patrician conservativism with a strongly populist version. The latter *seems* committed to entrusting power to the mass on the assumption that its own values are those of the great majority. How much of this is hypocritical gamesmanship, how much sincere populist enthusiasm? Clearly, both factors come into play, which has caused problems for the electoral ambitions of conservatives on a national scale. The chief lesson, however, is that "conservativism" as a viable political movement cannot function as the public voice of classical Hobbesian conservatism, with its rigidly hierarchical and anti-populist dogmas.

On the other hand, the ideology of unrestricted democracy, as articulated in current political life, is beset by its own contradictions. The charge of "elitism" which the populist right continually hurls has a certain merit, if only because today's most impassioned egalitarian theorists are not connected to any effective mass movement, let alone to a functioning political party pledged to their values. They are, indeed, "theorists" in the most pejorative sense, preaching in the name of a congregation that hasn't bothered to show up for the sermon. Part of their dilemma is the anti-intellectualism endemic to American life, which has always despised arid theorizing and the appearance of superfluous abstraction in political thought.

Compounding this, in recent years university-based left-leaning intellectuals have fallen into the paradoxical habit of celebrating the virtue and wisdom of demotic cultures through a form of discourse that comes across as "elitist" in the worst possible sense. Anyone who has followed the fortunes of postmodern doctrine—always embraced in the name of radically democratic political goals, to be sure—will be aware of how hermetic and insular it is. Even worse, it seems to provoke its adherents to the most galling kind of snobbery—obnoxious, self-infatuated, and comically hierophantic. There can be few better vantage points to view blatant one-upmanship in action than the forums in which postmodern cultural theorists (all of them supposed radically egalitarian democrats) vie with each other in torrents of clotted prose to see who can be the most perversely obscure. Even in less-confining contexts, ideologies that see themselves as paragons of the democratic ethos tend to exhibit elitist attitudes in practice. Environmentalism, for instance, though it probably has a broader popular base than any other crusade more or less connected to the "left," is nonetheless notably antidemocratic in certain respects. Over and over again, spokespersons for the movement make clear that the only way to avert imminent ecological disaster is for the great mass of humanity (or, at least, Western humanity) to abandon its greedy, profligate ways in favor of the environmentalist covenant. At least in tone, radical environmentalists tend to come across as a self-anointed elect, demanding that the ignorant multitudes convert to their way of thinking forthwith. The fact that most of us continue, wisely or not, to "vote with our feet" in favor of automobiles and cheap airfares is unlikely to persuade these missionaries to question their assumptions or modify their values, no matter how greatly they are outnumbered. For them, the only true democracy is one in which their own postulates are firmly in the saddle.

Leaving aside the hypocrisy of either right or left, the point is that in contemporary society, purely democratic values are in perpetual contention with elitist ones, but the struggle between them is not always waged overtly between declared ideological enemies. It is also thoroughly embedded in the internal contradictions of the contending factions themselves. Oscillation between populist and hierarchical assumptions seems an unavoidable part of the game of thinking politically in the modern world. To say this is no more than to say that politics, even at its most idealistic, is not a good breeding ground for doctrinal consistency. Still, the fact that the contest is so muddled on the practical level is insufficient grounds for slighting its importance. It pervades everything because it is so fundamental. On the one hand, we have all, as citizens of Western society in the twentieth century, internalized a notion of fairness that is reluctant to deny anyone a voice in public debate and a measure of authority over public decisions. On the other hand, most who have thought about the matter in any depth acknowledge that insight and competence, as well as the kind of selflessness necessary to worthy citizenship, are by no means equally distributed throughout the population. The disparity cannot be redressed in any easy fashion, if it can be redressed at all.

In the context of this book, these tensions are important as they affect the role of science in the public sphere. The fundamental question is a simple one but excruciatingly difficult to answer convincingly: What do democracy, as a reigning assumption of political life, and science, as a means of investigating the natural world, owe to each other? How far should the rules and mores of one domain intrude upon the other? Given the open-ended majoritarianism inherent in democracy, how far should it yield, in various cases, to the authority of scientific knowledge obtained out of sight of the larger population by means inaccessible to their critical evaluation? Given the inherent elitism of science, which will endure as long as disparity of talent among individuals remains a fact of life, how far should it defer to the decrees of the larger democratic process? Must democracy be defended against its own worst propensities in order to found public debate on the most reliable possible knowledge? Should the passion for truth about the world which drives the best science yield, at certain times, to the fears and misgivings of a wider populace, even if these spring from imperfect knowledge or actual ignorance?

My main concern is for the fate of science in a society whose strongest ideological commitment is to democracy—democracy, defined, moreover, with strong overtones of populism and implicitly imprinted with the dictum that all opinions deserve to be heard out with equal respect, regardless of provenance. But, insofar as science is being used as a test case against which to judge the strengths and flaws of democracy as such, the proper fate of democracy is itself an issue. If there are irreconcilable conflicts between the democratic ethos and the ability of science to pursue and promulgate knowledge, and, moreover, to have that knowledge respected, then my first instinct is to say that it is democracy, rather than the scientific culture, that ought to be modified. This view, clearly, will not draw instant assent from everyone. But it is premature to draw such lines until we have garnered a fuller understanding of how science interacts with the existing mechanisms of political democracy.

The questions I have raised are not abstract. They are repeatedly evoked by real-life conflicts, as well as by theorists of the social context of science. Scientists, as well as engineers and physicians, are continually faced with demands from politicians and governmental organs. They are confronted as well by grass-roots groups that are even more clearly democratic in philosophy and practice. These demands range across a wide spectrum of issues, from the prohibition of cloning to the legitimation of alternative medical practice to closing down scientific facilities deemed to be hazardous. The easiest cases to evaluate are those in which most or all of the fault lies with the scientists and their institutions. The plant geneticists who helped the Brown and Williamson Company deepen the addiction of smokers by developing high-nicotine tobacco were clearly off the reservation, both morally and legally. Obviously, they are as far gone in depravity as the weapons experts who provide high-tech killing power to assorted dictatorships, or the ostensible expert witnesses who custom-tailor their expensive testimony to the needs of a paying client.[1]

This sort of thing is mere venality and corruption, easy to analyze morally, if not always so easy to defeat in practice. A more difficult and, alas, more common situation arises when competent scientific opinion finds itself in conflict with strongly held community opinion that is articulated through political action or the judicial process. In such cases, as I have insisted throughout, the presumption of epistemological superiority must go to the scientists. Absent some showing of actual falsification or fraud, there is no court entitled to overrule this supposition, at least on the philosophical plane. But it is a fact of life that this seemingly straightforward (and, admittedly, rather smug) rule of thumb provokes a great deal of uneasiness, even on the part of people generally well-disposed toward science. It offends their sense of fairness and their notion that each dispute must be settled on a freshly leveled playing field where, initially, all claimants are to have equal standing. They realize that such an ideal is impossible to achieve in practice, and that in actual cases, scientific opinion must be given due respect. Still, they are haunted by the wish that the disjunction between expert and layman were not so absolute. They would prefer to be able to interrogate presumptive scientific expertise, and to insist that it explain its conclusions in terms they can fully grasp. In short, they object not so much to science, as such, as to the gulf of incomprehensibility that yawns between the professional and the rest of the population. Most scientists are quite eager to reduce this gap. Many of them—certainly those with academic teaching responsibilities—have considerable practice in doing so. But the process has definite limits, even when attempted in perfect good faith. Inevitably, most laymen are able to perceive far less of the logic behind expert opinions than they wish. Consequently, these opinions tend to strike them as summary judgments *de haut en bas*. Therefore, they provoke resentment, and sow distrust as much as reassurance. The end result is that the triumph of scientific accuracy (if that is indeed what takes place) is not really accepted unconditionally as a triumph of justice.

The situation is complicated, moreover, by the fact that conflicts between scientists and other social groups are often only incidentally about factual matters and the competence to decide these accurately. Instead, they are conflicts about moral values or about the right of a community to define itself in opposition to what orthodox sources of knowledge, including science, claim to know. Here, the facts that scientists can lay out, even if convincing by all conventional standards of reliability, are insufficient to assuage the psychological unease that lies at the root of the dispute. The question raised by these episodes is an ancient one, as old, at least, as Sophocles: Does truth have an absolute right to be heard and understood as such, even at the cost of social peace? My own deep feeling—evident, I trust, throughout this book—is that truth does, indeed, have such rights. Moreover, it is clear that I regard the natural sciences, imperfect as they may be, as by far the best instruments for discerning such truth as may be accessible to human beings. This, obviously, is a very strong claim, but logic, philosophy, and history are, I think, equal to the task of upholding it.

The real problem, however, is that many sincere people, including those who have little trouble conceding the epistemological point, draw back at the blatant implication that science, on account of its factual accuracy, simply has the right-of-way over all contending points of view. The question they raise is whether the relentless insistence on truth seeking that lies at the core of the scientific enterprise is always the path least injurious to the social health of the community.[2] Might not a greater tolerance for fiction, or myth, or refutable pretense buy more peace of mind for most of us than the single-minded quest to root out error or fantasy wherever it lurks? If the answer to this question is positive, might it not be the duty of a democratic system to restrict science in some respects, to give less than absolute backing to its claims, to restrict the sphere in which it holds sway, in order to meet the desire of ethnic or religious groups for tranquillity? One could argue—indeed it has been argued—that such a social compact would not constitute a rejection of science or even mitigate its ability to pursue most of the questions that scientists find important.[3] Neither would it entail rejection of the material and social benefits that obviously accrue from a base of scientific knowledge. Rather, it would merely free citizens to accept as much or as little of the scientific worldview as comports with their own sense of fitness, without being made to view themselves as exiles from modernity and enlightenment. It would remove much of the onus from unconventional belief systems simply by abolishing the sense that one particular belief system—that of scientists—is officially empowered to sit in judgment of the rest of them. This, according to its proponents, would not be a regime of rampant irrationalism, merely an arrangement under which orthodox science would have to be more gentle and diplomatic, and rather less arrogant, in pressing its claims.

One such proponent is the sociologist Sharon Traweek, who has spent most of her career studying the folkways of high-energy physics laboratories. She puts the case in a rather high-keyed *cri de coeur*:

> European arts no longer set the world standard; the "human condition" is no longer defined in Europe or North America alone. Beauty, truth, and logic have multiplied and dispersed. Some have just begun to notice that the same has happened with science. What is the name of that obsession for singularity and unity, for an order that does not divide, for a world of symbiotic union, for a world that begins and ends with an indissoluble ego? What is the name of the rage against a world of particular plurals? Is it like the rage that some felt against a heliocentric universe or the rage that others felt against a Darwinian world? Why should there be only one way to think well, only one way to have fun with our minds? Why is mental monogamy required? Are we still fighting about monotheism, Manichaean fallacies, and Albigensian heresies?[4]

As it stands, this peroration is rather overwrought, to say the least. Relativism is to our cultural avant garde what absinthe and laudanum were to a previous fin de siècle crop of Bohemians, and Traweek gleefully revels in it. Moreover, given the proliferation of ads for phone-in psychics on TV, it would not seem, all in all, that she has very much to complain about.

But let us look beneath the postmodernist theatrics of statements like this to see if they really embody a serious critique of the role of science, as a system of ideas, in a pluralist democracy. Ignoring the epistemological grand-standing of such flights of rhetoric, we may tease out a more sober question. On what grounds should we think it healthy that science (and, perforce, scientists) sit in judgment of the creeds, myths, dreams, fantasies, and speculations of the great mass of nonscientists. With what warrant does a democratic system, one which, moreover, is a crazy-quilt of ethnicities, sects, and doctrines, anoint one particular community—the scientists—with the authority to declare any of the others ridiculous in part or whole? This need not be taken as an epistemological question, in the strict sense. Its main concern is the conditions under which tolerance, intellectual liberty, and the freedom to elect one's own belief system can flourish. Even granting that science holds all the epistemological trumps when the cards are dealt out at the philosopher's table, why should it have the right to flourish them in every public arena, irrespective of the concrete questions at stake or the psychological costs to those who face derision? Oughtn't we to distinguish between cases where adherence to scientific standards of truth is vital in order to avert death and suffering, and those where little more than the vanity of scientists seems to be involved?

Throughout this book, I have continually argued from the postulate that science must automatically be declared the outright winner whenever it comes into conflict with contradictory belief systems. But have I truly reckoned with the possible social harm, in particular the denigration of the idea of universal participation in public discourse, that might be implicit in my credo? In elevating science to a position of such categorical authority, am I not, in effect, making second- (or third- or fourth-) class citizens of everyone else?

Let me play devil's advocate for a few paragraphs with respect to my own declared position. I confess that I find it nearly impossible to adopt, even *arguendo*, the relativist philosophical position, or anything like it. It is too philosophically hollow and collapses into self-contradiction at the lightest touch. Instead, I shall merely try to find, in concrete contexts, the best case that can be made on the basis of those moral and political precepts that, in my view, inspire postmodern science critics to uphold epistemological relativism, or at least to pretend to do so. If I am right that it is moral conviction rather than philosophical flightiness that is largely responsible for the current academic enthusiasm for relativism, I shall do little to diminish the persuasive force of the argument by simply discarding the relativism and placing all the weight on ethical and political considerations. If there is any case to be made in this way, it must be made in the actual world, not on the basis of "theory"—relativist, postmodernist, or otherwise. I shall therefore refer to actual confrontations, in particular some of those I have cited earlier in the book. The issue will not be whether orthodox science is "right"—I frankly take that as a given—but whether and how far that kind of truth warrants authority in a societal context.

Let us first revisit the situation brought about by the opposition of Native American militants to the work of scientific (mostly white) archaeologists, especially the kind of work that involves unearthing and examining ancient relics, including, in some cases, human remains. As I noted earlier, Indians and their sympathizers have obtained favorable legislation that allows tribal authorities to hinder and sabotage research of this kind. They have often exercised this right, to the point where scholarly and museum collections have been dismantled, archaeologists have been prohibited from doing work on their finds, and research careers have been suspended or ruined outright.[5] Moreover, the main motive for hostility to archaeology seems to be a desire to sustain a flattering "creation myth," one that links indigenous groups with their tribal lands on a mythic chronological scale that extends back to the legendary dawn of time. On what grounds might the archaeologists be obliged to abandon their presumptive right to investigate the historical truth about the peopling of North America in order to accommodate Indian fables?

Most Indian sympathizers (assuming, of course, they don't share the activists' declared commitment to the literal truth of the myths) will immediately point to the historical horrors visited on Indian peoples by the rapacity of whites. They will go on, what is more, to emphasize the wretched situation in which most nonassimilated Native Americans still find themselves. In Indian communities, crime and alcoholism are rife, education is rudimentary, unemployment high, infectious illness endemic, violence frequent, and early death common. Beyond material deprivation and social disorientation is a sense of anomie and placelessness, the feeling that today's Indians are merely the living wraiths of cultures that were assaulted, robbed, dispossessed, and nearly dissolved by the heedless greed of European interlopers. What is missing is any sense of being anchored in an ancient, enduring, and dignified community—in short, the kind of consciousness that stems from pride in one's ethnic identity and which associates that identity with competence, power, and a permanent and honored place in the world.

Militant Native American nationalism promises to repair that deficiency. Its primary aim is to overcome the reigning sense of futility and unworthiness that afflicts Indians who compare their situation with the power, authority, prosperity, and accomplishment of mainstream "American" culture. An important step in achieving this is to wrest some of that authority from the mainstream, particularly the authority to recount the "true" version of Native American history. This may necessitate a conflict with the vaunted science that evolved in the bosom of European culture. Indeed, the high regard in which the Western tradition holds science greatly amplifies the positive psychological and moral effect of defying and nullifying it, at least as it impinges on the Indians' sense of identity and historical rootedness.

As against this, there are countervailing claims of archaeologists and their scholarly allies. What do these amount to? Surely, the overall status of science, as such, is not in question. Archaeology and ethnography are marginal subjects that contribute little, if anything, to the core of scientific knowledge. Biologists, chemists, physicists, and mathematicians will not miss a beat merely because paleo-Indian grave sites and kitchen middens are no longer being excavated. North American archaeology is basically little more than a recondite hobby pursued by a handful of privileged enthusiasts, most of them white. Other than their private satisfaction, not much is at stake in its continuance. True, they claim to be sympathetic to Indian ways and assert that their findings actually provide solid factual grounds for Indians to take pride in their ancestors' achievements. But shouldn't Indians be the ones to judge these claims? Aren't they the final authority on what constitutes a source of Indian pride? Even if what archaeology brings to light does constitute a valid source of Indian self-affirmation, why not leave it for another generation to discover this—one which is not so devoured by its own bitterness and sense of irreparable loss. Moreover, what, really, do the archaeologists stand to lose, beyond a bit of their inappropriate self-certainty? At worst, they won't be able to publish a handful of papers in obscure journals and might have to find some other part of the country or the world to study. Where is the real harm in that? Mustn't archaeologists' ability to scratch the itch of private curiosity yield to the urgent psychic needs of precarious Indian communities to have control over their own sense of history?[6]

Of course, this argument begs certain questions: Are the Indian militants who make life miserable for North American archaeologists really as representative of their tribes and communities as they claim to be? And do these psychological theories of the relation between ethnic pride and the elimination of social pathology really hold water, or are they merely the sentimental efflux of unreflective pity? I am willing, however, to put aside such objections. I allow the argument I have just outlined the full force of its own assumptions. Does it then constitute a case so strong that it compels the deference of a democratic society? By this I mean a society that takes seriously the right to a dignified social identity and which is imbued, moreover, with a sense of justice recognizing the urgency of the claims of the historically dispossessed. Before attempting an answer, I want to scrutinize some other claims that run, more or less, along the same lines. The case for "Afrocentric" education, for instance, exhibits many of the same features. It addresses a similar evil—the historical confiscation of meaningful social identity and pride in one's ancestry, and therefore in one's own people. The fact that the victims are descendants of unwilling immigrants rather than dispossessed natives is an inessential distinction. Again, I allow the hypothesis that to repair psyches so damaged, to restore them to a plateau of pride and self-confidence that will enable sound social functioning, can only be done on a social or collective rather than on an individual basis. Further, I concede, as a debating point, that in this instance Afrocentric theorists are largely on the right track when it comes to devising appropriate remedial mechanisms for black youth. The question, then, is whether the pseudoscientific myths, and even the fake science, that infuses so much of the Afrocentric curriculum at this stage would in itself constitute a sufficient reason for refusing to implement it.[7] Can a program whose results, hypothetically, are so salutary and so necessary to our collective social health at this point in history, be disallowed? What damage would it do, for instance, to impute mathematical discoveries to imaginary Egyptians rather than real Greeks, if the result is to make it easier for large numbers of black children to devote their energies to learning mathematics and science? Indeed, what particular damage will be done if, in the course of developing this kind of esprit, rather extravagant theories of the scientific achievements of the Egyptians or the Dogon are handed out as fact? Tall tales are hardly unknown in education, after all. Generations of white American children were solemnly assured by teachers and parents of the literal truth of patriotic whoppers. It is hard to believe that their overall education was compromised or their character degraded because of this imposture. Once more, the argument can be made that the rectification of these fancies should be left to another generation of American blacks, one where deep alienation from the larger society is no longer commonplace, and which can take for granted the intellectual self-confidence of its youth.

Perhaps I concede too much to Afrocentrism—a rather extravagant doctrine—even for argument's sake. But the same essential points could be made about other schemes for multicultural education that are less overblown in their rhetoric, yet which still propound ideas that are scientifically dubious in the name of building pride and confidence in youngsters whom American culture has habitually denied a fair shake.[8] The same essential question that echoes, however histrionically, in Traweek's tirade recurs again here. Why must every organ of our social structure be compelled to genuflect repeatedly and unconditionally to the views of the scientific community, no matter what immediate psychic harm is done to local populations? Why must we forgo the potential benefits that might result from deviating on occasion from the epistemic puritanism imposed by single-minded loyalty to scientific truth?

This, moreover, is an argument that could well be made from the cultural right as well as the cultural left. The case of creationism—or, slightly less abrasively, "intelligent design theory"—in biology provides a fascinating instance. Few academic liberals will automatically accept the creationist constituency as an "oppressed class" comparable to Native Americans or blacks, although black fundamentalists, in fact, constitute a significant percentage, while the stratum from which white supporters are drawn can scarcely be placed at the top of the socioeconomic ladder. Indeed, the complex of attitudes and beliefs, creationism included, that forms the fundamentalist creed could well be accounted as a collective defense mechanism against the hard hand of contemporary capitalist culture. Irrespective of whether such facts will win the sympathy of most professors, the argument can certainly be made that this community, like any other, is entitled to celebrate its own notions of what gives dignity and meaning to life. Religious conservatives often do form communities in the narrow geographic sense as well as doctrinally. Religion is frequently a factor in determining where people choose to live, as well as whom they will choose as friends and associates. Especially in areas where they predominate numerically, why shouldn't conservatives' views at least be heard in the public schools, even when science scorns them? It will not do merely to ignore or disparage the psychological justification for this view. Creationists, too, have a philosophy of life, even if it is not one that is well represented in academic philosophy departments. Having that philosophy respected, recognized by the larger society as a reasonable core around which to build one's life, is a necessary component of dignity and self-respect.

One token of that recognition, then, would be a system of public education that did not relentlessly insist on insulting a central tenet of those beliefs while obdurately refusing to let anything be said in its defense. On what basis, after all, are the local schools of a community obliged to defer to the doctrines of a psychologically remote and rather smug mandarinate? What political consensus set that kind of pecking order in place? The exclusion of even mild and expurgated forms of creationist doctrine from the schools reposes merely on one Constitutional provision, the Establishment Clause of the First Amendment. But this interpretation is really a novel formulation without much warrant in the historical practice of the century and a half following the adoption of the Bill of Rights, a period in which Christianity (in its Protestant version) was tightly integrated into the pedagogy of most schools. At least in areas where there is substantial doctrinal unanimity among the local populace, oughtn't that doctrine be allowed some small opportunity to confront the repugnant doctrine that the state insists on imposing on schoolchildren? Doesn't the state's refusal to countenance such challenge degrade and diminish the very notion of democratic citizenship? Doesn't it dissolve, rather than build, social solidarity and a habit of mutual respect? And, finally, why should the right of veto be accorded to science and its apologists? Do the scientists really have anything at stake other than the gratification of knowing that they have the muscle to shove their doctrines into every schoolroom in the nation, and to shove all contending doctrines out? This is not the advancement of science (which is hardly threatened, nowadays, by creationism in any form) but only self-adulation on the part of scientists. What is the political and social warrant—which, remember, is quite a different thing from epistemology in the abstract—for such a peculiar arrangement? Even science itself is the loser thereby, since respect and support for science is crushed out in a population that is usually inclined to support it.

Readers will note that the three cases argued so far all draw upon the same essential logic. They insist that a principal obligation of a democratic society is to recognize the rights of its constituent groups as groups, as well as those of individuals. The first two exercises follow the current doctrine of the fashionable left, while the third urges a course supported almost exclusively by the right.[9] But both accept implicitly the idea that, for many if not all people, membership in a democratic polity is mediated by membership in a primary community bound by loyalties and commitments more deeply felt and intimate than abstract agreement with a particular constitutional system.[10] This argument is often advanced from the left as "identity politics" and the sanctity of "local knowledge."[11] Conservatives adopt a different vocabulary, as well as a different spectrum of tribes and folkways to champion, but their point of view echoes that of the multicultural left to a remarkable degree both in the formal logic of its argument and in its emotional appeal. The cases at hand also display the convergence of important segments of both right and left in their respective views of the social role of knowledge. The core of the doctrine is that abstract accuracy or reliability is not an ultimate virtue. It can be trumped by the need to avoid the social harm that can be wrought by insisting that communities abandon their sustaining myths merely because science believes it has proved them false. On this view, viable democracy can only exist on the basis of the consensual association of communities held together internally by deep and even irrational emotional ties. Thus, each group must walk somewhat warily in the presence of all the others, taking particular care not to insult any faction's articles of faith.

By this token, even the scientific community must play by the same rules. Science is, in the social sense, just another minority held together by the consensus of its members, although it is a particularly convenient one to have around. Notwithstanding its noteworthy cleverness, it must eschew the temptation to debunk publicly the dearly held creeds of other groups. Even more, it must abandon any hope of using the organs of government to advertise its superiority. That would serve no purpose other than to gratify the vanity and swell the prestige of scientists. Science, on this view, must learn diplomacy, tact, and discretion. It must, above all, develop a modus vivendi even with groups that, for reasons *they* find adequate or compelling, choose a doctrinal path that in some respects turns away from what scientists claim to have learned.

This brand of communitarianism is, in some ways, skeptical, if not antithetical toward the individualism sanctioned by most of the Enlightenment thinkers we are supposed to revere. It rejects the view that society is merely a constellation of self-defining human atoms. It qualifies the notion of "individual rights," at least to the extent that it sees aggregations of individuals—communities of feeling, doctrine, and interest—as entities in themselves, deserving of rights of their own. It might be argued that the historical consensus in this country has come round to this way of construing democracy. When both right and left argue, albeit in different lexicons, for the same underlying view of what constitutes fairness within the social organism, isn't that evidence that this view reflects widespread agreement?

It is not necessary, however, to endorse this kind of communitarian view of how society should be structured in order to reject the idea that the political system has an unqualified obligation to uphold the superiority of science as a knowledge system. The uncompromising individualist also has problems with allowing such a role to government, even if he scorns the idea that communities based on doctrinal solidarity have any special rights. Indeed, to the extent that individualism takes on the adamancy of absolute libertarianism, the idea that science has any specific right to count as our "official" standard of knowledge seems even weaker than it does from the communitarian point of view. The libertarian creed more or less insists "every thinking being for himself," so that every aspect of a worldview, every lineament of a philosophy, must be created anew within each individual. Thus, while anyone is free to be as enthusiastic as he chooses over science, and even to exert every effort to win others to the same viewpoint, no one is entitled to conscript the machinery of the state in order to anoint science with special authority.

This, in fact, is the social logic behind the relative passivity of the state and the legal system in the face of the proliferation of pseudoscientific health nostrums. True, the FDA tries, from time to time, to rein in the most outrageous shills for miracle cures—but the process is long, tedious, and, apart from some cases where the trail of dead bodies is long, usually rather ineffectual. Every once in awhile, health quacks do get prosecuted. But in our litigious era, the courts pose greater danger to their critics, since, for the well-heeled quacks, libel suits constitute a relatively cheap and easy way to quash outspoken protest. The fact that the market for health care has become a free-for-all that provides a good living for cranks and easy pickings for outright charlatans is an instance of the triumph of the libertarian ethos in one area where, until a few decades ago, the paternalism of the medical establishment largely reigned. This power, however, has wilted, not because conventional medicine has fallen short in any conventional sense—it has advanced at an enormous pace—but because the culture has fallen out of love with anything that smacks of conventional wisdom, and, even more, because we are now immutably committed to the creed that each person is the best judge of his own interests. We are more than ever wedded to the idea that everything falls within the realm of private action, and that no one, certainly not the state, has the right to interfere with the decisions of an individual as to how to get through life and deal with its difficulties.

The percolation of absolute libertarianism into the cultural subsoil hasn't been completely uniform and irreversible—we are probably less latitudinarian about the use of drugs than we were twenty-five years ago—but on the whole, the idea of being second-guessed by the state on matters of private choice seems more repugnant than ever. In particular, science has become a paper tiger when it comes to trying to discipline the health and nutrition market. This doesn't mean that the conclusions of science are automatically discounted. Rather, it reflects the common opinion that no authority, however august, should be allowed to interfere with the sovereignty of individual choice. More and more, autonomy is prized over prudent deference to authority. "It works for me!" is the byword of the day, with no requirement that efficacy ought to be double-checked by anything beyond raw subjectivity.

The philosophical basis for all this is reasonably clear and consistent. If a culture is to prize individualism, then it must surrender the right to superintend individual choices. The law and the state have no business looking over anyone's shoulder in matters of private concern. The right to freedom is, necessarily, the right to make one's own mistakes, no matter how grave. In particular, no matter how one admires the success of science, there is no way to compel anyone else to admire it, or even to take it seriously.[12] Democracy, on this view, requires that the coercive power of the state over actions, and certainly over ideas, be held to a minimum. Anything else would be a standing invitation to statism and the rule of conformity.

Again, we see in this attitude a curious convergence of left and right. The left deploys and defends these ideas with respect to certain kinds of behavior: homosexuality, birth control and abortion, provocative artistic expression, and radical criticism of the existing order.[13] By the same token, the right is stridently individualist in tone when it comes to the availability of guns, the abolition of highway speed limits, the use of off-road vehicles, unrestricted campaign contributions, and, naturally, untrammeled capitalism. It is interesting to note that in the area of health care, both right and left unite to defend the prerogatives of alternative practitioners, using rhetoric that is surprisingly similar in tone. Here, the bohemian individualism of the left finds common cause with the freebooter individualism of the right. Neither countenances the intrusion of a scientific establishment, however wise, into the arena, and both celebrate the epistemological self-sufficiency of anyone who wants to claim it. Even where the intransigence of parents in clinging to unconventional therapies for their children has brought about tragedy, defenders of the right to make this kind of decision have come from both right and left. To put it another way, a film like *Lorenzo's Oil*, which depicted a gutsy family defying medical wisdom by finding, on their own, an effective treatment for a supposedly hopelessly ill child, was enthusiastically cheered by both ends of the political spectrum.[14] Again, one may infer that the de facto fusion of seemingly incompatible ideological positions over this issue reflects a political—and moral—consensus that permeates the society as a whole.

Here, it becomes obvious that a certain amount of self-contradiction must afflict both right and left. The communitarian view of things is obviously at odds with unlimited individualism in many ways. The two are nearly antithetical. Is this because there is more than one "right" and more than one "left"? Frankly, I don't believe this accurately explains the situation. I think it is more correct to say that on some issues, the right, generally speaking, takes the "communitarian" point of view, on others, the "libertarian." The right thinks like a communitarian in condemning sexual nonconformity and like a libertarian in condemning gun control. Similarly, the left is libertarian on gay rights and communitarian in promulgating speech codes on campus. As is usually the case in human affairs, consistency is rarely a major preoccupation of any enthusiast. One notes, however, that in both libertarian and communitarian modes, neither the left nor the right is very willing to speak out for the authority of science.

In my view, neither the communitarian nor the libertarian critique of science can be sustained, whether articulated as right-wing theory, left-wing theory, or neither. They fail not only because of insufficient regard for the role of accuracy as the standard of public discourse, but also because they take an undemanding view of what constitutes a democratic polity. If there is a failing in the common and, indeed, the academic view of democracy, it is that it views democracy as essentially a default state of society and political systems, one that comes into being, almost by definition, once antidemocratic practices and institutions are abolished. The price of universal veneration of democracy is that the notion is never really examined, let alone subjected to a sustained and suspicious critique. Because it has no competitors, at least at the level of publicly acceptable rhetoric, it has no clear definition. Right and left disagree volubly and passionately about the obstacles that stand in the way of the desirable state of affairs that democracy supposedly has the potential to achieve. The right insists that it is the bloated machinery of the state and its bureaucratic organs, along with the excessive leverage accorded to left-leaning elites and their pet ideas. In the conservative version, these stand in the way of a popular will that, left to express its true intent, will enthusiastically endorse traditional social values, conventional religious piety and its place in public culture, individual responsibility, and, of course, the uninhibited working of the free market. The left version locates the hard core of resistance to democracy in the domination of a plutocratic code of values, as well as the power, overt and hidden, of those who actually control substantial wealth. To this list of villains, today's leftist usually appends a catalog of "hegemonic" ideas such as racism, male supremacy, and antagonism toward unconventional sexual proclivities. If all these were abolished, or at least curtailed, it is assumed that democracy would swiftly gravitate toward substantial equality in the economic sphere (and thus in the realm of power). Sexual and racial equality—indeed, the disintegration of the illusory categories that sustain inequality—will follow as well.

My personal view is that, at bottom, the left sees things more accurately, although in academic circles its vision is heavily fogged by the overbearing pieties of identity politics, as well as the intellectual bloat of postmodernism.[15] Also, like any leftist, I am deeply suspicious of the right's ostensible populism. I think it is a ploy, in the final analysis, a tactic for getting the most mileage out of the demagoguery of Rush Limbaugh and Pat Buchanan (to name two currently prominent personages I find unenchanting). I assume that when all the masks are off, the true visage of the right is fiercely authoritarian. It is also obviously true that the left has often celebrated the triumph of despotism when it arrived under the banner of "socialism," but this, I think, is more a matter of lack of acumen than of hypocrisy (at least in this country). For the moment however, I am content to put these animadversions to one side, if only for the sake of argument, and to postulate the sincerity of the right as well as the left in endorsing the sovereignty of popular judgment enacted through democratic institutions. This leaves the opposed extremes—and most centrists as well—committed to essentially similar intellectual models. The universal myth is that once false idols are overthrown and false prophets discredited, once the barriers erected by obnoxious ideology and illegitimate hoarding of power are thrown aside, the ideal democratic citizen will stand forth everywhere, exercising the natural wisdom that will, at long last, bring forth the true fruits of democracy.

This is a myth, and a pernicious one, because it fails to reckon with the demands that democracy makes of its citizenry. Let me make clear that I'm not referring to any present statutory requirements that the United States (or any other conventionally democratic state) formally imposes on its citizens. What I am saying is that in order for democracy, as a real system of political and civil relations, to realize the hopes of its most fervent champions, ordinary people—that is to say people in the mean, people on average—will have to become a great deal better than they are, in this country and everywhere else. Democratic ideology is a gamble—perhaps a desperate gamble—that human beings can, by and large, rise to an unprecedentedly high level of moral—and intellectual—functioning. For most democratic theorists, it is a rather unconscious gamble, or, in some cases, one whose problematical aspects are swept aside by the expectation that a *deus ex machina* will automatically appear on the scene once the various villains who stand in the way of democracy are undone. Looked at coldly, however, it is a depressingly long chance to which we are committed as proponents of democracy.

Neither history, nor psychology, nor what we can observe of contemporary mass culture offers us much in the way of immediate assurance. Even if there is no conspiracy of vicious or wrongheaded people at the top currently interdicting the practice of true democracy, the best we can say is that we are in an interregnum, a probationary period where we must hope that some overall trend will emerge to augment the proportion of the population capable of living up to the ethical and intellectual demands of democratic citizenship. Our present *agora* is aswarm with fashions, beliefs, and opinions. Ideas and serious reflection are in short supply.[16] As far as discussion of science is concerned, it has been noted, quite correctly, that "Hostile politicians in democracies are restrained in the degree to which they attack science, in part because of the prestige science has and in part because science and its technological offspring have proven so important to advanced economies. But that restraint is in the context of using what are simply different criteria for making public decisions than scientists usually expect."[17]

These are the kinds of observations that instantly elicit accusations of elitism. To the left, they disparage the ability of the dispossessed masses to take power and the instruments of justice into their own hands. The right sees them as yet another effete rationalization for circumventing the common sense and sound moral instincts of the majority that upholds traditional values. The view expressed is, indeed, elitist in that it identifies the health of our society with a body of views and capacities currently shared by a tiny minority. But it is markedly anti-elitist to the extent that it sees these qualities as within reach of the vast majority, given a sufficiently astute system of nurturance and education. Elitist or not, it is an unpopular view. The left, over the past few decades, has been engaged in a wild scramble to find more and more pretexts for celebrating the merely demotic, especially as it is found in the cultures of those officially designated as oppressed. The distinction between high culture and low, between popular and sophisticated art, has been systematically reviled, and volumes are devoted to showing that the distinction is an arbitrary barrier erected in the interests of power. There are more than a few humanities departments these days where a publicly declared preference for Mozart over rap music is a ticket to ideological exile. The simplistic multiculturalism that has become an article of faith for the left is further evidence of how little respect is accorded depth and complexity.[18]

Conservatism, in its current American incarnation, has also fled from any emphasis on the traditional hierarchy of intellectual and cultural values. Country-and-western music is pretty much the conservative's equivalent of rap. It would, moreover, be hard to find an important conservative spokesperson who would champion the resurrection of a reading knowledge of Latin as an undergraduate requirement, or show the slightest public disapproval of the fraternity-and-football subculture that still dominates most campuses, efforts of some "politically correct" rightists to the contrary notwithstanding. If the kind of cultural traditionalism that used to be called "conservative" lurks anywhere in the odd corners of the current political right, it is certainly never called to the hustings. Anti-intellectualism is more powerful on the right than it ever was, and conservative "intellectuals," unlike their postmodernist or multiculturalist or radically feminist antagonists, don't even seem to be interested in concocting a theoretical justification for it.

The critiques of popular opinion articulated by Emerson or Mark Twain or Henry Adams or H. L. Mencken (to name only American exemplars) are further than ever from the discourse of practical politics. A sweep through dozens of cable TV channels reveals a pluralist desert, largely bereft of ideas, arguments of substance, thoughtful analysis, or wit. Even more discouraging, things hardly get any better when the "alternative media" are factored in. Political rhetoric, from all parts of the spectrum, seems to be one vast effort to disparage all the qualities of mind and spirit, all the perceptiveness and curiosity, that might make a truly democratic mass culture possible. From Louis Farrakhan to Catharine MacKinnon to Pat Robertson, we prefer our political godlings to be donkey-headed. We don't expect them to think much, merely to talk loudly, and we certainly don't expect to think hard about what they say.

What this all comes down to is that understanding what it would mean to attain authentic democracy requires our society to engage in a lengthy and strenuous educational process. It would be enormously complicated to describe all the mechanisms of that process, even if we could now envision them. Certainly, the current work is no place to do so. One thing that must be said, though, is that such a mission definitely has an epistemological component. Among the Augean tasks facing those who want to build democracy on steady foundations is the widespread cultivation of those intellectual qualities that enable people to distinguish reliable knowledge and solid information from their opposites. It hardly needs saying that in the current culture, the appetite for delusion is vast and the defenses against it are weak, which is why faith healers, UFO entrepreneurs, and psychic-hotline operators prosper. This brings into plain sight why the role of science in our formal and informal political arrangements has important consequences—for science, of course, but also for the democratic prospect.

Science is the one area of human experience that constitutes, on the whole, a vast, almost unqualified, epistemological success. There is no justification for being coy or diffident on this point. It is as absolute a fact as anything history yields. Science is our only unambiguously encouraging model of how evidence is to be garnered and weighed, how theories are to be formulated, tested, and, where necessary, modified or even rejected. Science is the one human practice whose claims, when well confirmed according to the internal logic of the field itself, are entitled to outface all contenders and to stand alone as the factual basis of decision in numerous areas of public concern. This is not to make the "scientistic" claim that all kinds of intellectual inquiry must strive to model themselves on science in general or on one particular science. It is, however, to assert that any scholarly discipline or mode of investigation that falls outside science is perforce less entitled to claim sweeping authority in the public sphere.

Of course there are "borderline" cases, especially in the social sciences, where much of the methodological and confirmatory machinery native to science has been (or could be) put in place. But these are ambiguous precisely because their predictive and explanatory reach still falls far short of what is normal for the natural sciences. Democracy therefore owes science a special place of honor not only because of the material rewards society harvests from scientific work, but also because it provides a standard of accuracy, reliability, and stability to which other branches of knowledge must ideally aspire. Science belongs at the top of the ladder not because scientists are cleverer or more diligent than other thinkers, but because science represents an intellectual apex whose standards deserve to be deferred to. It is precisely because successful democracy needs a successful means of filtering evidence and theories that the political culture of democracy must acknowledge that science has created such a methodology, and that it is without counterpart in other areas of experience. To heap this kind of flattery on science is simply to recognize the vital role of logic and sound evidentiary principles in human affairs. This is what gives science its special social status as our chief instrument for dealing with a vast array of practical problems.[19] Not to have that methodology respected for its accuracy and objectivity[20] is to court disaster.

These observations are vital to refute the positions I introduced earlier as debating points. Consider first the "communitarian" arguments for refraining from imposing scientific findings upon subcultures or communities of belief that find them distasteful or disorienting. In rebuttal, we have to bear in mind that the "pluralism" so celebrated in American culture is a conditional, not an absolute value. Clearly, by tradition and law in most democratic countries, such subcultures have the right to take form and to endure without suffering direct hostility or interference from the state.[21] But government has no positive duty to screen these communities from ideas that make them queasy, particularly when such ideas have scientific warrant. In fact, if there is any general guideline that ought to govern state action in this area, it is that individuals, most definitely including nominal members of idiosyncratic subcultures, have a right to the full spectrum of notions and opinions that circulates in society at large. They are especially entitled to learn those things that, for compelling reasons, are regarded as best confirmed. Well-accepted scientific theories definitely fall within this category.

State-sponsored education has a positive obligation, not only to present these ideas, but to declare explicitly that they are vouched for by sound methodology. Education must inculcate epistemological wisdom, as well as deliver sound content. In learning about evolution, say, the student not only learns what the theory asserts, and the evidence for it, but that it is backed by a method for evaluating evidence that is, so far as we know, without peer. A crucial part of the lesson is that evolution is generally accepted for the same reasons that other scientific findings are generally accepted. It does not pose an exceptional case. By the same token, there is a positive duty not to present crank theories and pseudoscience as though they were entitled to comparable respect. This clearly applies to creationism in all its naked and disguised versions. Again, the guiding principle is that education has epistemological responsibilities.

The same arguments apply, virtually unchanged, to Afrocentrism and other seductive ethnic myths. To let these slide by unchallenged is not generosity but rather a piece of condescension that, at bottom, patronizes the students being "encouraged" as incapable of attaining mature standards of thought and judgment. The argument for suppressing archaeology in order to salve the pride of Indian militants is undercut by the same logic. Education has a wider meaning than formal schoolwork. The state has a duty to abet this more general process, which certainly includes research into all sorts of matters, along with public display of the results. To suppress such work in the name of preventing hurt feelings is to enthrone sentimentality and, even worse, expedience and political calculation as epistemological principles. This is the opposite of the lesson we need to learn in building a mature democratic culture.

Moreover, there need be no scruple about the possibility of seducing young people or anyone else away from the articles of faith that bind them to the community in question. If democracy means anything at all, it means that individuals have the right to disassociate as much as to associate, and that hereditary notions of caste and status cannot be recognized officially. Even more, it means that every individual has the right to be acquainted with ideas that might very well provoke a break with ancestral creeds, especially when those ideas proceed from epistemic processes which society as a whole is obliged to respect. The eminent political philosopher Michael Walzer is on the mark: "Many of our important groups are in the sociological sense involuntary associations. . . . We don't decide to join, we are enrolled by our parents. . . . We can only decide to stay . . . or not. Solidarity is an experience before it is a choice. What is crucial for freedom is not that the choice come first. . . . What is crucial is that, if one is born into these or those groups, it is possible for one to get out."[22] As Walzer also points out, "Examples of social tyranny can be found in almost all the groups that make up civil society."[23] By this token, if the idiosyncratic notions of a particular cult are to be presented at all in the course of schooling, it must be made clear that they lack the evidentiary authority characteristic of science. The most important reason for teaching science, at the elementary level, is to give young people a sound sense of what science *is*, and that includes the intellectual parsimony that abjures superfluous hypotheses. It sabotages this task when we refrain from pointing out what science *isn't*. Perhaps science teachers should not go out of their way to be heresy hunters. But when pseudoscience insists on showing up at the schoolroom door, there is no dodging the obligation to show why it is defective from a scientific point of view. Communities of belief, whether religiously or ethnically based, have the right to exist in peace but do not have legitimate power to use state machinery to custom-build intellectual barriers around themselves.

The libertarian argument also fails, though not necessarily at the most basic level. It is certainly arguable that adults have the right to choose or refuse any treatments or regimens they desire. It does not therefore follow that the state must be neutral among all the conflicting claims, or that it must not allow privileges to one school of thought which it denies to others. The libertarian fallacy is that claims to personal freedom of action automatically trump every other principle that might be mooted. In fact, confronted with any libertarian claim, the state is obliged to reckon both social harm and the danger to individuals who are not competent, in the legal and psychological sense, to exercise the asserted liberties. Put practically, the danger of quackery, for instance, is not only to those who resort to it as informed adults with their eyes open (if not sharp). It also threatens children and those whose illness or incapacity places their fate in the hands of others. Here, the state definitely is obliged to intrude its epistemological judgment, which to all intents has to be that of orthodox science. More abstractly, the notion that every individual has both the right and the duty to construct his or her own episteme from the ground up, and that the state may not play favorites among these, seems wildly idealistic and, in the end, fatuous. To state it once more, the endorsement of epistemological hierarchies is one of the functions for which a state exists. It performs in this manner, for example, when it creates a court system and lays down rules for its operation. One may not, for instance, walk into a courtroom and, on the basis of deeply held personal belief, demand trial by water, not even when all parties agree! The acceptance of the privileged role of science is equally within the state's competence and, at this point in history, constitutes a positive duty. Finally, one must observe that it is legitimate—visionary and perhaps even "elitist," but nonetheless legitimate—to hope for the emergence, in the far future, of a citizenry far more immune to credulity, quackery, and the antiscientific tall tales of cults, than is now the case. Democracy, as an enduring principle, may eventually depend on this, and it is proper for a political system that aims for democracy, always avoiding undue coercion, to encourage such a trend.

In the light of all this, what should we make of calls to "democratize" science itself? This demand has been raised by a number of intellectuals, some of them (but only a minority) scientists. It is not embodied in a popular movement. Indeed, the number of people who have even heard of the idea is small, and the number of actual supporters, minute. There are connections to movements that do have a modicum of support, mostly in relation to health and environmental issues, and presumably the theorists of democratization hope to weld this loose amalgam of causes into a more unified lobby centered on the idea of bringing science, as such, under greater popular control. A recent article by sociologists David H. Guston and Kenneth Kenniston, sympathizers of the movement, points out the chief concern:

> [T]ension between democratic politics and scientific practice arises from the fact that democratic processes and goals are largely incompatible with scientific processes and goals. One might call this exclusionary tension, because the requirements for membership in decision making within science are more exclusive—that is, being a scientist or expert—than for membership in democratic decision making in general. Democratic decision making constantly seeks to encourage and expand participation; scientific decision making limits it. There is a risk that science may oppose democratic decisions that deviate from or deny some scientifically defined truth.[24]

Guston and Kenniston, though relative moderates within the spectrum of those calling for democratization, clearly believe that in clashes between scientific knowledge and the decisions of ostensibly democratic political mechanisms, the onus falls on scientists who do not defer to mass opinion but insist on the primacy of scientific insight. It is also clear that they make the unexamined assumption that increasing the number and scope of participants in "decision making" automatically makes democracy deeper and stronger and is therefore a praiseworthy development. I think this is a facile assumption and, what is more, a very wrongheaded one. It is precisely because it is such a common one, especially among liberal thinkers, that it has to be confronted with the antithetical idea that, in some ethical sense, if not as a matter of law, citizenship has to be earned. True citizenship requires the development of a capacity for reflection as well as a base of generally accurate knowledge about the world. The kind of participation that Guston and Kenniston condone arises, in many cases, within a population lacking these capacities or possessing them only in rudimentary form.

Sympathizers with democratization will, on the whole, emphatically reject any notion of a "scientific literacy test" as the precondition of participation in decisions linked to science, but that is precisely what I think is necessary, in a metaphoric if not literal sense.[25] This is not a matter of actually limiting the franchise or restricting anyone's right to speak out on any issue. Rather, it is a call for the cultivation of a certain mood among the public and within the political elite. What I would like to see is a good deal of skepticism about populism and naive majoritarianism and a frank recognition that competence and knowledge do, in fact, vary greatly from one person to the next. This requires acknowledging that the risk is not that "scientists may oppose democratic decisions" that thumb their nose at scientific truth, but that populist impulses that take small account of scientific truth may, from time to time, conscript a majority in one ostensibly democratic forum or another. On this issue, the reflexive egalitarianism that Guston and Kenniston hint at, and which is asserted much more aggressively by more radical thinkers, is not only mistaken in theory but appallingly dangerous in practice. One need only look at examples of the kind of "expanded participation" the democratizers endorse in order to be made uneasy by the notion. J. Michael Bishop, a Nobel laureate in medicine, drives home the point anecdotally:

> For more than five years now . . . the University of California, San Francisco, has waged a costly battle for the right to perform biomedical research in a residential area. For all intents and purposes, the university has lost. The opponents were our neighbors, who argued that we are dangerous beyond tolerance; that we exude toxic wastes, infectious pathogens, and radioactivity; that we put at risk the lives and limbs of all who come within reach. . . . One agitated citizen suggested in a public forum that the manipulation of recombinant DNA at the university had engendered the AIDS virus; another declared on television her outrage that "those people are bringing DNA into my neighborhood."[26]

Under the reign of the participatory ethic advocated by the democratizers, it is hard to find any grounds to impeach this "community" decision. Under the reign of common sense, it is hard not to be horrified by it. One can argue for more diplomacy and approachability on the part of the research community, more patience and a greater effort to reach out and educate people out of their misconceptions. One might also warn scientists in a position like that of the biomedical researchers at UCSF against arrogance and high-handedness. But the fact is that developing the kind of informed public awareness that would be immune to the alarmism Bishop describes is, at best, a long-term project.[27]

Snobbish as it sounds to point it out, the people to whom such an effort must be addressed tend to dote on confessional talk shows and tabloid sleaze. The prospect for implanting a capacity for mature reflection on scientific issues in such a population is decidedly bleak. The sad fact is that democracy cannot now be measured by the capacity of popular opinion such as Bishop describes to enforce its judgments, but rather by the capacity of regulatory mechanisms to deflect such populism-in-the-raw. In saying this, I am distending the usual notion of democracy so that it incorporates some responsibility to logic, evidence, and truth that goes beyond mere nose counting. This move introduces dangers of its own, no doubt, but I am willing to risk them. Ultimately, one must reckon with the hard fact that this is a society in which the *National Enquirer* outsells *Scientific American* by an absurdly large margin.

The history of attempts to mobilize public opinion around scientific and technical issues is replete with examples of wrongheaded enthusiasm and blind dogmatism. This is not to say that all such movements invariably take the wrong side of the issue or veer into error, but that pattern is by now so familiar that one must ask whether something systematic is at work. I have already cited some of these examples, including the campaigns against power-line radiation and the massive lawsuits against the corporations and physicians involved in the cosmetic use of silicone. In the same vein, we may list organized opposition to microwave communication antennas or preservative food irradiation, or genetically engineered crops, as well as organized support for "alternative health care" in various forms.[28] Some of these campaigns arose from serious scientific questions, some were delusory to begin with, but all have evolved into dogmatic sects beyond the reach of contradictory evidence. They are largely comprised of people whose capacity to understand the scientific points at issue is minimal. Even the anti–nuclear power movement, though it continues to address grave and unsettling questions, increasingly resembles a quasi-religious organization, where the absolute unacceptability of atomic energy is an article of faith forever immune to empirical challenge.[29] It is clear that all these movements exist at least as much to gratify their adherents' desire for spiritual fulfillment as to make responsible and thoughtful criticisms of technology and medicine. If they are prototypes of what democratization of science is supposed to bring us, then the case for such democratization looks terribly weak.

It is further weakened when one looks at the philosophical musings of some of its best-known spokesmen. The most bizarre of these tie "democratization" to epistemological issues, insisting that the failure of science to incorporate appropriate political views damages the reliability of its findings. Feminist critics such as Sandra Harding and (more moderately) Helen Longino insist on this point, though their claims are based more on ideological acts of faith than on pertinent arguments.[30] So do a number of would-be sociologists and cultural analysts of science, with equally tenuous logic.[31]

In most of these instances, the desire to broaden participation in science is compounded with other political motivations, chiefly to dim the glowing image of science as it now exists, because science is historically the product of a white, Eurocentric, imperialist and male supremacist culture.

The mélange of theories and exhortations that thus emerges has little philosophical force and offers small hope for any kind of worthwhile political change, even on its own terms. However, even the more moderate apostles of democratization, those who avoid baroque epistemological flourishes, frequently celebrate populism in science without due consideration of its pitfalls, even to the extent of railing against scientific orthodoxy in the name of demotic enthusiasms. Richard Sclove, author of *Democracy and Technology* and head of the Loka Institute (an advocacy group promoting a supposedly more democratic science) exemplifies this. A recent paper of his urged that science be funded and directed in the interest of ordinary people, rather than the corporate elite (a fair enough goal). But it put forward, as an example of this strategy, a vastly increased outlay for research in "alternative" health care.[32] Specifically, Sclove wants the Office of Alternative Medicine to be greatly enlarged, with the aim of legitimizing a flock of novel theories and treatments. His rationale? Such nostrums are credited by a large percentage of the population.[33] This kind of logic, in which ultimate wisdom reposes in demotic belief, is unfortunately characteristic of the science democratization movement, and diagnostic of its weaknesses. In the end, many of its advocates seem more interested in subverting the reputation of standard science and its ostensible "elite" than in basing public policy on the soundest possible knowledge.

Another factor frequently neglected by the would-be democratizers is the possibility that the reforms they urge may actually backfire against the political ideals they espouse. Most of the call for greater democracy comes, as one would expect, from people who don't hesitate to describe themselves as leftist. Their direct experience is with various movements concerned with health and technology issues where the leadership is basically leftist and where the rhetoric reflects that tradition. But it is far from clear that this is the only possible orientation for a pressure group bent on confronting science. Moreover, even within the existing movements, much of the rank-and-file support comes from issue-oriented individuals whose enthusiasm for a broader left-wing program is at best lukewarm. It is quite possible that if the political mechanisms for which the democratizers lobby are ever put in place, it will be their enemies on the right who will benefit most.

If legislation and policy ultimately accord more power over science to popular opinion, allowing it a greater voice in determining the direction science is to take, and even the theories it is to anoint as legitimate, there is no guarantee that the populace will come down on the side of progressive ideas. It might well call for more military research and less expenditure on

AIDS research. It might support the kind of psychological theorizing that disparages the assumptions and goals of feminism and the gay rights movement. It might well prefer psychics and quacks to physicists and physicians. Above all, as experience should demonstrate to even the blithest of democratizers, it might well enhance the already considerable power of creationists and other fundamentalists to discredit and silence uncomfortable scientific ideas.

But misgivings about unconstrained populism and naive majoritarianism do not play much of a role in the thinking of leftist science and technology critics. These theorists continue to place their faith in the idea that there exists a vast progressive majority out there awaiting only the right kind of exhortation to catalyze it into a sweeping force for desired reform. They expect that the tide of opinion forming behind environmental and technology-related causes will also take up all the other banners of the contemporary left. This assumption has not played out particularly well in the context of any other political question, and there is little reason to believe that it is any more realistic when specifically applied to science, technology, and medicine.

It is understandable and even desirable that prophets and visionaries imagine a day when the highest intellectual achievements of our culture are largely accessible to the vast majority. In such a society, the case for democratic oversight of science would be quite strong. Indeed, it probably would not even have to be made, for it is hard to imagine that such a society could come into existence without more or less automatically bringing huge numbers of people into the "scientific community." It is still more admirable, however, to create practical mechanisms that stand a chance of achieving even a modest advance in the direction of this impressive goal. As experience with the intractability of reform in science education shows, such a task is enormously difficult. As the intense popular obsession with cults, fads, cranks, and pseudoscience shows, the demographic raw material for this intellectual upgrade is, to put it kindly, unprepossessing. What we need for the present is not a science that is "democratized" according to some idealistic formulation, but a science that can honorably serve our present, very imperfect, democracy, and which can contribute to the daunting task of perfecting it.[34] Under the circumstances, the best kind of idealism is necessarily temperate, hardheaded, and extremely skeptical about the instantaneous perfectability of humankind. It is an idealism that should take science, despite its manifest imperfections, as an example, not a target.

FOURTEEN

Authority

A common assumption among scholars who study the relation between science and society is that science is closely tied to the social forms and ideational constructs of the culture that produced it. This postulate governs sociological as well as historical work. At one level, it is tautologous. Modern science was built up within the confines of one relatively uniform culture—that of the Christian West—by groups of people actively engaged with one another on a level that has to be described as "social." Moreover, much of this activity had the condign approval of the political and social elite—thus the *Royal* Society and the *National* Academy of Sciences—and, indeed, closer to our own time, the support of governments. These days, as a matter of course, the modern state underwrites scientific projects with political prestige as well as funding. In return, politicians bask in the reflected glory of scientific achievement, which even now can stir a jaded public with pictures from the surface of Mars or the depths of the oceans. The practice of science is knit into the economic fabric at every level. The costly apparatus of science, ranging from universities to research labs to space probes, is one of the burdens our social system takes on without excessive complaint. Scholarship as a vocation, especially in connection with higher education, is having a hard time of it right now for a variety of reasons, but if disaster befalls, then, rightly or not, the humanists and social scientists will probably bear the brunt of it. The natural sciences will probably escape with light damage because they constitute an indispensable sinew of political and economic power, even when engaged in seemingly "pure" research. By one device or another, the hegemonic forces of this culture are going to find a way of supporting science, along with scientific medicine and technological programs of all sorts. Not to do so is suicidal, and doing so, despite the expense, rarely offends the sensibilities of the public, even in its most penny-pinching mood.

Yet if the previous chapters have demonstrated anything, it is that science—as a body of knowledge, as a method for investigating the world, and as a point of view about the ontological constitution of the world—meshes poorly, sometimes very poorly, with the rest of society's ideas, expectations, and institutions. The paradox is that while science is native to this culture, it is also alien. It is a presence that shapes our social life in countless ways, yet appears ineluctably mysterious as it does so:

> It follows that the process of discovery, the inner fire of the scientific enterprise, cannot be communicated effectively to the citizen who doesn't already know a substantial amount of science. Only when he possesses some of the content of science can he grasp its living culture. Then he can understand how scientific knowledge is validated and how best to make judgments on his own accord. Graphs and "margins of error" make sense to him. He can explain them to others. Controls, multiple competing hypotheses, and disconfirmation become habits of thought. Accounts in newsmagazines are read with an engrained reserve, and scientists are viewed less as savants than as the artists and lucky conjurers they are in fact. Moral-tinged controversies are weighed with close attention to testable reality in the physical world. Of course, these abilities are very limited today.[1]

If polls are any guide, science is one of the few aspects of our culture that still appears to command near-unanimous respect, but this respect is booby-trapped by the widespread incomprehension that lumps all sorts of dubious belief systems, from astrology to cancer quackery, together with legitimate science.[2] Science suffuses our economy with a flood of technological goodies, and makes it possible for most of us to contemplate with complacent satisfaction a prospective life span that would have required extensive good luck only a few generations ago. But the obscurity of the processes that generate these treasures almost invariably stirs up an undercurrent of unease among the beneficiaries. Stephen J. Gould observes: "The idea that science is monolithic, incomprehensible, soulless, and basically bad for us forms the core of a central paradox of our times: Science has become least popular and most feared at the height of its influence and intrinsic interweaving with our daily lives and activities—least pursued and cherished when most essential to the core of education for all thinking people."[3]

But Gould only depicts the paradox that nestles within a greater paradox. For, as many measures of public opinion continue repeatedly to demonstrate, science retains its status as one of the most honored and respected of human vocations. Scientists are still relatively unsullied culture heroes. Notwithstanding, Gould is on target. Distrust of science is also at an unprecedented pitch, especially among the nominally well educated. How are we to understand these contradictory currents? Science, we may note, is an honored guest when it brings cheerful news, but the welcome curdles when it challenges wishful thinking. Society fawns upon science when science enriches it and flatters its ambitions, but turns hostile when science looks askance at its treasured illusions. The public that is rapt to hear about the age of the universe from the Hubble space telescope is furious to learn of the age of the Shroud of Turin as revealed by radioisotopes.[4] The public image of science is continually whipsawed between two poles of popular feeling. The status of science as the acknowledged repository of invaluable truths is secure only insofar as those truths are comfortable.

The interface between science and the "rest of society" is a tumultuous and complicated zone, crisscrossed by alliances and enmities. Apparent amity may flare into distrust when the work of scientists unexpectedly touches a nerve. Recently, we have seen this at work in the controversies over cloning and over biotechnology in general. Beyond the concrete threats rightly or wrongly perceived by the general public, there lurks a deeper issue: the material, indeed, mechanistic, basis of life and all biological processes, and therefore of those processes, like thought, emotion, and belief, that most of us like to believe are self-willed tokens of irreducible individuality. The downfall of the superconducting supercollider is more subtle but even more instructive. On the surface, and to some extent in fact, it was the result of questions being raised about costs and priorities, and about prospects for practical exploitation of the knowledge to be gained. But it also revealed a deeper nervousness about science and scientists. As more and more people—intellectuals as well as politicians—became aware of the motivation of the physics community for building the device, they became increasingly uneasy and alienated. The reason was, basically, that as the apotheosis of "curiosity-driven science," the supercollider was seen as the adjunct of a hermetic and hopelessly obscure philosophical odyssey. It was as if Congress had been asked to underwrite a $12 billion sanctuary so that a tiny, exclusive contemplative sect could meditate at its ease. The physicists seeking answers about the validity and limitations of the "standard model" stood forth as a secretive freemasonry of mages, arrogant, aloof, and contemptuous of the demand for comprehensible explanations. They seemed to be in quest of a kind of esoteric wisdom that could never radiate beyond a tight inner circle, and which would always strike outsiders as impenetrable wizardry. The high-energy physics community appeared (because it *is*) the most rarefied of insider cultures. Understandably, this awakened resentment among those permanently frozen out—rather more than 90 percent of the species.[5] As it happens, most of the arguments against the supercollider were weak, even on their own terms. The project was not ridiculously costly, measured against its timescale. Certainly, compared to the space-station boondoggle, it was a model of parsimony. Although its purposes may have appeared hermetic, it is not unreasonable—quite the contrary—to think that the basic understanding that might have come from it would have eventually made its way into technological innovation.

Despite its great cost, and the occasional obscurity of its immediate goals, research science is anything but a freeloader. If anyone still takes the Marxian notion of "surplus value" seriously, scientists and engineers are the most exploited workers who ever lived. Science is harnessed ineluctably to all the machinery of the modern state and the modern economy, as even the most fanatic devotees of pure knowledge for its own sake concede. So, despite the overweening self-assurance of scientists and their immutable sense that, in some ways, they stand atop the ladder of human achievement, science is the bondservant of the encompassing society all the same. But there is a striking psychological reversal. If science is a servant, then the servant bears the mien, the gown, and, above all, the book of Prospero, while the beneficiary of the enchantment is reduced, most often, to mute incomprehension or surly resentment toward the source of his comfort, riches, and power.[6] Even the wealthiest and most astute chieftains of business and finance are to some extent abashed by the fact that at the heart of their enterprises there are secret goings-on that only their scientist-underlings can grasp. Indeed, the mystique of our current culture hero, Bill Gates, doesn't merely arise from his outlandish net worth. A good part of it comes from the fact that he is an authentic technonerd whose empire was bootstrapped into greatness from his early success in writing clever computer code. He qualifies, therefore, as an authentic philosopher-king, as our culture has transmuted this category.

Others aspire to that title. Some of them are scientists who are prepared to claim that their expert knowledge of plasmids or quarks can be extrapolated to comprehensive understanding of social and political questions. This tribe is, however, rather small and scattered. More common, for whatever reason, is the intellectual affiliated with what used to be called the humanities who sets himself up as prophet and lawgiver in the realm of social ethics. It is hard to account for this phenomenon completely, but it is safe to state that it arose in large part from the hard fate of the would-be "revolutionary" politics of the '60s counterculture. This managed to survive after a fashion, but only as the owlish postmodernism that, in convoy with academic feminism and identity politics, constitutes the Established Church of the contemporary university.[7] The quirks and furbelows of "political correctness" are hardly a major concern here. But I note that one of the chief themes of postmodern trends in literary studies, cultural anthropology, and so forth, is the unreliability, culture-boundedness, and dispensability of science.[8]

What is at work here is not dispassionate acceptance of the consequences of favored philosophical axioms. Rather, it is the distress and unease science generally provokes among nonscientists, articulated in a style appropriate, in its word-clotted fashion, to the academic community. Even more than the general public, the brain-proud humanists who bestride our campuses are chagrined by the idea that they are dependent on, and even controlled by, a corpus of ideas that they find alien and impenetrable. The patchwork of doubtful research programs and quasi-scholarly methods called "science studies" crystallizes this resentment in particularly strong form. Here, the attempt to devalue science and to relativize its knowledge as the effusion of one limited subculture is the very core of the "discipline."[9] Clearly, science studies would not have been so successful in making its niche in the academic world without the strong sympathy of a community of humanists and social scientists who favor the enterprise because it promises to cut science down to size.[10]

There is, of course, nothing new in the discomfort of intellectuals with a science to which they can make no contribution and which consistently defeats their attempts to understand it. The titanic Goethe provides the classical example of an incontestable genius—a polymath, in fact—whose frustration over his own limitations in absorbing the science of his day led to bitter discontent.[11] But by and large most humanist scholars have, until recently, accepted the supremacy of science on its own ground, even as its increasing technicality carried it further and further out of their ken. Their tone, in speaking about science, was more melancholy than hostile. A recent example of this attitude comes from critic and scholar Clive James (who indeed cites the cautionary example of Goethe):

> The difference between me and the microbiologists, of course, is that they know what they're talking about, whereas I know only how to talk. It is a difference basic to the life of the mind in our time—a time that can usefully be thought of as going back to Goethe, who didn't like Newton's theories about color. Goethe had good humanist reasons for his dislike but didn't have the math to back them up. Science was already off on its own; there were already two cultures. It could be said—it should be said, in my view—that only one of these, the unscientific one, is really a culture, since the mark of a culture is to accumulate quality, whereas science merely advances knowledge. But my view is part of the unscientific culture, and has no weight in the scientific one, which settles questions within itself, marshalling evidence powerful enough to flatten cities and bore holes in steel with drills of light.[12]

James, however, will be regarded as a defeatist by the new breed of humanist. Wielding the magic word "epistemology" as though it were an irresistible talisman, they delve into the jumble-shop of postmodernist aphorisms to concoct the charms that will ransom them from the incomprehension science seems to impose. A thousand rationales—feminist, anthropological, sociological, linguistic, ecological, even esthetic—are flung about to justify the notion of science as merely the blind faith of a monomaniacal tribe. (Although, paradoxically, science is also supposed to be the cutting instrument of raw political power—quite an accomplishment for what is asserted to be a collective illusion.) Taken individually, these arguments are hopelessly thin. Taken together, they form a tangle of inconsistencies, and are thus even less convincing. What they testify to, beyond general political discontent with the ruling institutions of contemporary life, is a desperate longing to redress the imbalance that prevails in the world of professional thinkers. This arises from the dichotomy that insists that science matters greatly because it tells us how the world works, whereas the work of humanists and many social thinkers matters hardly at all because it only tells us how their minds work. Even worse, it implies that science has the automatic right-of-way over any other mode of thought. Once it claims competence on a range of questions, everything "unscientific" must yield. The claims of science are serious and compelling, the claims of other modes of thought, no matter how charming or engaging, merely whimsical or illusory. This is not the sort of assertion that needs to be made by individual scientists, or recorded in formal manifestos. It is by now engraved in our cultural experience and ratified by the practice of our political and social institutions, at least when they are desperate. When something really important is on the line—the outcome of trade wars, shooting wars, or weather emergencies—it is the scientists who get called out to come up with the appropriate magic wand.

Whatever might be said for the special talents of humanistic thinkers and social philosophers, in practice the outside world pretty much shrugs them off. Even worse, the spirit that has governed their disciplines has tended to look at the question of "truth" with a resigned shrug. Grace of expression and moral passion have traditionally counted for more than the evidentiary protocols that the sciences relentlessly apply. At some level, these disciplines have implicitly abdicated. They have declared that they are not to be taken seriously as the sciences are taken seriously, that what a scholar asserts of a poem by Keats or a fresco by Cimabue is merely part of the cultural atmosphere, not a hard-nosed proposition on whose truth important questions will turn. But there is only so much of this devaluation that intellectuals can stand, particularly when they hold themselves out as the bearers of liberating political ideals. Sooner or later, the idea that the methodology of science constitutes the epistemological gold standard, against which every important claim must be measured, ceases to be merely irritating and becomes insupportable. The resulting explosion unleashes a rain of epistemological fantasies, along with the shimmering illusion that science has somehow been dethroned.

Academic life in the United States and elsewhere has been afflicted with this kind of uproar for a decade or more. Some of our most highly reputed thinkers have joined the insurrection. They have become mongers of paradox. At a time when science is unprecedentedly successful in redeeming the promises it made ten or twenty or thirty years ago, it is pronounced unreliable and flighty. As the various branches of science find themselves less and less isolated from one another, conceptually and practically, the new critiques of science declare it to be fragmented, disunited, riven with mutual incomprehensibility. Whether one likes it or not, science now stands more firmly than ever as the arbiter of proposed world views and the censor of palpable impossibilities. More than ever, philosophy, properly speaking, has to answer to it. In response, the postmodern humanist declares science inconsistent, hallucinatory, a mere text among texts.[13]

I think an end is in sight to this particular barracks uprising. The weapons it wields are simply too feeble and the apparent bloodlust is too diluted by self-doubt. Also, much to their amazement, the epistemological Jacobins have found that, in the cut and thrust of academic feuding, scientists are quite capable of hitting back with real ferocity and a much surer sense of the strengths of their own case and the weaknesses of their critics. Academics who rail or snipe at science are rather like well-brought-up children who have made a deliberate decision to misbehave and outrage their elders on some solemn occasion. They are terribly self-conscious and jittery about the whole business, and gnawed by the suspicion that they might lose their nerve and fail to go through with the thing.[14] When confronted with scientists' hard stares, they fidget and prevaricate and look as though they would really prefer to be elsewhere. They have been caught out in something naughty ("transgressive" is the favored word these days) and find that the requisite boldness isn't part of their makeup. Scientists find it easy to win these showdowns.[15] Not only does science hold all the aces, intellectually speaking, but its scholarly opponents, since they are not constitutionally stupid, soon realize how things stand and begin to think about cutting their losses. Things will probably cool off, to the extent that literary critics and devotees of cultural studies will stop saying provocative and silly things about science (in public at least).

Yet, quite apart from all this contention—the "science wars" to give it its frequently used if ungraceful title—the long-standing culture of the university is, as I have noted, now in serious trouble. University life is likely, for a host of reasons, to become stingier, meaner, less commodious. But, as I have also observed, university scientists, as the mainstay of the overall research culture, are simply too valuable to the outside world to alienate, demoralize, or banish. They are willing to work for the substantially lower salaries of academic positions in return for the freedom and autonomy this affords. That gone, they will decamp in droves for the more remunerative pastures of business and industry, with a sigh of regret at having to abandon much of their "curiosity-driven" work, but knowing that serious intellectual challenges still abound. It is unlikely, therefore, that Procrustean schemes of university "reform" will be impressed on scientists with anything like full force, especially if corporations and business consortia, acting on reasoned self-interest, step in as patrons of science departments. Though they won't get off scot-free, scientists will be offered a separate peace on far more generous terms than their colleagues in the humanities who will bear the brunt of increased regimentation and loss of power. Hierarchy will then replace collegiality within faculties, with most scientists somewhere near the top tier. If this prognosis is correct, one almost certain consequence is renewed antagonism toward scientists on the part of other intellectuals. If the current rationales don't endure or reappear, some new ones will doubtless be concocted. No other outcome is possible if the primacy of science as the true intellectual engine of our culture becomes so starkly engraved in academic life.

As a campus fashion, anti-science hasn't been completely harmless. It dilutes the already minimal experience of science available to students not majoring in the subject. It wins a few (though not many) converts to its slogans.[16] It has spread a bit of anti-science paranoia around the campus.[17] It has begun to breed a class of self-described "policy experts" who might cause mischief until they are firmly refuted.[18] It can also seduce a few bright but naive students away from scientific careers, in the name of feminism or some similar piece of ostentatious political high-mindedness. As one repentant academic feminist puts it:

> Groups such as the Association for Women in Science are sincerely attempting to dismantle the barriers that women still face. But others, in which I am sorry to include all too many of my fellow feminist philosophers of science, are really more interested in undermining the epistemological authority of science and making it subservient to their own political agenda than in making science a truly inclusive discipline.[19]

Most unforgivably, in league with the apostles of "identity politics" it can beguile minority students with extravagant praise of the pseudoscience practiced by the supposedly ancestral culture.[20] Most seriously, it can, through its influence on pedagogical theory, obstruct and delay desperately needed reforms in elementary and secondary science education.[21] In sum, therefore, the end of anti-science restiveness on campus will solve some real problems and curtail some galling nonsense. But it will not do very much to alleviate the uncomfortable position in which science finds itself with regard to the larger society.

The populist form of anti-science is a much more robust creature than its academic cousin. This version—versions, rather, because it is protean as well as hardy—is likely to endure and to re-create itself in a variety of contexts without any foreseeable end. There is very little of the anti-science case articulated by humanities professors which cannot be found in some commonplace maxim of a world largely indifferent to professors of any kind. The popular version of anti-science not only propounds the same themes as the academic version, it does so with greater brevity and honesty. Lacking the baroque verbosity of the "theorists," it simply gets to the heart of the matter more quickly and accurately. If the contrast points to anything, it is to the greater intellectual self-confidence of the supposedly unlettered masses. If there is a crisis of the authority and legitimacy of science—and I would argue that such a crisis has been slowly and incrementally building—then it is centered upon the grotesque mismatch between the scientific worldview, on one hand, and notions deeply and inextricably embedded in mass culture. Even to call them notions can be a misnomer, for they are often poorly articulated or completely inexplicit. They take the form of individual preferences, propensities, and stubborn habits. Their antiscientific character may not advertise itself on the surface but becomes visible in the context of concrete issues and debates.

I have, I think, outlined the sources of these misgivings—a view of the world that is stubbornly teleological and anthropocentric, a propensity to form elaborate and self-sustaining belief systems that fill an emotional vacuum, a sentimental preference for the anti-intellectual and supposedly instinctive side of life. Such predilections are further sustained by the cultural norms and institutional practices that help define contemporary culture. These include an easy indulgence of demotic preferences, including the demand for a world untroubled by vigorous debate, demanding thought, deep ideas, and precise, eloquent language. Jean Bethke Elshtain, one of our shrewdest social critics, expresses deep foreboding about a society in which the very notion of authority has disappeared from the political vocabulary:

> And what happens then, once everything is up for grabs and persons place no confidence in authoritative norms and claims? We know Tocqueville's response: at that juncture we're on a fast track toward democratic despotism, a world in which we have a hard time accepting that anyone—or any authority—has a legitimate claim to constrain another's actions or, indeed, to call us to action. . . . Authority and liberty are not opposites, they are twins. If we kill one, we lose the other. For legitimate authority is about standards, distinctions, plural institutions, norms, claims, and counterclaims—all of which are under attack today.[22]

The notion that hard work, sometimes desperately hard work, is a prerequisite for true citizenship, and for the authority that comes with it, has all but disappeared in the intellectual train wreck that constitutes contemporary political thought, whether of the right or the left.[23] The hidden assumption is that if democracy distributes rights equally, it must, by some unspecified alchemy, arrange for talent, insight, and judgment to be allotted with comparable fairness:

> This ideology begins with one of the great liberal ideas that defined this country: equality is not something attained or purchased or inherited. It is simply possessed, universally at birth. Failures to realize this ideal have done nothing to weaken its strength: when we stand before the law as citizens, we are supposed to stand equally.
>
>   But as de Toqueville realized, such a revolutionary idea could not be applied in the political realm without also resonating in all other aspects of American life. In fact, it has now become the defining aspect of American culture. We are so loath to make distinctions (except to correct earlier distinctions) that we now doubt whether any are worthwhile or even possible. We ask democ-

racy to be the moral, spiritual and esthetic compass by which our culture evolves.[24]

One corollary among many is the assumption that the hard work of learning science can be left to scientists, while the authority to dispute or undercut the claims of science nonetheless remains with a general population that has evaded such work.

This notion is reinforced by a system of education that wanders blindly in search of an effective method for teaching science at any level. Its negative consequences are substantially enhanced by the timidity of science educators in the face of religious and ethnic groups who insist on blurring or compromising the intellectual authority of science in order to leave room for their treasured myths. They are amplified by a tradition in the press and the mass media which distorts and crops the image of science in order to make it mesh with the desires and preconceptions of a mass audience. They are strengthened still more by a governmental and legal system that is edgy, uncertain, and frequently wrongheaded in dealing with scientific questions. Moreover, the intellectual shortcomings of politicians, jurists, and bureaucrats exist in peculiar resonance with the would-be intellectual avant garde. Gerald Holton, pondering the weakened image of science, remarks: "[T]here is another, an ideologically based, movement, which is also changing the environment for science. It comes from a very different direction [i.e., as compared with standard Capitol Hill politics], yet in an unplanned way it happens to intersect with the politically based one. Its purpose is to question the moral authority of science altogether."[25] The resultant of all these circumstances is a huge corpus of social habit that is sunk fast into the tissue of the culture and has the potential to resist, and even to blockade, scientific judgment as it impinges on social, political, economic, and environmental questions.[26]

This was made all too concrete in the respective fates of the superconducting supercollider and the space station project. Scientific opinion supported the former by a substantial majority, and opposed the latter by an even larger margin. Yet the politics of the day, formed by a complicated matrix of public opinion, economic pressure, diplomatic rationale, and media imagery, led to a completely opposite result. Of course, even the most perfect political arrangements can lead to mistakes and misfires, on occasion, but it isn't merely the final outcome of this contest that reflects badly on the public's grasp of science. The point is that the entire debate produced evidence, from across the spectrum of educational background, ethnic affiliation, and political ideology, of pervasive failure to understand any of the scientific issues involved, particularly the profound difference between the fundamental investigations undertaken by high-energy physicists and the one-shot engineering stunts proposed for the space station. Both enemies and supposed "friends" of science conflated the two. The worst of the rhetoric denounced the ambitions of the physicists to see ever more deeply as impious and blasphemous as well as costly. It suggested not only a superstitious fear of offending numinous powers by penetrating too far into their realm, but, even more, a sullen, envious suspicion of knowledge itself on the part of those unable to make much sense of it. Though the wisdom of funding this or that project was the ostensible focus of the debate, what it really disclosed was not differences in policy or perspective, but rather the enormous gap in worldview between scientists and laypersons, the latter even including many people who think of themselves as science supporters. One observer, ironically commenting on the ambiguous majesty of science, notes:

> The rhetoric of science has, however, accumulated such prestige since the seventeenth century that it has become difficult not to believe that through science the world is finally telling us what it's really like. That if the world could speak for itself it would speak science. Indeed it can sometimes seem as if scientific descriptions are not made by people at all, especially when these descriptions involve accounts of our cosmic irrelevance. Science often suggests that the most brilliant thing about us is that we invented science. There is something ironic about such man-made accounts of our own freedom and redundancy. It is like God proving that God is dead.[27]

The natural reaction, even of the well-educated nonscientist (and of politicians, well educated or not) is to try to devise some pretext for leveling the epistemological playing field, so as to taste, at least occasionally, the pleasure of chastening scientific hubris.

I return again to my unapologetic founding assumptions. By and large, scientists know what they're doing and, over the long (and not-so-long) haul tend to get things right. By and large, critics of science, whatever their ideological roots, tend to get things wrong. No doubt, hard digging and ingenuity can come up with some counterexamples, and the cleverest of the critics try to claim paradigmatic status for these. None of them are sufficient, in my judgment, to dislodge my basic point: all in all, science has no peers and no challengers. In matters of public interest where science has some bearing, and where it is vital to get the most accurate possible picture of how things work, science is—or ought to be—the trump card. This is not to deny the possibility that the issues reach areas where scientific knowledge is incomplete and fragmentary, and where scientific opinion is itself seriously divided. But in those cases, most scientists in the relevant area will be reasonably frank about the uncertainties, and any competent inquiry will bring them to light. In any event, as we have seen, even scientific unanimity will be flouted when other social and political forces contravene it. As one analyst puts it, "Never 'on top,' . . . they [scientists] have not developed a skill for direct or indirect rule."[28] Those who portray science as an imperial cultural presence ascribe far too much social force to scientific authority. If they are right, why are we still debating whether and how evolution is to be taught in the biology classroom?

It is, in my view, vital to change this situation and to give science a social authority commensurate with its astonishing success in living up to its own ambitions. The corollary is that we have to ignore or reject its rivals. To speak of any knowledge-claim as "unscientific" is to disparage it; the fact that a proposition has no scientific warrant ought to count against it. Claims thus tainted include religion and much of what passes for philosophy. This is not a principle gladly received in a culture where the ability even to judge claims of scientific validity is so appallingly rare. It dispossesses the vast majority of people and sets them on the sidelines of many vital debates. It confers the right to judge some of their most passionately held beliefs upon strangers who are intellectually and temperamentally quite different from them. It is a severe, even dismissive, verdict on the ability of most people to think accurately. In short, it is hardly a principle that can hope to endear itself to a culture whose central myth is that competence is found everywhere and that one person's opinion is as good as another's. But it is a principle that we neglect or fudge only at our great peril.

In saying this I court accusations of "scientism," the political doctrine that only the scientifically trained are intellectually mature enough to take political responsibility, and that the methods and attitudes of scientific and technological work can be transferred, largely intact, to the instruments of political governance. This idea, usually identified with the "positivism" of Auguste Comte and his followers, is universally reviled these days, and I have no immediate desire to resurrect it.[29] Even the most ebullient of the proponents of science take special care to repudiate "scientism," while science's critics resort to the word continually as an all-purpose term of reproof. Perhaps one can argue that a doctrine so unanimously despised, so subject to automatic condemnation from all sides, probably has virtues that have been deeply obscured by history. Some sober and objective thinker ought perhaps to resurrect "scientism," at least for the sake of argument, to see what points might honestly be made in its favor. But that is a project for an imaginary future.[30] Let us agree, without reservation, that there is no reason to believe that scientists, engineers, and physicians are ipso facto equipped with superior political judgment, or that they have a more reliable sense of which ideals and goals this society ought to pursue. Let us here adopt the democratic postulate that, over the great range of purely political and social ideas, at least, no group possesses or should be allotted special competence. Even agreeing that political discernment is not uniformly distributed in the population, let us concede that no particular vocation or economic situation gives one a better chance at it, no more so than membership in a particular race, ethnicity, or sex. My point is that even so, science, as a system of knowledge and as a body of techniques for getting things done, deserves a high order of respect from the institutions of our society, a presumptive, if not absolute, immunity to being overruled by at least the coarser and more volatile aspects of the democratic process. Science, in short, ought to be insulated from the impulsiveness of vulgar majoritarianism and populism. It ought to be hedged by a substantial barrier of respect, even amounting to deference, from the quarrelsomeness, pugnacity, and one-upmanship of the political game.

There is an analogy here to the notion of constitutional rights and human rights generally. The doctrine of innate and inalienable rights has fortunately evolved into a sacred juridical realm. It stands above the fractiousness of immediate political quarrels and is warded against the intrusion of popular clamor by all sorts of explicit and unspoken safeguards. This is a stringently limited analogy, however. It neither provides clues for the explicit political arrangements that ought to be made on behalf of science, nor for the tacit understandings that will be needed if such institutionalization is to go beyond cold formality. Even worse, it is no guide to how these arrangements could be put into place, in a practical sense, nor how the necessary understandings might be inculcated in a vast and vastly varied population. To cite one very obvious distinction, the notion of rights tenders a common promise to all citizens. Even those who are dismayed by some immediate application of the doctrine usually temper their outrage by reflecting that a time may well arise when they will desperately need the shelter of inviolable rights. But if our civil and political culture is to carve out a special niche for science, the process will necessarily be perceived as conferring a special civic status on a tiny, privileged caste of citizens, a feature, no matter how prudent and defensible, that will automatically generate considerable popular antagonism. Moreover, the key actions will have to be taken by a political class within which direct knowledge of science is sparse, and firsthand understanding of the idiosyncrasies of the scientific subculture is essentially nonexistent.

A peculiar fact is that in a society where science and technology have assumed premier importance in so many facets of economic and social life, and where they undergird the enormous diplomatic and political power of the most dominant nation ever to exist on earth, scientists, even engineers and physicians, are virtually absent from the vocation of serious politics. At most, they play a part in the actions of special-interest groups concerned with the environment, public health, and so forth. But few are party activists at even the most local level, and it is rare to find a candidate running for public office who has a scientific background, let alone an ongoing commitment to a scientific career. Thus, the prospects for shoring up the social influence of science, and amplifying its political voice in appropriate areas, are, if not bleak, then extremely hypothetical. They are contingent upon intermediate developments—like an enormous improvement in the quality of science education for the nonspecialist—that are still far away.

What I advocate has to be argued for, finally, in a particularly unfamiliar language. I suggest that an intellectual elite be recognized as such by the informal tenets of our collective life and, to some extent, by formalized legal machinery. I insist, moreover, that the fruit of their work be accorded a special status, not only when it is received with goodwill by most of the population, but even—in fact, especially—when it disconcerts and shocks popular opinion. In short, I am in favor of authority—authority recognized, empowered, and perhaps even institutionalized. Concomitantly, I implicitly seek to legitimize order and hierarchy as well. The ambient society must recognize the necessity for order, hierarchy, and authority, both within science and in the interactions of science with the wider culture, if science is to function well in its own terms and maximize its usefulness to everyone else.

Notwithstanding the horror evoked by these words in contemporary social discourse, honesty compels their use. But these ideas transgress certain assumptions about social life that form the core of our political culture. Our "left" and liberal traditions are emphatically egalitarian, as well as antiauthoritarian. Our "conservative" traditions, rooted in the model of untrammeled enterprise and the ascendancy of the self-made man, are also deeply libertarian and antihierarchical. The traditional model of European conservatism, with its continuing invocations of stability, legitimacy, and the immutability of social station, is at best a pallid ghost in the councils of the American right.

Paradoxes abound here. I am celebrating authority, hierarchy, and order on behalf of an institution—science—that prospers amazingly because of its intense commitment to meritocratic values and its consistent openness to talent wherever found. These values, in mutant form, pervade our culture, yet the result is at least partly inimical to science. "Openness to talent," in the wider society, means openness to all new ideas, no matter how inane or impossible. "Meritocracy" means celebration of whoever succeeds by whatever means, including demagoguery or arrant hucksterism. Likewise, the liberty to criticize received ideas and to press inquiry without limit, essential to the scientific culture, has its demotic counterpart in the ethos that labels every crank a Galileo, and every wild speculation an idea worth considering.

Despite the dark imaginings of some of science's more wistful critics, this suite of concepts—authority, order, hierarchy—is almost never mentioned, and certainly never praised, in the course of educating a professional scientist. If there is a pervasively idealized personality type, it is that of the dreamer, rebel, and iconoclast—in short, Galileo again, though much better understood than in the popular version. But, rightly understood and divested of their odious connotations, these notions are as vital to the health and longevity of science as the libertarian and individualist values that everyone is eager to celebrate. It is some consolation that, in the context of science, these ominous terms must be construed so that the notions of power and subordination they invoke exist more in the realm of ideas than in the relationships of people (although they can never be entirely absent from the latter). Authority is rooted in the cogency of theories and the rigors they must endure to pass into the canon. Order means the interrelation of theories, in particular their mutual consistency and their power to illuminate one another and to bring unity to our overall conception of the universe. Hierarchy means recognizing that not all ideas are equally well founded, that highly confirmed theories trump speculative ones, that the imaginative impulse itself must be tempered and disciplined by the explanatory power and self-consistency of theories already in place. It is merely the structural outcome of the skepticism that is a necessary counterweight to imagination. All of these are key to the realization that, while Galileo was a rebel, not all rebels—only a tiny fraction—are Galileos.

The attitude that social institutions ought to take toward science is not a servile one, but merely realistic. It would accept what seems obvious to dispassionate inquiry: science is far better at describing the world than any contender, and far more successful in reaching its declared goal than any other systematic body of knowledge. Consider, for a moment, how much sense it makes to accept the assumptions about social and individual behavior encoded in our Constitution as lasting, unimpeachable truths, while subjecting the natural sciences to every possible quibble when they enter the public sphere. Our society must devise a method of accepting the authority of well-confirmed scientific knowledge smoothly and ungrudgingly. If such a mechanism could be put into effect, it would inevitably discomfit all sorts of factions and special interests, both those defended by current critics of science on the left, and those, more used to getting their own way, on which the right typically dotes. But it would steer us away from a multitude of disasters and toward a more mature apprehension of how to accommodate our policies and political visions to a recalcitrant reality. Our culture needs to learn how to distinguish among theories claiming scientific status, to judge which should be taken at face value, which heavily discounted, and which merely rejected outright. As Karl Popper remarks, science education for the non-expert ought to produce people "who can distinguish between a charlatan and an expert" (though he is quite aware of the obstacles science educators face).[31]

Finally, we have to realize that the focused unity of science holds important epistemological lessons for everyone else. The system of assumptions on which we ground our policies and planning must be continually monitored and inspected for accuracy and consistency, which will mean, among other things, checking it against relevant scientific findings. Whether politics and the achievement of social consensus are ever going to be true "sciences" is open to severe doubt, but it seems to me that the virtues of anarchy and caprice in public affairs are much exaggerated, and that the model provided by science for gathering and analyzing knowledge, and for comparing it with experience, is worthy of at least some emulation. The point is that science is almost as important in providing a model for organizing, structuring, and pruning cognitive systems of "nonscientific" knowledge as in supplying society with reliable information from within its own specialties. All this requires a reorientation, if not an utter rethinking, of the attitudes that inform our most sprightly politics, on the right as well as on the left.

Western culture, and, more specifically, American culture, are imbued with the notion of continual upheaval and change, the continual process of discarding the outworn and instituting the novel. We build up our social and economic institutions as a palimpsest of "breakthroughs," the latest one always covering up a bit of its predecessor. It is not a bad model; indeed, science might well be pointed to as a prototype, and thus as evidence that it is a good model. But its dialectical opposite has, perhaps, been too much neglected. Maybe what we need is a more studied appreciation of stability and continuity, and more attention given to the virtues of an equilibrium that only shifts slowly rather than being displaced with stunning abruptness at unpredictable intervals. Paradoxically, science, in its cumulative qualities and in its constant need to harmonize new perspectives with classical bodies of knowledge, provides a model for this as well. At any rate, the notion of "deference," so strange and even revolting to American sensibilities, will have to infiltrate the culture, if only with respect to powerful ideas rather than to their authors and advocates. It seems to me, finally, that this is a necessary condition if science is to do its best for us, to open the gates of various Edens, as well as to ward off a host of incipient disasters. The demotic pugnacity and irreverence that are so much a feature of the American cultural landscape, and which energize the singular virtues of our way of life, also have great corrosive power. When it is allied to the massive ignorance, confusion, and even hostility, that characterizes popular attitudes toward science, and put into the service of the daydreams of cults and ideologues, it is a dangerous engine indeed. Far better, then, to temper it with some modest notion of due respect, with the idea—the factual idea—that there are some areas of knowledge where authority counts for something.

To say it once more, as a quasi-political project, reorienting society's relation with science to produce even a modest increase in the perceived authority of scientific knowledge is desperately difficult. As Miroslav Holub rightly observes, "Definitions of wisdom that exclude science, at this stage in the planet's history, are just lazy, persistent metaphors."[32] But to incorporate this insight into the common assumptions of our culture is a daunting task. Where to start? Even that is obscure. The notion of setting apart a restricted class of Americans to sit in judgment on all the others is spooky and obnoxious, particularly when the claims of unimpeded individualism are advanced to oppose it. "An open society . . . and a democracy cannot flourish if science becomes the exclusive possession of a closed set of specialists."[33] This is a valid insight, yet it must be reconciled with the pressing need to erect barriers against epistemological anarchy. Perhaps we will have to devise a new vocabulary, to euphemize "authority" and "order," before we can even fairly begin. Also, there is the question of where such a crusade is to fit into the panoply of current politics. So many apparently more exigent matters stir the public temperament that a debate on the epistemological virtues of science and the proper scope of its knowledge and competence is unlikely to get much of a hearing.

Great tensions afflict our current political mood. They are centered on notions of wealth and class. Basically, this is a nation in which much of the effective control of things belongs to a small, hegemonic class of people with embarrassingly large stores of money. This, baldly stated, invites the charge of "vulgar Marxism," but vulgar or not, Marxist or not, it seems to me inarguably true. At the other end of the spectrum, there is a rather larger class of the truly dispossessed, the people who, if they know where their next meal is coming from, are certainly unsure of where next week's paycheck is coming from. Between lies the amorphous mass of the "middle class," riven by conflicting fears and allegiances, oscillating continually between ideological poles. On the one hand, their material interests, in terms of job security, health care, environmental and workplace safety, educational opportunities for their children, and a secure old age, depend on wresting a good deal of wealth out of the hands of the narrow plutocracy. Yet they feel little sympathy with the underclass, are disdainful of its mores (racism has something to do with this) and frightened of being contaminated by them. Moreover, they are out of patience with the poor for a variety of economic and cultural reasons and reluctant to make any sacrifice for their uplift. They like the libertarian ethos trumpeted by the right but lack the financial means to take advantage of it. At the same time, many of them—particularly women— are quick to take note that the right flies from the implications of its own libertarian propaganda when dealing with issues like abortion rights. Yet their own libertarian enthusiasm cools when it is the "cultural left" that takes the most ostentatious advantage of the mooted liberties through movements such as "gay rights." They are suspicious of state authority, yet fearful of social disorder welling up from below. They are somewhat swayed by the argument from the left that their most intractable enemies are those at the top of the tree, but feel an envy for the wealthy that sometimes amounts to admiration. The tug-of-war that constitutes American electoral politics has been fought over their electoral loyalty for three decades or more, and the outcome, in terms of ultimate equilibrium, is far from certain.

The point of this exegesis of the current American political situation—I apologize for its length—is that any serious, systematic consideration of the social role of science is unlikely to occur at the level of instrumental politics. There are too many other items on the agenda, most of them far more urgent than even well-known issues like science education, science funding, and the role of scientific expertise in courts and hearing rooms. Scientific issues only get a smidgen of attention if they can be put in an apocalyptic light. Perhaps that is a good thing, if we assume that any attempt to marshal public opinion behind the institutional authority of science might well backfire. We had a hint of this in some of the contention over the failed attempt to create a national health-care system: an inordinate amount of lobbying was devoted to putting "alternative treatment" under the umbrella of the proposed coverage. But, in any case, the surface of American politics is hardly so placid that one can expect anything like a stately debate over even a fraction of the issues I have raised. The best we might expect is that some cause like tort reform might run interference for an effort to make courts more responsible to scientific evidentiary standards.[34] Issues of the public accountability of science—and the public accountability *to* science— are likely to be introduced, if at all, piecemeal or through the back door, as sideshows related to issues deemed more compelling.

But is it even possible, through legal and regulatory measures or merely through some new attunement of our cultural mood, to augment the status of science and respect for its judgments? There is, I submit, a precedent, at least in the American system, though an ambiguous, even dubious one. Consider the Federal Reserve Board. This is supposedly a regulatory organ of government, created by legislation and, in theory, liable to be abolished or to have its policies overruled by legislative action. Yet in practical fact, it amounts to an extraconstitutional "fourth" branch of government and is in some ways the most powerful one. Its control over the pace of economic activity and, in effect, the employment rate affects millions of lives and careers, though most people have only a vague notion of what the Federal Reserve Board is or does, if even that. The Fed can exercise control instantaneously, on the whim of a handful of appointed officials, without debate, without recourse or appeal, and even, for the most part, without being subject to judicial review. If Congress, or the administration, made a serious attempt to reverse one of its judgments, a storm of indignation would be unleashed from "responsible" financial circles, decrying the attempt at "political interference" with a supposedly apolitical institution. An attempt to abolish it or seriously restrict its freedom of action would, practically speaking, instantly terminate the career of any politician foolhardy enough to endorse the notion. There is no doubt that, in the community of movers and shakers, the pronouncements of the chairman of the Fed get much more serious attention than anything a representative, senator, or president has to say.

I am not, of course, suggesting the establishment of a "Federal Science Board" with powers and immunities analogous to those of the FRB. For one thing, as a loyal, if hard-bitten, socialist, I seriously challenge the Fed's ostensible political neutrality. Without question, it is an arm of the plutocracy and will do grievous damage to the lives and prospects of millions of working people, rather than roil the sensibilities of the banking and investment community. One might well suspect that the hypothetical Science Board, if comparably empowered, would be similarly tilted in terms of the interests it was committed to serve. My only point, then, is that, despite the democratic veneer of our formal political structure, it is perfectly possible to create essentially sub rosa loci of enormous political and administrative power that are utterly insulated from what most people think of as the political process. Perhaps—and I emphasize that this is extremely tentative—a modest step in that direction with respect to the policy organs concerned with science would actually be beneficial. This is, of course, sheer heresy to pure democratic sentiment—though, as I have pointed out, it is at worst a minuscule affront compared to the gross reality of the Federal Reserve Board. But, as I have continued to insist throughout this book, the flaws of naive majoritarianism and blithe populism have great potential to cause harm, especially in situations where public opinion is for one reason or another disinclined to pay heed to scientific fact.

Conceivably, such an apparatus could actually arise in America if a sufficiently powerful faction of the "unofficial" permanent government, that is to say, the reigning plutocracy, came to regard the slippage of scientific authority as a serious threat to the kind of governance they prefer. They might, in effect, do the right thing for the wrong reasons. The intermittent struggle over tort reform provides an example. Here, would-be reformers, most of them representing huge corporations, certainly want to exclude a lot of fake science from judicial procedures, if only out of selfish motives.

Ideally, one hopes for something better, for a public that, becoming aware of its own limitations, is willing to surrender a bit of its presumptive authority in return for the demonstrable consistency of experts and scholars. This would require a good deal of diplomacy on the part of the science community, as well as a softening of the mood of suspicion and alarm that frequently grips popular feeling when scientific issues are raised. To illustrate, it would require the kind of public mood that would, hypothetically, be at least willing to take a hard look at the possible resurrection of nuclear power if the consensus of scientists and engineers moved strongly in that direction. Such a reform assumes the emergence of a cultural consensus that there are such things as quackery and scientific imposture, and that honest scientists have both the capability and the responsibility to challenge or even ban them. This could not, in my view, occur without a corresponding act of renunciation on the part of scientists. They would have to foreswear all political ambitions, in an institutional sense, and credibly abjure any idea that they have a right to interfere in areas beyond the immediate competence of science. They would have to acknowledge that, in the large, science is responsible to democratic institutions. That is to say, the surrounding society has the right to decide, on grounds that go beyond scientific judgments, what kinds of research it wants to give high priority to, what kinds of projects are important, what kinds of social policies science should serve in the broad sense. Even missteps like the supercollider and space-station decisions could be accepted with reasonably good grace if they were clearly transient aberrations of a system of democratic guidance of science that was for the most part sound. Another element would be a credible pledge from the scientific community to deliver its judgments on scientific questions in a way that abolishes any suspicion of undue financial or institutional influence. This might indeed require more internal disciplinary machinery within science than presently exists, though it is not clear how formal or rigorous this would have to be. At any rate, secretiveness would have no part in any of these deliberations. Science must not only be open, it must be seen to be open.

Critics of science, especially those steeped in the fatuities of recent "science studies" doctrine, might be skeptical that such an arrangement could or should be worked out. Their preference, so far as I can see, is for the kind of "consensus" building that reflects the idea that there is no truth but socially negotiated truth. Therefore, they think that scientific truth can be whittled away as part of the negotiating process, with no prejudice to the real interests of the community. Such is the fruit of philosophical relativism! These critics, rarely found outside the academic hothouse, are rather easy to disarm in intellectual struggles. Doing so benefits the cause of serious scholarship, negates the untoward consequences of academic fads, and generally limits the harm that perverse doctrine can inflict on the outside world. But this damage is small in any case, and forestalling it is only a tiny part of a vastly more difficult project. All of us, scientists and nonscientists alike, must ultimately create and sustain a society and a culture that is mature enough and brave enough to handle the gifts—and the uncomfortable truths—that science affords. The enormousness of this task is frightening. But I should like to believe that it is also inspiring.

Notes

INTRODUCTION The Rule of Opinion and the Fate of Ideas

1. "Without democratic control, there can be no earthly reason why any government should not use its political and economic power for purposes very different from the protection of the freedom of its citizens" (K. Popper, *The Open Society and Its Enemies*, vol. 2, 127).

2. "As long as science is the search for truth, it will be the rational critical discussion between competing theories, and the rational critical discussion of the revolutionary theory" (K. Popper, *The Myth of the Framework*, 58).

3. A rare exception to the near-universal concurrence of Western intellectuals that democracy is indispensable to the development and continued existence of decent societies may be found in R. D. Kaplan, "Was Democracy Just a Moment?" Kaplan, a true disciple of Thomas Hobbes, takes the highly unpopular view that in most places at most times, anarchy and social dissolution pose the greatest threat, and that a beneficent autocracy will probably be necessary to insure a minimum of civil decency. Kaplan's recommendations do not apply only to the "Third World"; he suggests that even the highly industrialized democracies of the contemporary West may find themselves sliding, piecemeal, into de facto oligarchic and authoritarian systems.

4. Recall that Alfred Russel Wallace, codiscoverer, with Darwin, of evolution by natural selection, reverted to a teleological—indeed, a religious—view of nature because he could find no satisfactory evolutionary explanation for the extraordinary power and versatility of the human intellect, which, as Wallace saw it, far exceeds what any environment would have demanded of it for mere survival.

5. It has been argued by some recent historians of science of the "social constructionist" school that, far from being inexorable, scientific results (that is, the consensus of the scientific community) are contingent on social and ideological factors. S. Shapin and S. Schaffer's *Leviathan and the Air-Pump*, for instance, argues that but for the disposition of contemporary political forces, the "experimental" science of Robert Boyle and the Royal Society might have lost out to a "Hobbesian" science eschewing experiment in favor of deductive rationalism. Despite the enthusiasm for this thesis among theorists predisposed to overrate the "social," I think the evidence both for Shapin and Schaffer's particular claims about seventeenth-century science and for the broader thesis is vanishingly slight. A more interesting case is made in J. Cushing's *Quantum Mechanics: Historical Contingency and the Copenhagen Hegemony* that the "complementarity" view of quantum mechanics urged by Niels Bohr and Werner Heisenberg grew out of idiosyncratic philosophical commitments and was not inherent in the actual formulation of the theory itself. Moreover, notes Cushing, the wide acceptance of the so-called Copenhagen interpretation would not have occurred if

certain mathematical analyses, accessible even in the 1920s, had been carried out and made known. I note, however, that Cushing's observation really applies to the philosophical interpretation of the counterintuitive mysteries of quantum mechanics, and not to the so-called formalism that physicists, irrespective of metaphysical predisposition, use routinely to make predictions about observable phenomena. The latter, I claim, exemplifies an "inexorable" scientific theory.

6. See D. Hollinger, *Science, Jews, and Secular Culture* (chapter 5: "The Defense of Democracy and Robert K. Merton's Formulation of the Scientific Ethos"), for an account, centered on the work of R. K. Merton, of the way in which many intellectuals of a previous generation came to regard the practice of science and the belief in democracy as closely linked aspects of society. See also Y. Ezrahi, *The Descent of Icarus.*

7. A depressing example in our own day is provided by the Russian mathematician Igor Shaferevitch, a strident nationalist, anti-Semite, and reactionary, despite his superb scientific credentials. See M. Gessen, "The Anti-Sakharov."

8. The celebrated science historian I. B. Cohen argues (*Science and the Founding Fathers*) that close familiarity with eighteenth-century science strongly influenced the political thinking not only of Thomas Jefferson and Benjamin Franklin, but of several other statesmen who were prominent in designing the American system of government. See also S. Doron and R. B. Bernstein, "Exploring the Age of Experiment in Government," for a review of Cohen's book.

9. M. C. Jacob, "Reflections on the Ideological Meanings of Western Science from Boyle and Newton to the Postmodernists," 339.

10. H. L. Miller, "Science and Dissent in Post-Mao China," 4.

11. M. Nanda, "The Science Wars in India," 80–81. See also Nanda, "The Science Question in Postcolonial Feminism," "The Epistemic Charity of the Social Constructivist Critics of Science and Why the Third World Should Refuse the Offer," and "Modern Science and the Progressive Agenda." For a contrasting view, see S. Visvanathan, "A Celebration of Difference," which sees Indian science as a threat to Indian democracy. Visvanathan, an anthropologist, takes a rather romantic view of demotic wisdom as against formal science in the Western mode.

12. The negative example I have in mind—and here I anticipate pain and outrage from his admirers—is that of Richard Wagner. I have always felt that this composer's well-attested callous greed, selfishness, and egomania are stamped indelibly on his music and somehow render it less than truly great, despite its masterful inventiveness and searing emotional power. By contrast, although Beethoven's most honest biographers show us a man of stunted emotional development, whose vicious prejudices wrecked the lives of a number of his unoffending relatives, there's something about *his* music—for want of a better term, its nobility—that leads me to impute a like quality to his character, however imperfectly his actions in everyday life reflect it.

Some readers will doubtless regard these observations as moonshine; others will wonder why I seem compelled to illustrate the point with dead German musicians.

13. H. L. Miller, "Science and Dissent in Post-Mao China," 18–19.

14. See, e.g., P. R. Gross and N. Levitt, *Higher Superstition*, "Knocking Science for Fun and Profit," and "The Natural Sciences: Trouble Ahead? Yes"; N. Levitt and P. R. Gross, "Academic Anti-Science"; and N. Levitt, "Vetenskapens pessimism, antivetenskapens nihilism" and "Knowledge, Knowingness, and Reality."

15. See S. Weinberg, "Physics and History," for a brief but pointed defense of the "Whig" approach to the history of science (Weinberg actually invokes the term). B. L. Silver, *The Ascent of Science*, is a recent overview of the development of science that also takes a Whiggish approach. (See D. Goodstein's review, "On the Shoulders of Giants.")

16. Various other names are current: science and technology studies, sociology of scientific knowledge, cultural studies of science, and (more narrowly) feminist epistemology. Much the same current of sentiment swirls beneath all of these variants.

17. See D. Bell, "Desperately Seeking Redemption," for a caustic view of the popularization and commodification of supposed American Indian and Australian aboriginal shamanic lore in contemporary Western popular culture.

18. E. O. Wilson, "Scientists, Scholars, Knaves and Fools," 6. See also Wilson, "Back from Chaos," for a defense of the ethos of the Enlightenment as the enduring key to sustaining our culture ethically, as well as materially.

19. It is often alleged that notions like this one represent "conventional wisdom," with the implication that it is thereby deficient or shortsighted. In my view, it would be more accurate to say that the attitude that holds the valorization of scientific knowledge to be "conventional wisdom" is, itself, a strong example of conventional (and not terribly wise) wisdom. See J. M. Bishop, "Paradoxical Strife."

20. K. Popper, *The Myth of the Framework*, 200.

21. But see N. Lemann, "Rewarding the Best, Forgetting the Rest," for an intelligent critique of cold-hearted meritocracy. See also J. Bernstein, "The Merely Very Good," and D. Bromwich, "Democracy, Merit, and Presumptive Virtue."

22. I can, perhaps, illustrate this through my own experience as a math teacher. Often, when students in elementary college math courses (calculus, differential equations, linear algebra, and so forth) are confronted by a "difficult" problem (meaning, usually, that a certain sequence of operations each of which has supposedly been mastered individually, is called for), I have the sense that they know "what" to do, but that their actual performance is inhibited by the lack of an instinctive sense of "how things flow," so they are locked in continual struggle with their own doubts about whether the whole business will really work out (rather like a novice driver learning how to parallel park). On the other hand, when I, as instructor, am asked to show the solution, I have what I suppose is the lamentable habit, especially when I am looking at the problem "cold" for the first time, of blithely sailing through the steps of the solution with nary a pause. I suspect that I fall into this way of doing things partly out of the near-sensual pleasure I derive by letting my "gut feeling for the subject" work its will unchecked. Often, the response of students is to look at me as though I were a supercilious stage magician who has just pulled the three of clubs out of someone's coat pocket. No doubt this is suspect pedagogy, but it is common enough.

I am not, by the way, boasting of extraordinary mathematical skills. Problems like these are, by the standards of professional mathematics, ridiculously elementary. The fact that students catch some sense of this is part of the pedagogical problem.

23. S. Jones, "The Set within the Skull," 13.

24. T. Ferris, "The Wrong Stuff."

25. Anti-evolutionism in America is primarily—but not exclusively—the province of fundamentalist Protestants. The Roman Catholic Church, thanks to John Paul II's October 1996 endorsement, now fully accepts the idea of the evolutionary origins of all species, including the human, though of course it insists that this is consistent with divine direction and control. On the other hand, the notorious anti-evolutionary television special *The Mysterious Origins of Man* (broadcast on NBC) was tailored to the beliefs of an American sect of Hindu origin. (See *National Center for Science Education Reports*, Winter 1995 and Fall 1996.) See, as well, for a curious example of "left-wing" anti-evolutionism, M. Robinson, "Darwinism." (N.B.: The similarity between Robinson's title and that of the current chapter is a simple, if striking, coincidence.)

26. It is noteworthy that in Europe, anti-evolutionist feeling is far weaker and almost entirely quiescent as regards public education. (See S. Coleman and L. Carlin,

"No Contest.") This probably has something to do with the fact that in the Western world aside from the United States (and, to a certain extent, Canada), the biblical-literalist tradition doesn't survive in the form of strong and viable fundamentalist churches. Furthermore, an explicit tradition of anti-clericalism (as opposed to pluralism) is present in many countries. It is nonetheless arguable that the existential unease embodied in American anti-evolutionism is nonetheless present in these places in other forms.

27. E. O. Wilson, "Resuming the Enlightenment Quest," 16.

28. Ibid., 16.

29. See S. Weinberg, "Reductionism Redux," for an astute defense of reductionism. Weinberg's noted book, *Dreams of a Final Theory*, also makes the same points. On the other hand, see T. Todorov, "Surrender to Nature" (a hostile review of E. O. Wilson, *Consilience*) for an example of the unease that can unhinge even an astute thinker when he fails to make a clear enough distinction between metaphysical and methodological reductionism.

30. Wilson prefers the perhaps more diplomatic term "consilience" to describe what I have termed materialism. The terms "materialism" and "reductionism" have the tendency to raise hackles, as much among sensitive secular humanists unattuned to science as among the devout. "Monism" is not so great an irritant, probably because it hasn't as much currency.

31. For some discussion of this fascinating, if highly technical, question, see D. Bohm and B. Hiley, *The Undivided Universe*; J. Cushing, *Quantum Mechanics*; D. Wick, *The Infamous Boundary*; S. Goldstein, "Review Article."

32. Here, once more, I stress that I am talking about reductionism as a mode of fundamental understanding of how the universe works, not as a tenet of methodology. To recur to the zoologist-zebra example, a zoologist would be explicitly antireductionist, in my sense, if he were to invoke some special "life-force," not even in principle reducible to ordinary chemistry and physics, to account for the properties of zebras (or any other organism). "Vitalism," as this doctrine is called, was incorporated into the views of many scientists until the end of the nineteenth century, but is now no more respectable than geocentric astronomy. See R. A. Frosch, "Reductionism and the Unity of Science."

33. Notwithstanding this observation, there have been a considerable number of scientists willing to embrace an antimaterialist view, often out of religious hopes and convictions. Most, nonetheless, accept materialism at least tacitly as the governing assumption of their actual scientific work. See K. D. Fezer, "Is Science's Naturalism Metaphysical or Methodological?"

34. See B. Latour, "La fin de science?" and T. Pinch and H. Collins, *The Golem*.

35. A. Hale, "Astronomer Hale Speaks Out about Poor Prospects for Science Careers." There is a further stinging irony in the fact that the discovery to which Hale owes his celebrity, the Hale-Bopp comet, will chiefly be remembered as the pretext for the mass suicide of the Heaven's Gate UFO cult. Hale, as one sees from news coverage of this bizarrre event, has already expressed his disgust at this prospect.

36. See, for instance, R. Sclove, *Democracy and Technology*. Sclove, to be fair, is chiefly concerned with democratizing specifically *technological* decisions and has little to say about pure scientific research. He avoids conflating technology with basic science—in this respect he is conservative and even unfashionable. Moreover, he doesn't seem to have much sympathy with postmodern challenges to the very notion of objective knowledge of a real world. He genuflects occasionally to the notion, favored by multiculturalists, of "local knowledges" supposedly as robust as conventional science, but only briefly and quietly.

37. See R. H. Brown, *Toward a Democratic Science,* for another plea for the "democratization" of science where ideology has been allotted too strong a role. Unfortunately for the logic of his argument, the courage of Brown's convictions derives largely from fashionable but flabby cultural constructivist theories concerning the nature of scientific discourse. Brown's pantheon (pp. 176–177) of stellar theorists includes Bruno Latour, Sandra Harding, Evelyn Fox Keller, Stanley Aronowitz, Donna Haraway, Alan Gross, Steve Fuller—in my view, a tribe one does well not to rely upon. (For critiques of these thinkers, see P. R. Gross and N. Levitt, *Higher Superstition;* P. R. Gross, "Author's Response"; N. Levitt, "Author's Response"; A. Sokal and J. Bricmont, *Impostures intellectuelles;* and S. Haack, "Misinterpretation and *The Rhetoric of Science.*" Also N. Koertge, ed., *A House Built on Sand* and J. Bricmont, "Science Studies—What's Wrong?"

38. R. Sclove, *Democracy and Technology.*

39. See, for instance, S. Harding, "Science Is Good to Think With."

40. I note a few specimen issues where organized "citizen" intervention in technological or health issues has most likely come down on the wrong side of the question: the supposed carcinogenic properties of low-frequency electromagnetic fields; the ostensible dangers of genetically engineered foods; the ostensible immunological problems caused by cosmetic silicone implants; the degree of danger to human health posed by environmental traces of dioxin. These will be discussed in greater detail in subsequent chapters.

41. Daniel Bell, commenting on the recent documentary film *Arguing the World,* reaffirms this famous credo, first formulated in his book *Cultural Contradictions of Capitalism.* (See D. Bell, "Of 'Trotskyites' and 'Trotskyists.'") Another meditation on the congruity of cultural conservatism, political liberalism, and economic egalitarianism can be found in R. Shattuck, "Education." Similarly, M. Lilla, "A Tale of Two Reactions," takes ironic note of the syncretism between left and right in certain key aspects of American life and culture.

A more lengthy exposition of where I stand politically (for those who care) may be found embedded in the novels of the much underrated (and pseudonymous) writer, K. C. Constantine, whose work, thanks to the peculiar limitations of publishers and editors, is usually exiled to the "mysteries" section of the bookstore. Here are some titles: *Cranks and Shadows, Bottom Liner Blues, Sunshine Enemies, Upon Some Midnights Clear, The Man Who Liked Slow Tomatoes.*

42. T. Ferris, "The Risks and Rewards of Popularizing Science," B6.

43. P. R. Gross and N. Levitt, as well as N. Levitt and P. R. Gross, and N. Levitt, as cited in n. 37.

44. See F. J. Dyson, "Is God in the Lab?" for a meditation on the very different religious views of two eminent physicists, R. Feynman (*The Meaning of It All*) and J. Polkinghorne (*Belief in God in an Age of Science*). According to recent surveys, the overwhelming majority of scientists are nonbelievers (see E. J. Larson and L. Witham, "Leading Scientists Still Reject God").

45. R. Dawkins, "An Apostle of Science."

ONE *Culture*

1. F. L. Byrne and A. T. Weaver, eds., *Haskell of Gettysburg,* 158. In order to keep intact my Civil War buff credentials, I must point out that the term "Pickett's Charge" is a common misnomer whose origin is probably the dominance of Virginians in creating early versions of many of our Civil War tales. (See C. Reardon, *Pickett's Charge in History and Memory.*) The assault was actually under the overall command of Lieutenant General James Longstreet and was carried out by one division from his own

corps—that of Pickett—a division from A. P. Hill's Corps, under James Pettigrew, and a "demi-division," also drawn from Hill's Corps, under Isaac Trimble. Subsequent to the initial effort, Confederate brigades under Lane and Wilcox tried to come up in support, but by that point the main force was broken and in retreat.

2. Haskell's diaries show his familiarity with Homer, at least in translation.

3. Haskell was a Dartmouth graduate, but he had grown up on a Vermont farm and was largely self-educated before entering college.

4. See T. Gitlin, "The Dumb-Down."

5. See, for instance, Grant's *Memoirs*, now regarded as an American prose masterpiece. Especially noteworthy are the dispatches and orders Grant composed during the war. When Matthew Arnold's largely approving review of the *Memoirs* briefly offered some mild criticism of the general's grammar and usage, Mark Twain (Grant's publisher, as well as his admirer) rose in wrath to denounce Arnold's own solecisms, and to declare the work a masterpiece of style as well as substance.

6. Frederick Douglass, July 5, 1852:

What, to the American slave, is your Fourth of July? I answer, a day that reveals to him, more than all other days in the year, the gross injustice and cruelty to which he is the constant victim. To him, your celebration is a sham; your boasted liberty, an unholy license; your national greatness, swelling vanity; your denunciations of tyrants, brass fronted impudence; your shouts of liberty and equality, hollow mockery; your prayers and hymns, your sermons and thanksgivings, with all your religious parade and solemnity, are to him mere bombast, fraud, deception, impiety, and hypocrisy—a thin veil to cover up crimes which would disgrace a nation of savages. There is not a nation on earth guilty of practices more shocking and bloody, than are the people of the United States, at this very hour.

Go where you may, search where you will, roam through all the monarchies and despotisms of the old world, travel through South America, search out every abuse, and when you have found the last, lay your facts by the side of the every-day practice of this nation, and you will say, with me, that for revolting barbarity and shameless hypocrisy, America reigns without equal.

7. A. Trollope, *North America*, 143.

8. Ibid.

9. Ibid., 121.

10. The literary scholar and educational theorist E. D. Hirsch, in *The Schools We Need and Why We Don't Have Them*, makes the interesting point that the egalitarian, liberal—indeed, left-wing—aims are best served, educationally, by a schooling grounded in traditional pedagogy and traditional subjects. Equally interesting is the sidebar that accompanies the article, which quotes a memoir by Avis Carlson of her girlhood in Kansas in the early 1900s, specifically, her grade-school education. It deserves quotation:

Recently I ran onto the questions which qualified me for my eighth grade diploma. The questions on that examination in that primitive, one-room school, taught by a person who had never attended a high school, positively daze me. The orthography quiz asked us to spell 20 words, including "abbreviated," "obscene," "elucidation," "assassination," and "animosity." . . . Two of arithmetic's ten questions asked us to find the interest on an 8–percent note for $900 running two years, two months, six days; and also to reduce three pecks, five quarts, one pint to bushels. . . . In reading we were required to tell what we knew of the writings of Thomas Jefferson, and for another of the ten questions to indicate the pronunciation and give the meanings of the following words: zenith, deviated, misconception, panegyric, Spartan, talisman. . . . Among geography's ten were these: "Name three important rivers of the U.S., three of Europe, three of Asia, three of South America, and three of Africa. As one of physiology's ten we were asked to "write 200 words on the evil effects of alcoholic beverages." . . . In history, we were to "give a brief account of the colleges, printing, and religion in the colonies prior to the American Revolution," "name the principal campaigns and military leaders of the Civil War," and to "name the principal political questions which have been advocated since the Civil War and the party which advocated each."

Ms. Carlson passed this exam shortly before she turned twelve.

I can't resist comparing this account with the remark with which a prominent scholar in a major American university begins his recent book on "culture, science, and technology": "This book is dedicated to all of the science teachers I never had. It could only have been written without them" (A. Ross, *Strange Weather*). I apologize for the redundancy of once more citing an item that I made much of in an earlier book.

11. The noted mathematician James Sylvester came to the University of Virginia in the early 1840s, in part to escape the anti-Semitism that held him back in England. Unfortunately, the young cavaliers in his charge did not see eye to eye with him on the matter of grading standards, and their violent threats soon obliged him to flee for his safety. More than thirty years later, he returned to this country to become one of the leading figures at Johns Hopkins, which had just been founded on the model of the German system. Thereby, he became perhaps the most important founder of American mathematical research.

12. A "Golem" according to Pinch and Collins.

13. In fact, in recent months there have been moves toward significant increase in congressional science appropriations, but (as of the spring of 1998) the issue remained in doubt.

14. C. P. Toumey, *Conjuring Science*, 26.

15. G. Holton, *Einstein, History, and Other Passions*, 41.

16. C. P. Toumey, *Conjuring Science*, 21.

17. See, for example, J. Nattier, "Buddhism Comes to Main Street."

18. Entirely by ironic coincidence, on the day I wrote this I received a postcard from an art gallery which somehow got my name on its mailing list. It was to announce a new one-man show entitled *Primus corpus: DNA License*. A sample of the artist's work was included. It consisted of a small plastic bag stamped "Universal Notice—Only One—Original Human" containing some hair trimmings. Anyone struck dumb by this achievement is unlikely, I daresay, to be among the admirers of this chapter.

19. See R. Hughes, *The Culture of Complaint*, for an interesting meditation on these issues.

20. See W. Sampson, "Antiscience Trends in the Rise of the 'Alternative Medicine' Movement."

21. See, for instance, P. R. Gross and N. Levitt, *Higher Superstition* (pp. 89–92) on the well-known cultural studies celebrity, Andrew Ross.

22. See M. L. Perutz, "The Pioneer Defended." See also P. R. Gross and N. Levitt, *Higher Superstition*, for an unadmiring account of some of the most celebrated work in science studies, and N. Koertge, ed., *A House Built on Sand*, for a collection of sharp critiques of the field.

23. See P. R. Gross and N. Levitt, *Higher Superstition*.

24. In respect to postmodern theory (in its various manifestations) and its relation to science, see P. R. Gross and N. Levitt, *Higher Superstition*; G. Holton, *Einstein, History, and Other Passions*; and N. Koertge, ed., *A House Built on Sand*. See also A. Sokal, "A Plea for Reason, Evidence, and Logic"; K. Gottfried and K. G. Wilson, "Science as a Cultural Construct"; L. Evans, "Should We Care about Science 'Studies?'"; N. D. Mermin, "The Science of Science"; and A. Sokal and J. Bricmont, *Impostures intellectuelles*.

25. I have explored this theme at greater length elsewhere. See N. Levitt, "The End of Science, the Central Dogma of Science Studies, Monsieur Jordain, and Uncle Vanya."

26. This is not a hyperbolic figure of rhetoric, but rather, the literal truth. See P. R. Gross and N. Levitt, *Higher Superstition*, and N. Koertge, ed., *A House Built on Sand*.

27. Consider, for instance, the appearance of books like V. Deloria's *Red Earth, White Lies*, which celebrates Native American origin myths as literal truths and scornfully dismisses (with language and dogma clearly borrowed from the stockpile of postmodernist clichés) any scientific account of the peopling of the Western Hemisphere. See also M. Gladwell, "A Matter of Gravity," and E. Eichman, "The End of the Affair."

28. For instance, in the Kenyon affair at San Francisco State University, the faculty senate of this largely "progressive" institution condoned the introduction of "intelligent design" ideology into an *elementary* biology course by Professor Dean Kenyon, a devout Christian.

тwo *Mathematics*

1. H. Konig, "Notes on the Twentieth Century," 99.

2. A metaphorically appropriate example recently turned up in the news: a manufacturer of paper products has begun to produce toilet paper bearing an intricate design technically known as a Penrose nonperiodic tiling of the plane. Mathematician and philosopher Roger Penrose, who invented this pattern as an answer to a deep mathematical question, has filed an infringement suit against the firm. See S. Mirsky, "The Emperor's New Toilet Paper," and E. Eakin, "Paper Trail."

3. In fact, it would have a height equal to its diameter. Since the area is minimal, it would chill more slowly than any other cylinder, but it would also warm up more slowly.

4. Here, let me give the standard definition of limit; I omit some of the contextual background, such as saying what "$f(x)$" is supposed to represent, in order to stress the logical form of the statement, which is what causes all the difficulty:

We say that the limit of $f(x)$ as x approaches a is equal to L if and only if: for every $\varepsilon > 0$, there exists a $\delta > 0$ such that $0 < |x - a| < \delta$ implies $|f(x) - L| < \varepsilon$.

5. W. C. Dowling, *Jameson, Althusser, Marx*, 26.

6. D. Hoffman, review of J. Horgan, *The End of Science*, 263. Other commentators, including me, have noted Horgan's antipathy toward mathematical thinking. See N. Levitt, "*The End of Science*, the Central Dogma of Science Studies, Monsieur Jourdain, and Uncle Vanya."

7. Sometimes this is called the "Monty Hall Problem" in honor of the host of *Let's Make a Deal*, a game show that featured guesswork about which door hid a valuable prize.

8. With such a strategy, the expected winning percentage is

$$[2/3(X) + 1/3(Y)]\% = [1/3((100 + X)]\%,$$

which is clearly no greater than two-thirds. So the best strategy, unambiguously, is to switch every time. Those readers who insist that precognition, clairvoyance, or ESP will give rise to an even more efficient strategy are entitled, under the Constitution, to think so. I disagree.

9. M. Vos Savant, *Parade's Ask Marilyn*, 199–209.

10. Fermat was a seventeenth-century French jurist who was a devoted, and excellent, amateur mathematician. Together with Descartes, he deserves credit for the invention of analytic geometry. The conjecture, which Fermat claimed to have proved in a famous marginal note, has been known for the past century as "Fermat's Last Theorem," a misnomer, since it was unproved, unless one accepts Fermat's brief and extremely unlikely claim. Now it is a theorem (or rather a corollary to a powerful theorem of Wiles) and hence should be known as Wiles's Theorem.

11. In fact, there was a flaw in Wiles's original proof, which was repaired, with the aid of a collaborator, about a year later.

12. M. Vos Savant, *The World's Most Famous Math Problem.* I have referred to both the Three Doors Problem and to Ms. Vos Savant's book in an exchange with the sociologist Stanley Aronowitz over the value of "science studies." See N. Levitt, "The World's Highest IQ and Other Damned Souls."

13. Despite the "elementary" statement of the Fermat Conjecture, the work which proved it was highly technical and not in the slightest sense elementary. It's fair to say that among mathematicians, only specialists follow it with close understanding.

14. Some comical examples of this defensiveness (for such it seems to me) may be found in the work of cultural critics like L. Irigaray, *This Sex Which Is Not One*, and N. K. Hayles ("Masculine Channels and Feminine Flows").

15. J. Holt, "Let's Make a Deal"; M. Piatelli-Palmarini, *Inevitable Illusions.*

16. The example given by Holt concerns a hypothetical disease harbored by 500,000 Americans out of a total population of 250,000,000. A test for this disease is 99 percent accurate. If a randomly chosen individual tests positive for the disease, how concerned should that person be? The somewhat counterintuitive answer is that the odds are slightly better than five to one that there is no disease present. But this requires the assumption that the error rate is one percent for uninfected persons as well as infected, a point which Holt muddles.

17. R. S. Porter, "Inevitable Illusions." Here, I shall admit that I sent in such a letter myself, which ultimately was not published. This was the third time I got into such a wrangle with Holt over large or small inaccuracies in his articles about mathematics (see J. Holt, "The Newer Pardigm" and "Their Days Are Numbered," and the subsequent letters by N. Levitt, "Letter to the Editor" and "Sniper Attack"). I don't mean to impugn Holt's ability, overall; he is an interesting and able essayist. But he does seem to plunge into mathematical questions without sufficient consultation of experts who might help him avoid certain pitfalls.

18. See also N. Levitt, "Mathematics as the Stepchild of Contemporary Culture."

19. Paulos is also the author of *Innumeracy.*

20. A fascinating example of an intelligent layperson's attempt to analyze a highly technical problem in the absence of the technical—that is to say, mathematically grounded—knowledge usually thought necessary to such an analysis occurs in E. Scarry, "The Fall of TWA 800." Professor Scarry is a Harvard philosopher specializing in social and ethical issues. However, the article in question has nothing directly to do with moral values. Rather, it puts forward a purely technical hypothesis concerning the catastrophic explosion of TWA Flight 800 in 1996. Scarry's thesis is that a pulse of high-intensity electromagnetic radiation from a military plane or ship in the area might have triggered a sequence of events leading to the fuel-tank explosion that

brought down Flight 800. Clearly, Scarry has done an enormous amount of home-work and made a prodigious effort to learn about relevant technical matters. More-over, at least some of the scenarios she envisions are not inherently utterly implau-sible. Nonetheless, the effectiveness of her piece is undercut by her uncertain grasp of the physics involved, which renders her arguments vague and clumsy at crucial points. Scarry is certainly to be praised for the depth of her commitment and her sheer hard work. Nonetheless, condescending as it will seem, a very conventional assess-ment seems justified: if a physicist or engineer with direct professional knowledge of electronic warfare—radar, electronic countermeasurers, and so forth—and civilian avi-onics had written the piece, it would almost certainly have been clearer, more co-gent, better focused, more persuasive—and probably much briefer.

21. See, for instance, M. Angell, *Science on Trial*, for a patient examination of the consequences of the public's unfamiliarity with the statistical methodologies of medical research and clinical trials.

22. See G. Johnson, "The Unspeakable Things That Particles Do." This is a semi-humorous rumination on the impossibility of describing subtle phenomena in quan-tum mechanics to readers who have little mathematical background. The case in point is yet another confirmation of the so-called Einstein-Podolsky-Rosen effect in a re-cent experiment. This demonstrates that one particle can have an instantaneous ef-fect on a distant one. Einstein mooted this idea in order to demonstrate that, since "action at a distance" is impossible, there must be some flaw in the foundations of quantum mechanics. Subsequently, John Bell showed, via his famous theorem, that the paradox is even deeper than Einstein supposed. Unfortunately for Einstein's di-agnosis, experiment shows that the effect, even in Bell's more pointed version, is real, and that it is Einstein's critique, not quantum mechanics, that is in trouble. Ironi-cally, if Johnson had read further in the foundational literature, he might have seen that there is a plausible explanation for this phenomenon—one that is somewhat ac-cessible even to the mathematically naive—via David Bohm's notion of "nonlocal" particle mechanics. This is not to say that the issues can be accurately grasped with-out mathematics, merely that the matter needn't be as thoroughly unintuitive as Johnson makes out. See J. Cushing, *Quantum Mechanics*, for an account of Bohm's ideas and explanation of why they are still so little known, even in the physics com-munity.

THREE   *Teleology*

1. See E. O. Wilson, *Consilience*.

2. I write shortly after the opening of the film version of Sagan's novel, *Contact*. Ironically, the film traduces Sagan's own resolute unbelief by representing extrater-restrial intelligence as strongly suggestive of transcendence, or even divinity.

3. The role of the concept of "Nature" in conditioning our attitude toward sci-ence will be discussed at greater length in chapter 7.

4. It is a weary cliché of postmodernist thought that there is no such thing as a "natural" property; all attributes are "culturally constructed"; to hold otherwise is to be guilty of the enormous sin of essentialism. Of course, most ostensible post-modernists are (to their credit as human beings) passionate defenders of the "natu-ral" world against the encroachments of industrial civilization.

5. G. Holton, *Einstein, History, and Other Passions*, 13.

6. J. Barzun, *Science: The Glorious Entertainment*, 116.

7. The great resurgence of scientific cosmology began in the early 1960s with the

discovery by Arno Penzias and Robert Wilson of the "background black-body radiation" taken to be the signature of the so-called Big Bang.

8. S. Weinberg, *Dreams of a Final Theory*, 250.

9. See J. D. Barrow and F. J. Tipler, *The Anthropic Cosmological Principle*. But see also L. Smolin, *The Life of the Cosmos*, for a philosophically antagonistic view.

10. F. J. Tipler, *The Physics of Immortality*.

11. See, for instance, J. Polkinghorne, *Belief in God in an Age of Science*, or R. Jastrow, *God and the Astronomers*.

12. See E. C. Scott, "Antievolution and Creationism in the United States" and "Creationism, Ideology, and Science." For the contrasting British situation, see S. Coleman and L. Carlin, "No Contest."

13. E. J. Larson, *Summer for the Gods*. See, especially, the Introduction. See also D. Dennett, *Darwin's Dangerous Idea*.

14. Adam Sedgwick to Charles Darwin, November 24, 1859.

15. M. Robinson, *The Death of Adam*, 62. Ms. Robinson is a gifted writer, but an amazingly sloppy thinker, conflating evolution, natural selection, "Darwinism," and "Social Darwinism," undeterred by any intellectual qualms. She falls repeatedly into the genetic fallacy and adopts the odd tactic of trying to repudiate Darwin by citing the presumptively nasty ideas of Nietzsche and Freud. She seems to have convinced herself that to adhere to a Darwinian view of biological evolution (or its modern refinements) is necessarily to adopt the devil-take-the-hindmost social policies of the right. She never seems to have taken account of the fact that some of her proclaimed villains—S. J. Gould and R. Dawkins, for instance—pretty much share her ethical and political views (despite their own well-publicized differences over the nuances of evolutionary theory). Indeed, she seems utterly unaware that her religious, as well as political and social, views are very close to those of Alfred Russel Wallace, the coinventor of "Darwinism." I hesitate to label the work of an evidently compassionate and intelligent woman as "stupid," but it seems to me that no kindlier label applies. (The chapter "The Death of Adam" excoriating "Darwinism" also appears as a slightly different separate essay, "Consequences of Darwinism.") For other evidence of mistrust of biology on the left, see B. Ehrenreich and J. MacIntosh, "The New Creationism."

16. Unease about the doubtful ontological status of ethical and moral values within the scientific worldview is clearly one of the factors that motivates even sophisticated intellectuals. For an example, see T. Todorov's "Surrender to Nature," a caustic review of E. O. Wilson's *Consilience*.

17. D. Denby, "In Darwin's Wake," 59.

18. S. J. Gould, "Nonoverlapping Magisteria."

19. A similar point is made in T. Edis, "Relativistic Apologetics."

20. Of course, the notion of a God who commands earthquakes and hurricanes brings with it the ancient question of why a benevolent deity should permit evil, terror, and pain into the world. This will never be a dead issue. A columnist for the *New Republic* (James Wood, "Twister") recently fell to brooding over this very point.

21. Those unsettled by the ethical nihilism they perceive in orthodox evolutionary theory are hardly apt to be reassured by pieces like S. Pinker, "Why They Kill Their Newborns." Pinker asserts, rightly or wrongly, that there is a good deal of evolutionary logic to infanticide on the part of mothers, under certain circumstances. (For an account of Pinker's ambitious program to meld evolutionary theory with behavioral psychology, see J. Horgan, "Darwin on His Mind.") Attempts like that of E. O. Wilson ("The Biological Basis of Morality") to explicate human moral arrangements on the basis of "moral instincts" that coevolved with the rest of our physical and

intellectual equipment, while rejecting transcendental foundations for morality, will be equally unwelcome. (See, as well, F. Sulloway, "Darwinian Virtues," for an informative account of recent attempts to comprehend human moral sentiments on the basis of evolutionary theory.)

22. This does not prevent militant academic anti-foundationalists and anti-essentialists from denouncing biological categories. See, for example, J. M. Kaplan, "Problematizing Reifications and Naturalizations," or P. Thurtle, "The Creation of Genetic Identity." I have argued elsewhere (N. Levitt, "What's Post Is Quaalude") that beneath its skeptical veneer, "postmodernism" is relentlessly teleological in spirit.

23. A recent controversy over the teaching of evolution in North Carolina found a postmodernist (and presumably anti-foundationalist) scholar, Warren Nord of the University of North Carolina, jumping in on the side of the creationists. His rationale was multicultural: varied ethnicities, other than fundamentalist Christians, find evolutionary theory offensive, hence diversity demands that *everyone's* favorite creation story be taught in the public schools. (See M. Cartmill, "Oppressed by Evolution.") One cannot help but wonder, however, if even a staunch anti-foundationalist might not feel morally undermined by the Darwinian picture of a purely adventitious world.

24. See P. E. Johnson, *Reason in the Balance* and *Darwin on Trial*, as well as "What (If Anything) Hath God Wrought?" See also K. D. Fezer's review essay, "Is Science's Naturalism Metaphysical or Methodological?"; M. J. Behe, *Darwin's Black Box*; the reviews of Behe by P. R. Gross and K. R. Miller and S. M. Barr, "Debating Darwin"; and H. Allen Orr's "Darwin v. Intelligent Design (Again)," which thoroughly demolishes Behe's principal thesis.

25. See M. Gardner, "Intelligent Design and Philip Johnson," which gives a brief account of the "Intelligent Design" movement, as well as a specific rejoinder to Johnson. Another critique is to be found in E. C. Scott and R. M. West, "Again, Johnson Gets It Wrong." See also E. C. Scott and K. Padian, "The New Anti-Evolution—and What to Do about It."

26. R. T. Pennock, "Naturalism, Creationism, and the Meaning of Life," 10.

27. The metaphysical fog created by indiscriminate deployment of the term "nature" will be considered at greater length in chapter 7.

28. See N. Koertge, ed., *A House Built on Sand,* for an account of some curious doctrines of science studies.

29. F. J. Dyson, "Science in Trouble," 525. See also G. Easterbrook, "Science Sees the Light."

30.

The Sea of Faith
Was once, too, at the full, and round the earth's shore
Lay like the folds of a bright girdle furl'd;
But now I only hear
Its melancholy, long, withdrawing roar,
Retreating to the breath
Of the night-wind down the vast edges drear
And naked shingles of the world.
                    —Matthew Arnold, *Dover Beach*

C. Raymo, "Spiritually Homeless in the Cosmos," meditates on the persistence, in the educated classes of our own day, of feelings akin to those expressed by Arnold.

31. See C. N. Degler, *In Search of Human Nature,* for a history of the changing fortunes of biologism in social theory in America. See S. Goldberg, *The Inevitability of*

Patriarchy and *Why Men Rule*, for a closely argued, biology-based theory of social institutions that, right or wrong, flies in the face of contemporary liberal-humanist piety.

32. I am well aware of the deeply problematical nature of notions like "intelligence" and "race." I do not propose to explore this question here, but do note my own feeling that more confusion will be introduced into this kind of discussion by avoiding such terms than by using them in straightforward fashion.

33. Since the issue of hereditary group differences is such a touchy one, let me make clear the kinds of questions I think are at stake. It is not all that hard to poke holes in the arguments that Herrnstein and Murray make or cite, and it would certainly be foolish to regard these as settling the issue in any basic way. However, it is one thing to point out that an argument fails to establish what it intends to establish, and quite another to show that the intended conclusion is actually false. If a prosecuting attorney presents a sloppy, unconvincing case, it does not mean that the accused did not perpetrate the crime! This sort of reasoning—an invalid argument for A proves that A is false—is an elementary error in logic. Yet a surprising number of ostensible intellectuals fall into the trap in the case of the "race-and-intelligence" question. It is repeated endlessly in homilies denouncing *The Bell Curve* and similar work. The reasoning is that if some other factor than genetics *might* account for observed differences in test performance or whatever, then that other factor *must* account for it. In point of fact, it will take a substantial amount of research, and enormous ingenuity, to establish that there are no cognitive differences between the genetically distinct subgroups of humanity carelessly labeled "races"—if, in fact, that turns out to be the conclusion. F. Miele, "IQ in Review," provides a brief, dispassionate, and useful survey of the state of informed opinion on the IQ question.

34. Particularly illuminating is S. J. Gould's paper, "Human Equality Is a Contingent Fact of Evolution" (in *The Flamingo's Smile*), and V. Sarich's rejoinder, "In Defence of *The Bell Curve*." M. Schoofs, "The Myth of Race," repeats a number of now-familiar arguments against the possibility of important "racial" differences; although it is well informed on a number of scientific issues, it, too, evades the points raised by Sarich. See also P. Rabinow, in chapter 9 of *Essays on the Anthropology of Reason*, for an account of his efforts, in conjunction with a molecular biologist, to debunk Sarich's hereditarian views. This is a curious collaboration between Rabinow, an eminent postmodernist and Foucault scholar, and a working scientist, but what's interesting is how both men (at least in Rabinow's account) seem to share a methodology that comes down to assuming a priori there's something wrong with the hereditarian view, then rummaging about for anything that comes to hand that might be useful, at least rhetorically, in refuting that view.

35. M. Ruse, *Monad to Man*. Ruse's case for the covert or unconscious influence of teleological assumptions is strong in some cases, but rather forced in others. See the reviews of Ruse's book by P. E. Griffiths et al., "Onward and Upward." One confusing issue is that a notion such as "complexity" can be given a rather rigorous and objective definition, under which it is seen that evolution does, indeed, result in increasing complexity of organisms, perhaps inevitably. One must tread carefully, however, before identifying "increasing complexity" with "progress," let alone "purpose"—or accusing others of doing so.

FOUR *Credulity*

1. See S. A. Vyse, *Believing in Magic*; M. Shermer, *Why People Believe Weird Things*; and J. E. Alcock, "The Propensity to Believe," for general discussions of the psychology

of superstitious belief. See also S. Goldberg, "When Wish Replaces Thought," for more general discussion of unsupported belief.

2. D. Barry, *New York Times*, July 4, 1997, B3. The substance of the story is that in a continuing search for a woman who has been missing for four years, police, acting on instructions from a "psychic," are digging up an empty lot in Brooklyn. So far, they have unearthed the skeleton of a dog. The background material on the clairvoyant in question reads as follows:

> Ms. Allison, who works from her home in Nutley, N.J., said that she had worked on thousands of cases around the country over the last two decades, and had helped various police departments to solve hundreds of them.
>
> "My job is to find bodies and people who are missing," she said.
>
> But Ms. Allison said that she only works on criminal cases when a law enforcement agency requests her assistance in writing. "It has to be on official police department paper," she said. "That's the only way I work."
>
> One police official said that Ms. Allison had been retained by the missing woman's family to work on the case, but others confirmed that she was part of the official investigation. Ms. Allison said that she received a request for help in August 1994 from a detective with the New York transit police, which has since been merged with the Police Department.

There is no suggestion that any attempt has been made to find independent confirmation of the psychic's claims, let alone to investigate them in detail. It should be noted that in the past the *Times* has done similar stories, in a similar tone. None of these was marked by any noteworthy discovery on the basis of paranormal insight.

3. D. Barry, "The NYPD's Psychic Friend: When Technology Fails, Detectives Call on a New Jersey Woman's 'Visions.'" In fairness to the *New York Times*, one must point out that it is often appropriately skeptical. See, for instance, A. Harmon, "NASA Went to Mars for Rocks? Sure," the target of whose skepticism is the frenetic "skepticism" of a good part of the public concerning the reality (or the true aims) of the space program. Says Harmon, "But especially now, believing in aliens or Hollywood special effects may be more palatable than confronting the unsettling realities of what science is capable of."

4. Joe Nickell and James Randi, both associated with the skeptical organization CSICOP. I must point out here that I am a member of CSICOP's Committee on Media Responsibility, although I had no role in responding to the *Times* in the context of this story.

5. See M. Talbot, "America Image Disorder," for a sardonic account of the self-improvement-cum-spirituality cult originated by Marianne Williamson, which has carried its creator not only to wealth and celebrity, but won her entry into presidential retreats at Camp David. Needless to say, Williamson's chief villains are the spoilsports of the Enlightenment and the rationalistic and scientific worldview they celebrated.

6. C. McGinn, "Reason the Need," 41.

7. See P. R. Gross and N. Levitt, *Higher Superstition*, as well as "Natural Sciences: Trouble Ahead? Yes!" and N. Levitt and P. R. Gross, "Academic Anti-Science." See also N. Levitt, "The World's Highest IQ and Other Damned Souls" (together with the subsequent response to Stanley Aronowitz), "More Higher Superstitions," and "*The End of Science*, the Central Dogma of Science Studies, Monsieur Jourdain, and Uncle Vanya." Whatever the virtues of these efforts (and I think them substantial), excessive tactfulness is not among their vices.

8. As it turned out, the back-hoe hit water about eight feet down, just where the dowser said it would. However, inasmuch as, unannounced, I'd chosen that very same

spot on the basis of topography, I wasn't all that impressed. Alas for both the dowser and myself, the well dried up after a couple of years.

9. I would contrast this with what many regard as the equally perverse behavior of a jury in another infamous proceeding, the first trial of the policemen who savagely beat Rodney King. In the King case, I believe we saw a case of "jury nullification," plain and simple. The jurors had little doubt about the facts, but felt that the police were heavily provoked, King deserved what happened to him, and, moreover, they believed that the forces of anarchy and civil decay can only be held at bay by the kind of heavy-handed police procedure that occasionally results in such rough justice.

10. Of course, Freud was hardly resistant to the the charms of crank systems in general, having created the twentieth century's greatest example thereof. See F. Crews and F. Crews, *Memory Wars*; H. Eysenck, *Decline and Fall of the Freudian Empire*; and A. Grünbaum, *The Foundations of Psychoanalysis*.

11. T. J. Kuhn, *The Structure of Scientific Revolutions*.

12. See M. A. Churchill, "A Skeptic Living in Roswell," and N. Gerlich, "A Skeptic Crashes in Roswell."

13. For a fascinating analysis of the UFO cult, see F. Crews, "The Mindsnatchers." As Crews notes, polls reveal that an astonishing 64 percent of Americans believe that visiting aliens have made personal contact with earthlings. It is not clear, however, how deep or stubborn this belief is. See as well J. Leonard, "Culture Watch." An ironic testament to how deeply the notion of "alien abductions" is engrained in our common culture may be found in a poem of C. Dennis, "On the Bus to Utica."

14. S. A. Vyse, *Believing in Magic*, 51.

15. For a sober account, which places the brunt of the cynicism where it belongs, see W. J. Broad, "Air Force Details New Theory in U.F.O. Case."

16. J. McAndrew, *Case Closed*.

17. See W. Kaminer, "The Latest Fashion in Irrationality."

18. This text, from an Internet posting, was relayed to me by a sociologist of science who takes a jaundiced view of most postmodernist fads. I have omitted the names of some prospective contributors.

19. See A. Sokal and J. Bricmont, *Impostures intellectuelles*, for a caustic view of Deleuze, particularly with regard to his scientific pretensions.

20. S. Haack, "Concern for Truth," 58.

21. Ibid.

22. See B. Ortiz de Montellano, "Afrocentric Pseudoscience," and B. Ortiz de Montellano, G. Haslip-Viera, and W. Barbour, "They Were NOT Here before Columbus," as well as M. Lefkowitz and G. Maclean, eds., *Black Athena Revisited*, for informed critiques of university-based Afrocentrism. Also see D. Patai and N. Koertge, *Professing Feminism*.

23. See R. G. Walters, ed., *Scientific Authority and Twentieth-Century America*, for a compendium of essays expressing, in various ways, the resentment felt toward science by nonscientists. The late philosopher Thomas Kuhn seems to be the chief cult figure of this sect, a great injustice to Kuhn's memory.

24. Probably the best-known jeremiads are R. Kimball, *Tenured Radicals*, and D. D'Souza, *Illiberal Education*, both of which are interesting for some risible examples of academic folderol but compromised by their strident conservative agendas. D. Bromwich's *Politics by Other Means*, on the other hand, reproves many of the same tendencies from the perspective of the democratic left. J. M. Ellis, *Literature Lost*, is a recent and welcome addition to this genre, valuable for its etiology of current academic cults. As for satires, I recommend J. Hynes, *Publish and Perish*, and R. Grudin, *Book*.

25. See J. M. Ellis, *Against Deconstruction*. Also see J. Radcliffe-Richards, "Why Feminist Epistemology Isn't," and N. Koertge, "Feminist Epistemology."

26. A. Ross, *Strange Weather: Technoscience in an Age of Limits*.

27. See P. R. Gross and N. Levitt, *Higher Superstition* (chapter 4). The disdain is reciprocal; see *Social Text* 46/47, a compilation edited by Ross, the principal intent of which was to defend various versions of "science studies" from the animadversions of such as Gross and Levitt; unfortunately for Ross, the intent was largely subverted by the clever imposture "Transgressing the Boundaries," foisted upon Ross and company by my friend, physicist Alan Sokal. See also the Sokal-less book version, *Science Wars*, as well as *Social Text* 49.

28. An even more extreme work in this genre has recently appeared: J. Dean, *Aliens in America*. Professor Dean is keen to defend the community of fervent believers in UFO abductions on the grounds that their paranoia is an appropriate response to the racism, sexism, militarism, and all-around oppressiveness of American culture. If the author herself doesn't believe in UFO visitations, it is only because, as a postmodernist in good standing, she doesn't believe in the "truth" of anything. But clearly, her sympathy lies with the believers, while she has nothing but scorn for scientists and scientific rationality. For analyses of Dean's peculiar book, see F. Crews, "The Mindsnatchers," and J. Leonard, "Culture Watch."

29. D. Chopra, *Quantum Healing*.

30. M. Drosnin, *The Bible Code*.

31. See, for instance, M. Edmundson, "Save Sigmund Freud," for a typical example of the persistent nostalgia for Freudian "insight" which still inhabits the hearts of many intellectuals (Edmundson seems to be a professor of literature). For an explanation of this phenomenon (especially as it involves professors of literature), see F. Crews, "Freudian Suspicion versus Suspicion of Freud."

32. See A. Byrd, "Squaring the Circle," for an account of Thomas Hobbes's misadventures with mathematics and the notoriety they earned him.

33. My previous book (P. R. Gross and N. Levitt, *Higher Superstition*) addressed this phenomenon in detail. More generally, see J. M. Ellis, *Literature Lost*, for a caustic description of how scholarly standards in the academy have been displaced by fads which are justified by appeal to political agendas claimed (usually with exaggeration and often with outright untruth) to be progressive. R. Kimball, *Tenured Radicals*, makes similar points but has its own emphatic political agenda, which Ellis seems to avoid.

34. Even this is not as cut and dried as it seems. One recurrent theme in UFO mythology is that some of the cleverest ideas of modern technology were actually purloined from the wrecks of crashed flying saucers.

35. CUFOS, founded by astronomer J. Allen Hynek. Hynek was at one point a member of the panel assembled by the air force to scrutinize, and presumably debunk, UFO sightings. His motives for going over to the other side invite speculation. See I. Ridpath, "Interview with J. Allen Hynek."

FIVE *Technology*

1. I would have demurred even if the offer had come from Wozniak and Jobs who, at that moment, were busy inventing the Apple computer in a garage in the town of Los Altos, where I was living.

2. Indeed this happened; that industry was born, flourished, and died within a decade as desktop computers and software for them became commonplace. For an account of the process by which a pack of impecunious technonerds became billionaires and captains of industry, see B. Cringely, *Triumph of the Nerds* (video). As it now

happens, some industry leaders are pushing for a new generation of simplified "network computers" which will, in effect, act as dumb terminals that call upon the power of larger mainframes, via the Internet, for intensive computing chores. This would be, in effect, a reversion to the old "time-sharing" idea.

3. In fact, an IBM 650, together with its auxiliary devices, at Columbia's Watson Labs, circa 1959.

4. There is another irony in that Microsoft, the proprietor of what came to be called PC-DOS (and then MS-DOS) hadn't developed this software, either, but merely acquired it for what turned out to be an incredible bargain price from another firm, whose acronym for the system was, in fact, Q-DOS—that is, "quick and dirty operating system."

5. As is well known, the admiration of the most technoliterate segment of the market hasn't been able to save Apple Computer, with its slick MacIntosh series, from a similar decline. It is interesting to contemplate the "no-win" situation of the personal computer industry's giants. The designs for IBM's PC and its successors became dominant because they were essentially unpatentable (allowing IBM's shoestring-operation competitors to produce equivalent machines for ridiculously low prices) and featured "open architecture" (allowing all sorts of "third-party" manufacturers to turn out modifications and peripheral devices). IBM figured wrongly that the prestige of its brand name would offset the price advantage of its cut-rate rivals; lacking patent protection, it lost both sales and royalties. By contrast, Apple fanatically protected the proprietary status of its technology; although that kept clones and unlicensed peripherals out of the market, it also kept prices high, compared to IBM clones, and greatly constricted the degree to which its systems could be "customized" by individual users. Thus Apple suffered a drastic loss in market share; the final nail in its coffin was the appearance of Microsoft's Windows graphic interface for PCs. Windows was a somewhat inferior realization of the graphic interface idea that was at the heart of the MacIntosh, but it was good enough for the purpose of matching the operational simplicity of the Mac. Apple tried manfully to keep Windows out of the market on grounds of patent infringement, but ultimately Microsoft prevailed in the courts. As of this writing, it's not clear that Apple can survive. Just recently, Microsoft, whose relations with IBM have gone sour, has come forth as the prospective rescuer of Apple, but the future remains cloudy. (See J. Heilemann, "The Perceptionist.")

6. Microsoft's aggressive attempts to make up for its early neglect of the Internet has brought down the wrath of the Justice Department, which alleges that Microsoft has crossed the line into unfair restraint of trade. As of this date, major antitrust suits loom, with the Justice Department to be joined by a number of state governments. See J. Quittner and M. Slatalla, *Speeding the Net*, for a lively account of how "Web browser" software evolved from the plaything of professional computer scientists to the central fixture in a war between giant social forces.

7. See K. Auletta, "The Microsoft Provocateur," as well as N. Myhrvold, "The Dawn of Technomania."

8. Michael Wolff, quoted in J. Surowiecki, "Flame Wars," 32.

9. See H. Petroski, "Development and Research," for an analysis that emphasizes the "nonlinear" relationship between science and technology. For background, see D. Cardwell, *The Norton History of Technology*.

10. J. H. Ausubel, *Rails and Snails and Debate over Goals for Science*, 18.

11. My own primary affiliation is with the Gettysburg Discussion Group.

12. For a meditation on the unpredictability of the evolution of technology and its social consequences, see N. Myhrvold, "The Dawn of Technomania." Myhrvold, a theoretical physicist by training, is Microsoft's resident guru and futurologist. On the

other hand, see M. Minsky, "Technology and Culture," and A. Penzias, "Technology and the Rest of Culture," for enthusiastic, even gleeful visions of the delights in store for us as technology unfolds. Minsky, of course, is one of the world's great experts on artificial intelligence—indeed, on models of mental function itself—while Nobel laureate Penzias is codiscoverer of the cosmological "background radiation" that persuaded most astronomers and physicists of the essential correctness of the Big Bang cosmological theory. On the other hand, for a somewhat skeptical view of the economic utility of the computer and information technology explosion, see A. S. Blinder and R. E. Quant, "The Computer and the Economy."

13. See R. L. Hirsch et al., "Fusion Research with a Future." See also W. M. Stacey, "The ITER Decision and U.S. Fusion R&D."

14. On the other hand, the technology of light planes has stagnated, and the production of such aircraft has practically ground to a halt. As James Gleick explains, "Small-scale aviation has thus become one of the peculiar corners of applied science—along with some kinds of biotechnology, pharmaceuticals, pesticides, and birth control—in which the ever-present threat of litigation suppresses the good along with the bad" (J. Gleick, "Legal Eagles," 50). Tragically, shortly after this was written, Gleick himself was badly injured, and his son killed, when the high-tech light plane he was piloting crashed.

15. Consider, for instance, the idea of an external-combustion, closed-cycle turbine engine that could, without producing noxious fumes, run on pretty much any fuel—gasoline, kerosene, or alcohol.

16. In fact, battery technology may be beside the point, since fuel cells, which burn hydrogen catalytically, can probably replace batteries. On the prospects for electric cars (whether battery or fuel-cell powered), see J. Sutherland, "Unplug the Car and Let's Go!" Moreover, the U.S. Department of Energy has recently announced (October 21, 1997) that a newly developed fuel-cell system which uses gasoline, rather than hydrogen itself, as a primary fuel, shows considerable promise. See also A. DePalma, "Ford Joins in a Global Alliance to Develop Fuel-Cell Auto Engines."

17. The repeating rifles and carbines of the Civil War era fired a cartridge much like a modern rifle round, except for its black powder propellent. Ammunition of this kind would probably have been harder for the South to manufacture in quantity than the weapon itself.

18. A good history of the attempts to arm Union troops with rapid-firing rifles, and the resistance to those attempts, can be found in R. V. Bruce, *Lincoln and the Tools of War.*

19. Another noteworthy fact about the Civil War, reflecting a similar backwardness, was that neither army tried to keep accurate time on the battlefield. Technical difficulty wasn't the problem, since accurate timekeeping was an essential to accurate navigation and had long been mastered by both warships and merchantmen. (See D. Sobel, *Longitude.*) Nonetheless, despite what seems to us the obvious advantage of "synchronizing watches," especially to a large, spread-out army, the practice was unheard of.

20. H. Marcuse, *One-Dimensional Man,* 18. Among Marcuse's more prominent disciples is Theodore Roszak, whose *The Making of a Counter Culture,* an influential book of the late '60s, is largely a popularization of Marcuse's *One-Dimensional Man.* Roszak has since become a leading critic of technological society, whose allies include significant segments of the New Age subculture. See E. Mendelsohn, "The Politics of Pessimism," for a discussion of the critique of technology that emerged as part of the ethos of the New Left in the '60s, under the influence of Marcuse and others. See D. Sarewitz,

Frontiers of Illusion, for a contemporary instance of technopessimism. See also L. Marx, "The Idea of 'Technology' and Postmodern Pessimism."

21. J. Habermas, "The Analytical Theory of Science and Dialectics," 154.

22. "Modernity" is the term frequently chosen, although more recently some claim that the whole shebang has been transmogrified into "postmodernity."

23. See, for instance, A. Ross, "Science Backlash on Technoskeptics," for an especially simplistic version of the dogma that there is no such thing as disinterested science apart from greed-driven technology. See P. Forman, "Recent Science," for a somewhat more urbane, but still rather badly confused, version of the same thing. On the other hand, see P. Galison, "Three Laboratories," for a subtle and informed analysis of the convoluted interface between "pure science" and "technology" (although Galison's notion of technology is, implicitly, somewhat different from that employed in this chapter). Galison explicitly rejects the term "technoscience" (a coinage of B. Latour), which Ross dotes upon.

24. The central role of Bacon and Descartes in the demonology of most technophobic intellectuals is rather curious. Bacon was uncannily prescient about both the methodological adaptations necessary to produce true science and the power that science would unlock, but he was never, properly speaking, a scientist and is utterly absent from any of the literature that scientists encounter in their professional training. Descartes was, of course, a serious mathematician, but, from the point of view of science, a decidedly reactionary figure. His ideas on physics, though dominant for decades, were neither correct ideas nor particularly fruitful errors. His ideas on methodology stressed deduction from a priori insights over empiricism and the experimental method. His mind-body (or, perhaps, "spiritual-material") dualism has been thoroughly eclipsed by the monism that dominates scientific thought. Thus, the eagerness of many prominent "science critics" to lay the blame for the presumed evils of science on the doorstep of Cartesian "dualism" seems particularly misplaced.

25. N. Jardine and M. Frasca-Spada, "Splendours and Miseries of the Science Wars," 232.

26. A common move among "science critics," especially those with radical sympathies, is to point to booming bombs as particularly odious products of science-based technology, and thence to declare that since such horrors are born of scientific "objectivity," there must be something wrong with objectivity (or sometimes, that it oughtn't to be called "objectivity"). That this kind of logic involves a fundamental confusion of categories, a confounding of "is" and "ought," is trivially obvious. It is an instance of the triumph, all too frequent in contemporary academic life, of sentimentality over clear thinking.

27. No doubt large numbers of bureaucrats and politicians have convinced themselves that something called "high-energy physics" must be of great technological importance, at least to the military. No doubt there have also been high-energy physicists who have been careful to address such people in terms calculated not to dispel such delusions. In terms of politics and sociology, this is an interesting phenomenon, but it is unrelated to the nature of high-energy physics as a science. Within the so-called science studies community, there are scholars deeply proud of their political acumen who are thoroughly convinced that subatomic particle physics played an important role in the development of Cold War weaponry. This sheds light on the scientific sophistication of "science studies."

The foregoing observations do not, of course, deny the possibility that over the course of time the deep understanding of physical reality provided by contemporary high-energy physics (or, for that matter, cosmology) might lead to important technological innovations. Indeed, if we use a timescale of centuries, rather than years, I believe such consequences are close to inevitable. There is no predicting what they will be, however.

28. Many champions of avant garde sociology of science are quite eager to discern a large element of irrationality (or, at any rate, nonrationality) in the construction and acceptance of scientific theories. They greedily batten on any preferred example of such phenomena, which usually derive from sociological or historical "case studies." By and large, I think they have been chasing mirages. Such case studies are usually deeply compromised by the fanatical urge to debunk that marks their authors. The air tends to go out of them once they are subject to rigorous scrutiny. (See N. Koertge, ed., *A House Built on Sand.*)

It could be reasonably maintained that such irrationalism is sometimes present in fringe science or in highly immature protoscience. The "racial science" of the nineteenth century and the extreme eugenicism of the early twentieth afford examples. But this "science" is notable precisely for its fragility and ephemerality. On the other hand, some commentators are eager to point to the disputes and confusions, leading, in some cases, to outright mysticism, that swirl around foundational questions in quantum mechanics as evidence of irrationalism in scientific thought at even the highest, most "mathematical" levels. This is, indeed, an intriguing topic, but note that the puzzlement is concerned with the philosophical and ontological interpretation of predictive mechanisms that are themselves rock solid. Nobody within the physics community, irrespective of philosophical inclination, denies that the mathematical formalism at the heart of quantum mechanics is a supremely accurate device for describing how the world works. The question, a deeply vexed one for the moment, is how to get below the level of this formalism—if indeed this is possible—to obtain a description of the objective world at a more basic level. But the reasons for accepting the formalism and the modes of applying it, in "applied" as well as "pure" physics and chemistry, are supremely rational. See J. Cushing, *Quantum Mechanics*, for a fascinating discussion. See, as well, N. Levitt in *Physics Today* for a review of Cushing's book. See also D. Kleppner, "Physics and Common Nonsense," and S. Goldstein, "Quantum Philosophy," for brief but illuminating discussions.

29. The universal adoption of the QWERTY keyboard in English-speaking countries is often cited as the paradigmatic triumph of a bad idea. Another is the current multiplex-stereo system for FM broadcasts. I would also argue that the use of CD-ROMs on computers is another case of an inferior technology crowding out superior alternatives, in this case because CDs were already omnipresent as an audio medium, and so the basic equipment was easily available for adaptation to computers. (Come to think of it, the CD itself represents the triumph of a suboptimal, if not actually bad, technology; the sampling rate is too low and the discriminating power of the 16-bit sample too small for ideal sound recording. If the technology had been held back from the market for another few months while these parameters improved, consumers would now have a much better product.)

30. I apologize to all my fellow leftists for naming this phalanx of conservative thinkers as paragons of fundamental political acumen, but—alas—it seems to me that conservative fears have, over the years, given us sharper insight into the human condition than have our revolutionary hopes. It seems to me, finally, that if any optimistic thinkers are to arise to give us truly good news about the possibilities of fundamental social change, they will first have to deal—without evasion or sophistry—with the bad news reported by the pessimists I have cited.

31. E. Tenner, *Why Things Bite Back*, 271.

32. See Y. Ezrahi et al., eds., *Technology, Pessimism, and Postmodernism.*

33. L. Menand, "The Demise of Disciplinary Authority," 211.

34. There is, of course, the question of what has been responsible for this mood shift with its increasing valorization of hard-edged skepticism toward everything in sight. It could be argued that the failure of technology to live up to its utopian promises is one element. Perhaps; but I think it is a relatively minor one.

35. A. Harmon, "NASA Flew to Mars for Rocks?"

36. See E. Tenner, *Why Things Bite Back*, for a comprehensive account of unintended consequences.

37. See, for instance, H. M. Meade, "Dairy Gene," for an eloquent defense of using transgenic technology to produce substances urgently needed for medical use.

38. By the standards of many contemporary thinkers on "science, technology, and society," a book like Hackerman and Ashworth's *Conversations on the Uses of Science and Technology* seems outdated and naive. For that reason, it is a reliable guide to many issues. The dubious value of sophistication, by contrast, is revealed by a book like *Strange Weather*, by Andrew Ross, who is often held up as a very paragon of contemporary sophistication. (I have criticized Ross's book elsewhere; see P. R. Gross and N. Levitt, *Higher Superstition*.)

39. E. Tenner, *Why Things Bite Back*, 277.

six *Nature*

1. M. Douglas and A. B. Wildavsky, *Risk and Blame*, 127.

2. See E. B. Leacock, *Myths of Male Dominance*.

3. See S. Goldberg, *The Inevitability of Patriarchy* and *Why Men Rule*.

4. W.V.O. Quine, *From a Logical Point of View*, 1.

5. See the interesting remarks of M. Talbot ("The Perfectionist") on this point.

6. Curiously, many *soi-disant* postmodernists are eager to make the same point about the "unnatural" nature of the nature/non-nature distinction. They contend that the distinction is a social construction, and thus (as usual) the embodiment of forms of power and domination in linguistic convention. I would go that far with them on this narrow question, though of course I take vigorous issue with the more general postmodern dogma that there is no real, objective world out there (at least none known to us), merely texts and discourses. I also note that beneath the pose of skeptical sophistication, there lurks a propensity for old-fashioned, stiff-necked moralizing (usually on behalf of the victims, real and imagined, of white, male hegemony). As usual, all these resolute anti-foundationalists proceed as though their own fervent political and moral agendas had rock-solid foundations entirely immune to challenge. See J. Bennett and W. Chaloupka, eds., *In the Nature of Things*, as well as M. W. Lewis's review of the book.

7. An interesting example of the genre of nature-worshipping devotional literature (there is no other apt name for it) can be found in B. Krause's article, "What Does Western Music Have to Do with Nature?" Krause doesn't have too much to say about Western music as such (perhaps this is his sly way of making a point). Instead, he rhapsodizes about the music of hunter-gatherer cultures, which is actually embedded in the sound of "nature" itself. This essay is part of a special issue (vol. 2, no. 3) of the environmentalist journal *Terra Nova* (subtitle: *Nature & Culture*) which is concerned with nature and music. Oddly, the compact disk (*Music from Nature*) that accompanies the issue has snippets of Hildegard von Bingen—and Beethoven! For whatever reason, the mystical propensities of these German musicians warrant description of their work as "music from nature," so far as the editors of *Terra Nova* are concerned. I hope I do not offend the creators of this disk too much with my recalcitrance when

I report that, for me at least, this disk had an esthetic and moral effect opposite to that intended.

8. Chapter 4, discussion of the way in which "natural" is equated with "healthy."

9. S. R. Kellert, *Kinship to Mastery*, 83.

10. See E. O. Wilson, *The Diversity of Life*.

11. See B. Taylor, "Earth First! Fights Back."

12. In fact, this idea has gone beyond mere science fiction; see G. K. O'Neill, *The High Frontier*.

13. Recall the ill-fated "Biosphere" project, which was, in fact, an attempt to create a physically and emotionally sustainable microculture within a fully artificial environment. Unknown to most observers at the time, the project was the doing of a rather bizarre survivalist cult which intended it as an experiment toward the establishment of some similar community on Mars which was to serve as a refuge for privileged survivors of various earthly cataclysms. Among other things, the designation "Biosphere" was ironic—ironic because the colony was supposed to house a severely limited spectrum of life forms, whose interactions with each other were to be calibrated by human "biosystems" engineers.

14. S. R. Kellert, *Kinship to Mastery*, 3. Note that Kellert's argument is essentially "sociobiological"; yet because of the context, it will be accepted by many political "progressives" who regard sociobiology in its own right as anathema.

15. One example of extreme biophilia that certainly does cross the line into misanthropy can be found in those fundamentalist biophiliacs who deplore the extirpation of the smallpox virus from the face of the earth.

16. K. Lee, "Designer Mountains: The Ethics of Nanotechnology."

17. See B. C. Crandall, ed., *Nanotechnology*. See also W. Hively, "The Incredible Shrinking Finger Factory."

18. See M. W. Lewis, *Green Delusions*, for an account of the environmental promise of nanotechnology.

19. K. Lee, "Designer Mountains," 132.

20. Ibid., 134.

21. Some examples of the genre are A. Naess, *Ecology, Community, and Lifestyle*; V. Plumwood, *Feminism and the Mastery of Nature*; and V. Shiva, *Monocultures of the Mind*.

22. See M. W. Lewis, *Green Delusions*, as well as "Radical Environmental Philosophy and the Assault on Reason"; see also M. Bookchin, *Reenchanting Humanity*, and R. Denfeld, "Old Messages." The phenomenon of self-destructive antiscientific absolutism within the environmental movement is not limited to the United States. A distinguished Dutch medical scientist recently made the same point with regard to European eco-radicalism: "The environmental movement has been taken in tow by radicals who distrust scientific approaches to the environmental problems and as a consequence millions have been thrown away to the wrong projects. Of course this will be counterproductive and eventually this will lead to a backlash" (W. Lammers, personal communication).

23. But see V. Plumwood, "Being Prey."

24. B. N. Ames, "Does Current Cancer Risk Assessment Harm Health?" Ames's thesis is rebuked, but only partially and tentatively, in J. Wargo, *Our Children's Toxic Legacy*. See also M. Fumento, "Pesticides Are Not the Main Problem."

25. E. O. Wilson, *The Diversity of Life* (concluding chapter).

26. E. O. Wilson, quoted in D. Takacs, *The Idea of Biodiversity*, 252.

27. P. Brodeur, *The Great Power-Line Coverup*.

28. The chief reason that it seems unlikely that low-frequency radiation can have biological effects arises from the most fundamental equation of quantum theory,

$E = h\upsilon$, where E is the energy of the quantum, or photon, associated with the radiation, υ is the frequency of the radiation, and h is a universal constant of proportionality called "Planck's constant." The point is that the lower the frequency, the smaller the energy carried by a photon. An impinging photon can only trigger chemical change if its energy lies above a certain threshold. In the case of 60-cycle (in Europe, 50-cycle) household alternating current, the photon's energy is absurdly low.

29. See S. Schneider, *Laboratory Earth*.

30. Physicist Robert L. Park—certainly no friend to eco-alarmism—also concurs ("Scientists and Their Political Passions").

31. Quoted in P. R. Ehrlich and A. H. Ehrlich, *The Betrayal of Science and Reason*, 243.

32. It might be well to give some idea of the range of serious scientific opinion on the global warming issue. Stephen Schneider (*Laboratory Earth*) clearly believes that substantial warming is inevitable, with a probable increase in average global temperature on the order of 3°C by the middle of the twenty-first century. Thomas R. Karl et al., "The Coming Climate," assumes as much. However, a number of reputable climatologists, the best known of whom is probably Richard Lindzen, strongly disagree, and assert that a 1° rise is the worst we should expect. Schneider's view more nearly represents the current consensus, but some recently emerging evidence that the upper atmosphere (several thousand feet above sea level) is actually cooling, in defiance of most of the global warming models, has provided that consensus with serious food for thought (see F. Pearce, "Greenhouse Wars"). Moreover, some climatologists who accept the reality of anthropogenic greenhouse warming are unsure whether the effects will be serious enough to warrant costly measures (see W. K. Stevens, "Experts on Climate Change Ponder: How Urgent Is It?"). The uncertainty of the best available predictive methods are discussed in W. Stevens, "Computers Model World's Climate, but How Well?" Some "techno-optimists," including Jesse Ausubel ("The Liberation of the Environment" and "Environmental Trends"), think the problem is essentially self-correcting, given the tendencies inherent in the modern economy and its supporting technology. Nebojša Nakićenović ("Freeing Energy from Carbon") believes that the overall trend in fuel usage has been to "replace" carbon with hydrogen, and that this will greatly abate the projected increase in the greenhouse effect. A useful survey can be found in W. K. Stevens et al., "Global Warming."

33. See G. Easterbrook, "Hot and Not Bothered," for an account of the reluctance of all sides in the current controversy to devise practical measures for dealing with global warming, should it indeed prove inevitable. See also M. B. McElroy's sidebar to the Easterbrook piece, "Clean Machines."

34. See S. Nemecek, "Frankly, My Dear, I Don't Want a Dam."

35. The PBS documentary *Frontline: Nuclear Reaction* makes a compelling case that nuclear power has been inappropriately demonized and deserves serious reconsideration.

36. But see M. L. Wald, "Finding a Formula to Light the World but Guard the Bomb," which points out the possibility of building power reactors that will eliminate the possibility of the diversion of weapons-grade fissionable material. Of course, it is possible to use the radioactive materials generated by reactors as weapons of mass terrorism even without creating nuclear explosives.

37. G. Easterbrook, "Hot and Not Bothered," 21.

38. Similar misgivings impelled Europe's leading facility for high-energy physics to change its name from the "European Center for Nuclear Research" (Conseil Européen pour la Recherche Nucleaire in French) to the "European Laboratory for Particle Physics." Ironically, the organization is still universally known by the old acronym, CERN. See M. W. Browne, "International Language of Physics Ties Physicists' Tongues."

39. The only plausible danger from irradiated food—and it is not very plausible— is that the radiation will produce chemical "free radicals" which might cause cancer or other pathologies.

40. Author and activist Jeremy Rifkin is an interesting example. See G. Stix, "Profile: Jeremy Rifkin," as well as P. Gross and N. Levitt, *Higher Superstition*.

41. See Michael Shermer, "The Beautiful People Myth," for an analysis of the psychology behind this nostalgia for utopian societies that never were.

42. See M. W. Lewis, *Green Delusions*.

43. S. Schneider, *Laboratory Earth*, 115–154.

44. S. Schneider, spoken address.

45. See M. W. Lewis, "In Defense of Environmentalism," for a fair-minded review of *Betrayal of Science and Reason*, giving a good sense of its strengths and weaknesses.

46. See F. Miele, "Living without Limits: An Interview with Julian Simon." See also J. H. Cushman, "Industrial Group Plans to Battle Climate Treaty," for an account (leaked to environmental activists by a sympathetic insider) of plans by a consortium of major corporations, including Exxon and Chevron, to derail international agreements to limit carbon dioxide emissions.

47. M. Fumento, *Science under Siege* (dioxin chapter).

48. P. Ehrlich and A. Ehrlich, *The Betrayal of Science and Reason*, 167–170.

49. See as well P. R. Ehrlich et al., "No Middle Way on the Environment," for a further example of the strategy adopted by Ehrlich and his allies; the basic idea is to concede nothing to critics and to extract the gloomiest possible lesson from any given example, even if it means stretching a point to distortion.

50. See R. Shattuck, *Forbidden Knowledge*, for a sophisticated meditation on the possibility that the Faustian spirit of Western culture has carried us into territory we literally know more about than we should. This includes aspects of our scientific knowledge.

51. More recently, Japanese scientists claimed to have cloned cows from adult cells. See G. Kolata, "Cows Cloned in First Case Like Dolly's, Japanese Say." Even more recently, mice have been added to the list of clonable mammals. See J. B. Verrengia, "Hawaii Scientists Clone 50 Mice."

52. J. B. Elshtain, "Ewegenics." See also Elshtain, "Our Bodies, Our Clones," as well as L. R. Kass, "The Wisdom of Repugnance." Since I don't share Kass's repugnance, I have my doubts as to how wise it is. See, in addition, G. Kolata, "Scientists Face New Ethical Quandaries in Baby Making," which makes it fairly clear that, for good or ill (my personal view is that the threat of ill consequences is overblown), human reproductive technologies are likely to be put to use as soon as they become available, notwithstanding the subsequent repugnance of would-be moralists. See D. Shenk, "Biocapitalism," for a thoughtful piece that takes a middle course, realizing that biotechnology will quickly come to play a major role in human affairs, and that much of this will be for the good, but that disquieting questions remain.

53. My own point of view on Dolly and, beyond that, the possibility of creating human clones is that the ethical aspects of the affair have been greatly overblown. For a number of genetic and embryological reasons, it is clear that cloning does not produce a carbon copy of an adult. So many purely adventitious circumstances intervene between fertilization and birth that it is wrongheaded to regard the cloning process as the mere xeroxing of a fully formed organism. I assume that in the course of the coming decade, a human being will be cloned, most likely from a person with the wealth and the vanity to get the job done. I don't think anticloning laws will prevent this, even if they are widely adopted. I am not terribly perturbed at the prospect, and I certainly don't believe widespread social distress will result. Basically, it is

a stunt, and I think that once it has been done and replicated a few times, it will lose its fascination. No grave social or cultural consequences will ensue. Of course, this is merely my opinion.

54. The recent announcement by a physicist, Richard Seed, that he will offer human cloning services to the public has sparked considerable uproar. One observer, medical ethicist Willard Gaylin, claims that Seed's announcement triggered the "Frankenstein factor" in the public psyche. The most probable interpretation of Seed's announcement is that it amounts to little more than a publicity stunt. See G. Kolata, "Proposal for Human Cloning Draws Dismay and Disbelief."

55. See P. Bereano, "Don't Take Liberties with Our Genes," for a typical example of the genre.

56. Geneticist S. Jones has it about right, I think: "It [genetics] will become just another science, and as such, of no interest to most of those who gain from it. That day is closer than most of us realize" ("In the Genetic Toyshop," 16).

57. For a serious, thorough, and deeply informed discussion of the human and ethical consequences of burgeoning biotechnology, particularly as regards human genetics itself, see P. Kitcher, *The Lives to Come*. A healthy corrective to dystopian views of molecular biology may be found in W. Bodmer and R. McKie, *The Book of Man*, 257: "Far from worrying, we should relish the prospect before us. We are entering a golden era."

58. P. Admiraal et al., "Declaration in Defense of Cloning and the Integrity of Scientific Research," 11. The signatories of this declaration include Edward O. Wilson, along with such distinguished scientific, literary, and philosophical worthies as Sir Isaiah Berlin, Francis Crick, Richard Dawkins, Adolph Grünbaum, Sergei Kapitza, W. V. Quine, Simone Veil, and Kurt Vonnegut. Other defenses of cloning research and technology can be found in R. Dawkins, "Thinking Clearly about Clones," R. A. Lindsay, "Taboos without a Clue," and R. T. Hull, "No Fear." Other coolheaded perspectives appear in a special issue of *The Sciences* (September/October 1997) dedicated to the ethical and technical issues raised by Dolly. See B. Bilger, "Cell Block," T. J. Bouchard Jr., "Whenever the Twain Shall Meet," P. G. Brown, "What Hath Wilmut Wrought?," M. A. Di Berardino and R. G. McKinnell, "Backward Compatible," S. J. Gould, "Individuality," J. B. Gurdon, "The Birth of Cloning," R. Hart, A. Turturro, and J. Leakey, "Born Again," P. Kitcher, "Whose Self Is It Anyway?," H. M. Meade, "Dairy Gene," R. E. Michod, "What Good Is Sex?," and M. Z. Ribalow, "Take Two." See also R. Sapolsky, "A Gene for Nothing," for a primer on what is and is not "genetically" determined. Also see B. H. Kevles and D. J. Kevles for a critique in *Discover* of naive hereditarianism. It is of some ironic interest that some of the people raising the loudest alarms about Dolly are also among the most fervent opponents of hereditarianism, a noteworthy inconsistency. An exception may be found in M. Schoofs, "Fear and Wonder." A journalist who covers biomedical issues, Schoofs shares most of the cultural left's concerns about hereditarianism and the like, but seems to take the supposed menace of cloning in stride. G. Kolata, "On Cloning Humans, 'Never' Turns Swiftly into 'Why Not,'" points out that the alarmist rhetoric about the question seems rapidly to be fading, and that an increasing number of researchers are seriously considering the possibility. L. H. Tribe, "Second Thoughts on Cloning," is an op-ed piece (by an eminent legal scholar) that illustrates the point. See also G. Johnson, "Don't Worry, a Brain Still Can't Be Cloned," and "Ethical Fears Aside, Science Plunges On," as well as P. Kitcher, "Tall, Slender, Straight and Intelligent."

59. The news that a Swiss referendum (June 1998) has rejected (by a two-to-one vote) an attempt to ban transgenetic research in biotechnology comes as a relief, but not an unmixed blessing. (See also M. Specter, "Europe, Backing Trend in U.S., Blocks

Genetically Altered Food.") The fact that one-third of the voters in one of the most prosperous and best-educated countries in the world could be swayed by what are essentially medieval arguments is ominous. See M. Shields, "Swiss Vote Down Bid to Curb Genetic Research."

60. M. Douglas, *Purity and Danger*, 145.

61. M. Douglas and A. B. Wildavsky, *Risk and Blame*. Wildavsky was a noted neoconservative political and economic theorist whose last book, *But Is It True?*, aggressively challenged a spectrum of environmentalist pieties.

62. M. Douglas and A. B. Wildavsky, *Risk and Blame*, 169.

SEVEN *Ethnicity*

1. See G. Niebuhr, "Lutherans Bridge Old Divisions, but Decide to Keep One in Place," for an account of the decision of American Lutherans to affiliate, despite doctrinal differences, with the Presbyterians, the Reformed Church, and the United Church of Christ. A similar proposal with respect to the Episcopal Church barely failed, and will presumably succeed eventually, as is suggested by a follow-up story by the same reporter ("Lutherans Reconsider Episcopal Concordat").

2. See chapter 3, especially note 33.

3. See, for instance, the various books of Steven Shapin: *Leviathan and the Air-Pump* (with S. Schaffer), *A Social History of Truth*, and *The Scientific Revolution*. Shapin's recurrent point is that what we traditionally think of as a great advance in knowledge and in the art of acquiring knowledge is, in fact, a matter of evolution of social institutions, driven by social, economic, and political forces. Thus, the ascendency of the Royal Society and of Newtonian science at the end of the seventeenth century records not an amazing new chapter in our understanding of the physical universe, but rather a modus vivendi among contending social factions in England and Europe. Over the years, Shapin has moved away from a hard-edged "social constructivism" (which concedes nothing to the truthfulness of science), but he remains loath to acknowledge the way science is internally driven by its own logic and the demands of empirical accuracy. As a card-carrying Whig, I cannot share his reservations.

4. The key figures in this nihilistic view of human progress are Michel Foucault and Thomas Kuhn. I discount Foucault as being an interesting character, but an inaccurate historian and a muddled philosopher. Kuhn is another matter. Nowadays, it is impossible to talk about a scientific revolution of any kind without at least implicitly invoking his name and his startling book, *The Structure of Scientific Revolutions*. He was trained as a physicist and knew the details of science quite well. At the end of his life, he was plagued by the nonsensical dicta, especially the sneers at the very idea of objectivity, that were promulgated on his supposed authority. In my view, Kuhn was a thinker who raised interesting questions and stimulated valuable inquiry into the historical and social context of scientific work. He did much to banish shallow clichés from the discourse of the history of science, and to place in a problematical light many aspects of the development of scientific thought that had simply been glossed over. Yet in my view, Kuhn's answers to the questions he raised are simply not compelling. They are often wrong or misleadingly overstated. Seeking to banish simplistic and comfortable myths about scientific progress, he left in their place just-so stories that are more unsettling but hardly less simplistic. That he has become a totemic figure for the intellectuals of our age (even among those he would have scorned) says much more about our unfortunate age than about Kuhn. See P. R. Gross, "Bête Noire of the Science Worshippers."

5. I have previously alluded to my book with P. R. Gross, *Higher Superstition*, as a critique of the weaknesses of "social constructivism" and related schools of thought. See as well P. R. Gross and N. Levitt, "Academic Anti-Science" and "The Natural Science. Trouble Ahead? Yes," along with P. R. Gross, N. Levitt, and M. W. Lewis, eds., *The Flight from Science and Reason*. Other work on this theme includes: N. Koertge, ed., *A House Built on Sand*; G. Holton, *Science and Antiscience* and *Einstein, History, and Other Passions*; S. Weinberg, "Physics and History," *Dreams of a Final Theory*, and "Sokal's Hoax"; P. Boghossian, "What the Sokol Hoax Ought to Teach Us"; and M. L. Perutz, "The Pioneer Defended." J. Bricmont and A. Sokal in *Impostures intellectuelles* analyze the defective understanding of science manifest in much of recent "continental" philosophy.

6. "Whig history" is customarily denounced on the grounds that it is history written by the winners, in this case the triumphal institution of modern science. There is, however, a fallacy in this plaint: there is no logical reason why winners' history cannot be accurate history. Are we to believe that the history of World War II as written by unrepentant Nazis is more accurate than that written by American or British historians?

7. These include so-called feminist epistemology, social constructivism with its attendant relativistic philosophy of knowledge, "cultural" study of science, and, to an extent, "Afrocentrism," as applied to science.

8. M. Harris, "Post-Modern Anti-Scientism."

9. See D. Herschbach, "Imaginary Gardens with Real Toads," for a persuasive account of the deeply poetic nature of research science.

10. M. Harris, "Post-Modern Anti-Scientism."

11. See L. Kuznar, *Reclaiming a Scientific Anthropology*, and M. Cartmill, "Reinventing Anthropology," for critiques of this trend. See also R. Fox, "Anthropology and the 'Teddy Bear' Picnic," as well as "State of the Art/Science in Anthropology." One other traditionally scientific field that has succumbed, at least in part, to the blandishments of the postmodernists is clinical psychology; see B. Held, *Back to Reality*, "Constructivism in Psychotherapy," and "The Real Meaning of Constructivism." I note without surprise, but with considerable gratitude, that nothing similar has befallen the sciences traditionally thought of as "hard."

12. Harris, "Post-Modern Anti-Scientism."

13. L. A. Kuznar, *Reclaiming a Scientific Anthropology*, 183.

14. R. Fox, "Anthropology and the 'Teddy Bear' Picnic," 49.

15. C. Sartwell, "Science and Race in W.E.B. Du Bois." I note that this quote is from an abstract of a talk by Dr. Sartwell and might not reflect his exact words.

16. T. O'Meara, "Anthropology as Empirical Science," 354.

17. See D. Hollinger, *Science, Jews, and Secular Culture*, for an account of the way in which the secularization of high culture in America was intertwined with the rise to prominence of many Jewish intellectuals, both in and out of the sciences.

18. Tsushima was the 1905 battle in which the thoroughly modern Japanese fleet routed a larger, but technologically outdated, Russian enemy. This occurred barely forty years after Japan made the decision to throw aside the hide-bound, xenophobic Shogunate, and to adopt whatever it needed from the West, without embarrassment, in order to achieve political and technological parity.

19. India's recent plunge into an all-out nuclear weapons program is sad evidence of this fact.

20. "It is a well-known fact that the same anti-modernist intellectuals who have led the charge against science, have also vigorously critiqued all modern ideas, including

secularism, liberal democracy, industrialization, urbanization, etc. as Western imports, unsuitable for India's civilizational ethos. As many eminent Indian intellectuals . . . have pointed out, the uncompromising opposition of our nativists to all ideas modern implicitly, if not explicitly, legitimizes the agenda of the reactionary Hindu forces" (M. Nanda, "Reclaiming Modern Science for Third World Progressive Social Movements," 922). See also M. Nanda, "The Science Question in Postcolonial Feminism" and "The Epistemic Charity of the Social Constructivist Critics of Science and Why the Third World Should Refuse the Offer."

21. See W.V.O. Quine, "Commensurability and the Alien Mind"; see also O. Kenshur, "The Rhetoric of Incommensurability."

22. I have the private habit of referring to the postmodern doctrine that one culture's way of knowing may not be disparaged on the basis of what another culture's has to say as Boxer/Ghost Dancer epistemology. The Chinese Boxer societies *knew* that their magic practices protected them from the bullets of intruding foreign troops. Likewise, at about the same time, Native American "ghost dancers" *knew* through their own cultural logic that *their* magic nullified the bullets of the Federal Army. The Gatling guns and Maxim guns had different ideas, however, and these seem to have prevailed, incommensurability notwithstanding.

23. Fox, "Anthropology and the 'Teddy Bear' Picnic."

24. See, for instance, D. Gonzales, "The Origins of Indigenous Americans." The author of this letter to the *Chronicle of Higher Education* signs himself as an associate professor of English at Bemidji State University in Minnesota. Among his claims: "We have to set the record straight: Our oral history is trustworthy; we are from this continent; and we have been in the Americas for three ice ages, or more."

25. See M. Gladwell, "A Matter of Gravity," and E. Eichman, "The End of the Affair." Both these pieces report on the debate at New York University between the physicist Alan Sokal and defenders of the journal *Social Text* in the wake of the hoax article that Sokal palmed off on that publication. (See A. Sokal, "Transgressing the Boundaries.") What is interesting is that the session broke down in recriminations when Sokal raised the question of the relative accuracy of anthropologists' accounts of early settlers of the Americas, as against Native American creation myths. Many people in the audience defiantly refused to consider the question, since it invited the impermissible "privileging" of Western scientific ideas over aboriginal myth.

26. See K. L. Feder, "Indians and Archaeologists," for an account of this quarrel and its dire consequences for North American archaeology. Deloria's antiscientific stance is an important part of the story. One irritant in the situation is the fact that some of the oldest human skeletal remains found in North America seem to exhibit "Caucasian" morphology. See D. Preston, "Skin & Bones," for a brief account, accompanied by picture, of the reconstruction of the physiognomy of the controversial "Kennewick Man," a 9,000-year-old skeleton found in the state of Washington. The controversy stirred up by this reconstruction is reported in T. Egan, "Old Skull Gets White Looks, Stirring Dispute," and K. A. McDonald, "Researchers Battle for Access to a 9,300-Year-Old Skeleton." These stories also cover the attempts of Umatilla Indians to claim the Kennewick remains as "ancestral," and to bury them irretrievably. The role of the Army Corps of Engineers in sequestering the remains and possibly in destroying the archaeological value of the original site raises serious questions.

27. K. L. Feder, "Indians and Archaeologists," 79.

28. For a realistic critique of this myth, see M. W. Lewis, *Green Delusions*.

29. Ania Grobicki, cited in M. Cartmill, "Oppressed by Evolution," 81.

30. M. Dickison, "Maori Science." See chapter 4, n. 15, above, for an instance of

academic multiculturalist support for the idea of teaching various ethnic creation myths, as well as Christian ones, alongside evolutionary theory.

31. H. W. Verran and D. Turnbull, "Talking about Science and Other Knowledge Traditions in Australia."

32. "These black forms were not inherited from the Old World, the way the culture of immigrants was: they were decisively inventions of the New World, as American as apple pie" (K. A. Appiah, "The Multiculturalist Misunderstanding," 30).

33. C. L. Brace et al., "Clines and Clusters versus 'Race.'"

34. See P. R. Gross and N. Levitt, *Higher Superstition*.

35. I teach at Rutgers, the State University (New Brunswick, New Jersey). The fact that Ivan Van Sertiman is on the faculty as well (in the Africana Studies Department) may have something to do with the currency of his theories among students, and with the loyalty with which they defend them.

36. A. B. Powell and M. Frankenstein, eds., *Ethnomathematics*. This is published by the State University of New York Press as part of a series ostensibly dedicated to reforming mathematics education. Of this, more in a subsequent chapter.

37. A. B. Powell and M. Frankenstein, "Uncovering Distorted and Hidden History of Mathematical Knowledge," 57, n. 5.

38. M. C. Borba, "Ethnomathematics and Education," 266. Perhaps the right image to keep in mind when weighing this statement is that of an Egyptian or Sumerian scribe trying to keep track of bushels of wheat (including fractions of a bushel).

39. The academic left, in its current incarnation, is moralistic and hypercritical on many fronts, hence the currency of the term "politically incorrect." Amazingly, this intolerance does not extend to out-and-out fascists, providing they can be claimed as progenitors of postmodernism.

40. S. E. Anderson, "Worldmath Curriculum: Fighting Eurocentrism in Mathematics," 303.

41. M. Bernal, *Black Athena*. See M. Lefkowitz and G. Maclean, eds., *Black Athena Revisited*, for a collection of withering critiques of Bernal's claims.

42. M. Bernal, "Animadversions on the Origins of Western Science." This piece, originally published in *Isis*, is dissected by one of the contributors to *Black Athena Revisited*; see R. Palter, "*Black Athena*, Afrocentrism, and the History of Science."

43. See M. S. Frankel, "Multicultural Science," for an academic's plea on behalf of the notion.

44. The phrase is Meera Nanda's; I borrow it with gratitude.

45. The first explanatory panel in the display reads:

The Science of Prevention
Many traditional African societies believe the world to be permeated with a powerful vital force. Found in stones and other inanimate or vegetable matter, it is stronger in animals and still more powerful in man. This force can be trapped and used, for good or evil. What we might call "magic" is seen as a science, and one not to be wholly scorned.

46. An e-mail exchange with the editor of the series informs me that *Ethnomathematics* is selling quite well.

47. From the point of view of sheer lunacy, the worst paper in *Ethnomathematics* is one that claims that Navajo children are resistant to learning elementary mathematics because their culture is in some way too mathematically sophisticated; in particular, it is said to possess a near-mystical insight into topology. (See R. Pinxten, "Applications in the Teaching of Mathematics and the Sciences." The somewhat mystifying

title, left unaltered when this chapter was excerpted from a complete book, really means applications of *Navajo anthropology* to math and science teaching.) As a topologist by profession, I'm fairly confident in judging Professor Pinxten to be rather poorly informed.

I should also mention that the book contains B. Lumpkin's "Africa in the Mainstream of Mathematics History," which also appears in Van Sertima's *Blacks in Science, Ancient and Modern.* In the latter incarnation, Lumpkin's piece drew some skeptical attention from P. R. Gross and N. Levitt, *Higher Superstition.*

48. An interesting example is the notorious issue of the self-consciously political cultural-studies journal *Social Text,* which published physicist Alan Sokal's hoax article, "Transgressing the Boundaries." Note that many of the articles in the issue (those by editor Andrew Ross, Emily Martin, Sandra Harding, Hilary Rose, and Sharon Traweek) in varying ways try to make the case for supplanting, or at least supplementing, conventional science by "knowledges" that proceed from the putative special wisdom of the oppressed. In fairness, I should point out that another aim of this issue was to confront the supposedly reactionary tantrums of P. R. Gross and the present author. A measure of preemptive revenge came our way in the form of Sokal's joke, which was actually inspired by *Higher Superstition,* a fact of which I shall be eternally proud.

49. S. Fuller, "Does Science Put an End to History, or History to Science?" 40. The cited peroration occurs at the end of a piece that attempts, with indifferent success, to show that the Battle of Tsushima (i.e., the rapid success of post-Meiji Japan in acquiring a scientific and technological infrastructure) had a pronounced effect on Western philosophy of science. Someone ought to tell him about the *post hoc ergo propter hoc* fallacy. In any case, I am not all that worried about Fuller's predictions concerning the essence of tomorrow's science, since he seems to have a difficult time discerning the essence of today's—and yesterday's—science. As to his speculations on the new citizenry of the republic of science, I can assure him that the female citizens of the province of mathematics, though recently greatly increased in numbers, seem to look upon the subject pretty much as their male counterparts do.

50. See G. Bass, "Geek Chic," for an amusing and instructive account of how undergraduate culture at MIT has—and has not!—changed as the number of women students in that bastion of the technoscience culture approaches that of men. See also G. Sonnert and G. Holton, *Who Succeeds in Science* and *Gender Differences in Science Careers.*

51. Commentators on intellectual and political trends have noted the significant distinctions among "feminisms," particularly that between "equity" and "gender" feminism. (See C. H. Sommers, *Who Stole Feminism.*) Equity feminism is, by and large, the belief that all legal bars to women's social, economic, and political equality must be rooted out. It furthermore insists on the extirpation of the extra-legal assumptions and practices that exclude women from careers and positions of authority. Gender feminism, by contrast, is at once more radical and more mystical. It envisions women as inhabiting a culture fundamentally different from that of men, and embraces a politics that views the gap between the sexes as the fundamental, well-nigh eternal, dividing line in society, the site of perpetual warfare. This presumably can only end with the extirpation of all the cultural and cognitive categories historically imposed by "patriarchal" culture. There are some curious contradictions within gender feminism. On the one hand, in good postmodernist, "antiessentialist" fashion, it embraces the idea that "gender" is wholly a culturally constructed category, and thus that there are no significant differences between men and women other than those that have been arbitrarily created by the needs and forms of patriarchal culture. On the other

hand, in practice gender feminists behave like the most unrepentant of essentialists, endlessly praising supposed modes of thought behavior that are timelessly and perpetually characteristic of "women's culture."

At this point in history, most educated and ambitious women, even those who are in some respects political conservatives, emphatically endorse the ideas of equity feminism. However, the near-monopoly of gender feminism in academic venues and the pugnacity and extravagance of gender feminist rhetoric have led many of these women to decline the label "feminist."

52. There have been some important and useful changes in the ways in which science accommodates the personal lives of scientists; in particular, the rigorous "up or out" schedules imposed on young scientists have been modified, in many places, to give young women time to start families. Under the old regime, which kept new Ph.D.s chained to their lab benches for a decade without respite, this was rather difficult. It is now becoming much more common to allow women to take a few years off while still in their twenties and thirties to have children without penalizing them in terms of opportunities for advancement and tenure. In fact, this practice is spreading to men as well, in the recognition that they, too, have responsibilities as young parents.

53. It is a standard myth of feminist science critique that the presence of large numbers of women biologists, starting in the 1960s, was responsible for major revisions in the fundamental ideas of reproductive biology—supposedly, the model of the "passive" ovum was abandoned because of female insight. That this legend has been endlessly repeated does not, however, make it true. (See P. R. Gross, "Bashful Eggs, Macho Sperm and Tonypandy.")

54. See N. Koertge, "Are Feminists Alienating Women from the Sciences?" as well as D. Patai and N. Koertge, *Professing Feminism.*

55. There have nonetheless been cases where scientific projects *have* been derailed by irritated racial sensitivities. One instance that comes to mind is the archaeological investigation of the African Burial Ground in New York City. The white archaeologists and physical anthropologists who had done the initial excavation were prohibited from doing work on the materials they had retrieved and eventually had them confiscated and turned over to black researchers, in an episode somewhat reminiscent of the Native American sabotage of archaeology. In another famous case, scientists planning a conference on the genetic basis of the propensity toward violence saw their federal funding withdrawn and their meeting canceled when black politicians and some white sympathizers complained that the project was inherently racist.

EIGHT *Education*

1. The source for these assertions is R. M. May, "The Scientific Wealth of Nations." Although this represents just one survey and one methodology, it is doubtful that different approaches would change the results significantly.

2. D. Goodstein, "Scientific Elites and Scientific Illiterates." Goodstein, an eminent Caltech physicist, is also one of the most successful science educators in history through the series of filmed lectures in elementary physics which he created. It is regularly seen throughout the world.

3. See C. K. Yoon, "Evolutionary Biology Begins Tackling Public Doubts," for an account of the response that is only now emerging from professional evolutionary theorists.

4. This is not to say that official government policy on immigration is never onerous; on the contrary, it can be a severe headache, particularly for young scientists

who want to remain in the United States after getting their doctorates. The real point is that scientific institutions, academic and otherwise, make few distinctions between native-born Americans and freshly arrived outlanders.

5. The eminent chemist Roald Hoffmann argues that the scientific establishment in the United States ought not to lure able foreign scientists to permanent positions here and should encourage foreign Ph.D. students trained here to go back to their native countries. "We must restrict scientific immigration to the United States so as to improve the chances of scientific and technical development of the broadest range of countries" (R. Hoffmann, "Movement of the People," 387).

6. Rutgers, the State University, New Brunswick, N.J.

7. The NSF's recent efforts to develop the talents of youngsters have, at least in some cases, taken some strange turns. I learned from a junior faculty member of my department, of an NSF-sponsored summer program in which he recently taught. The idea was to provide enriched mathematics education for high school students (by teaching them about some aspects of population dynamics, as it happens). However, this program was not designed to give a boost to the most promising students; in fact, students who are doing excellent work in high school are specifically *barred* from enrolling. Rather, the idea seems to be to boost the skills of average students who are getting B's in their college math courses. In my opinion, such a program has virtually no chance of nurturing serious careers in science—at least not in mathematics. At most, it will help a handful of unremarkable students improve their precollege grades somewhat. This is not a justifiable use of NSF's scarce resources. It is, in fact, lunacy.

8. Again, I offer myself as an example (somewhat overstated, as it may be). My three college roommates were, like me, intensely serious math majors. Today, all of them are professional research mathematicians of the highest distinction (a status which, alas, I must decline to award myself). That might have been carrying things to excess, but it was in keeping with the spirit of the times.

9. L. E. Beyer and D. P. Liston, *Curriculum Conflict: Social Visions, Educational Agendas, and Progressive School Reform*, 143. The authors appeal to recent "critics" of science, including not only the much-misunderstood Thomas Kuhn, but those with a radical feminist social agenda such as Sandra Harding, Evelyn Fox Keller, and Ruth Bleier. Note that this book is published by the press of the school of education of Columbia University, one of the leading centers of pedagogical theory in the country.

10. The uncertainties of the National Academy of Sciences—at least of its president, Bruce Alberts—over educational issues are reflected in a *Business Week* story (J. Carey, "Everyone Knows $E = MC^2$—Now Who Can Explain It?") on the efficacy of standardized testing, particularly as it is used to determine admissions to graduate programs in the sciences. There is much to be said about the inadequacies and unfairness of these tests (and also something in their defense as rough screening devices). However, I also detect in Alberts's remarks a subtle desire to find shortcuts for redressing the shortage of women and blacks in science, a willingness, if you will, to let "democratic" values prevail over "elitist" ones. The writer of the story is largely sympathetic to Alberts's point of view, a fact of some interest since *Business Week* is not ordinarily associated with strong support of affirmative action.

11. G. Holton, "Science Education and the Sense of Self," 554. Holton gives a useful account of this bizarre episode.

12. D. E. Drew, *Aptitude Revisited*. See also G. Campbell, "Raising Expectations," a favorable review that enthusiastically endorses a categorical end to all "tracking."

13. See S. Brenna, "Hard Lessons," for a depressing account of the effects of antielitist dogma on New York City high schools. See also A. Kohn, "Only for My Kid" (rebutted by J. M. Rochester, "What's It All About, Alphie?").

14. W. Lammers, personal communication (September 9, 1997). Lammers is the former dean of the medical faculty at the University of Groningen.

15. Ibid.

16. Some readers will doubtless point to affirmative action policies in medical school admission as an instance of violation of meritocratic principles. Although I would not wish to deny that affirmative action raises some difficult philosophical and ethical considerations, I don't think that in the context of medical schools such practices are violently antimeritocratic. There may be some disparity between ordinary students and those admitted on minority-preference grounds, but in either case we are talking about people of considerable ability; it would not be remotely fair to say that intellectual standards have simply been cast aside.

17. S. V. Rosser, *Teaching the Majority*, 6.

18. Ibid.

19. Ibid.

20. The most risible of these contributions is probably M. A. Campbell and R. K. Campbell-Wright, "Toward a Feminist Algebra," which, in a prepublication version, was criticized (not without asperity and a certain malicious delight) by P. R. Gross and N. Levitt, *Higher Superstition* (chapter 5).

The most curious is K. Barad, "A Feminist Approach to Teaching Quantum Physics," which propounds the thesis that the notorious subtleties of that subject will in some way be made more accessible to women if greater emphasis is placed on the philosophical views of Niels Bohr, one of the great founders of the subject but a man with a decidedly mystical bent and an inclination to dogmatism on certain mathematical issues where dogmatism is far from warranted. Barad's chief theoretical prop is that Bohr's fondness for subjectivism is somehow consonant with (as usual) "women's ways of knowing," posited as antithetical to a presumably testosterone-drenched objectivism. My own view is that this is more than a little silly; whatever mysteries still obscure the foundations of quantum mechanics, they have little to do with sexual politics or with (largely nonexistent) sexual differences in cognitive repertoires for dealing with physics. I would also object, contra Barad, that the problem with Bohr's so-called Copenhagen philosophy is not that it has had too little influence on physicists and on the way quantum mechanics is taught, but that it has had too much. See, in this regard, S. Goldstein, "Quantum Philosophy" and "Review Article"; J. Cushing, *Quantum Mechanics* (as well as the review in *Physics Today* of same by N. Levitt); D. Wick, *The Infamous Boundary*; and D. Albert, *Quantum Mechanics and Experience* and "Bohm's Alternative to Quantum Mechanics."

21. S. L. Webb, "Female-Friendly Environmental Science," 206.

22. L. Williams, "The National Science Foundation's Systemic Initiatives."

23. At the moment, my teaching duties involve acquainting senior and junior mathematics majors at a better-than-average state university with "rigorous mathematics." It is an uphill struggle.

24. Circa 1957.

25. Something called the Division of Educational Systemic Reform gets credit for authorship of this document.

26. Regarding the gap between the cultures of professional mathematics and of the theory of mathematics education, R. A. Raimi and L. S. Braden astutely remark: "Professors of education, even mathematics education, are members of a 'second culture' almost as distinct from the world of mathematics as C. P. Snow's literary culture was from that of his friends the physicists" (*State Mathematics Standards*, 4).

27. A number of sincere and competent science educators, urging reform in the

name of "constructivism," use this rather plastic term in its most unobjectionable sense. That is, they reject mere "rote learning" in favor of the kind of active intellectual engagement that can "construct" the mental machinery necessary to understanding science. (See, for instance, R. Firenze, "Lamarck vs. Darwin," where "construction" means something very much like acquiring the capacity to make serious evaluations of scientific arguments.) However, for concision's sake, I will use the word mostly to denote the more radical, "postmodern" version.

28. A. Cromer, *Connected Knowledge*. It is amusing to note that Cromer's title is an arch riposte to constructivist theorists, for whom "connected knowledge" (with a meaning altogether different from Cromer's) is a talismanic term.

29. See ibid. for a trenchant critique of "social constructivism." See, as well, P. R. Gross and N. Levitt, *Higher Superstition*; S. R. Cole, "Voodoo Sociology"; A. Chalmers, *Science and Its Fabrication*; N. Koertge, ed., *A House Built on Sand*; S. Haack, "Towards a Sober Sociology of Science"; and L. Wolpert, *The Unnatural Nature of Science*. For less censorious but still negative evaluations of social constructivism, see M. R. Matthews, *Science Teaching*, and I. Hacking, "Taking Bad Arguments Seriously."

30. M. R. Matthews, *Science Teaching*, 144. The reference is to R. J. Osborne and P. Freyberg, *Learning in Science*. Matthews, especially in chapter 7, gives a useful account of how social constructivism merges with psychological constructivism in underwriting constructivist pedagogical practice. He offers a chilling sketch of the swift and virulent growth of constructivist theory of science education, and shows how much the doctrine of "social construction of science" had to do with it.

31. Holton, "Science Education and the Science of Self."

32. See M. M. Atwater, "Social Constructivism," for an enthusiastic recommendation of social constructivist philosophy as a vital component of multicultural science education. See also M. A. Boudourides, "Constructivism and Education."

33. See E. Rothstein, "The Subjective Underbelly of Hardheaded Math," for an account of the attempt to introduce an "ethnomathematics approach" into the math classroom. Rothstein, who has done graduate work in mathematics, takes an appropriately skeptical view of this innovation. Remarking on the efforts of Reuben Hersh, a sometime mathematician, to promote the ethnomathematics approach, Rothstein remarks: "Mathematics may be dependent on culture, but it is also independent of it. And so far, neither Mr. Hersh, nor any other champion of ethnomathematics has provided examples of mathematical facts that are culturally relative." This seems accurate.

34. D. Pomeroy, "Science across Cultures," 257. Quoted in Matthews, *Science Teaching*, 185. Matthews, especially in chapter 9, is well worth consulting for an account of how multiculturalism dovetails with the epistemological relativism promoted by the postmodernists and social constructivists.

35. Fairness requires me to point out that there have been proposals for multicultural approaches to science education that reflect a good deal of common sense, as well as a passion for social justice. See, for instance, C. C. Loving, "Cortes' Multicultural Empowerment Model and Generative Teaching and Learning in Science," which takes a dim view of ethnochauvinist nonsense and is equally skeptical of the postmodern and relativistic dogmatics often used to justify it. Contrast W. W. Cobern, "Defining 'Science' in a Multicultural World," which is irresolute. What is at issue, right now, is whether a commonsensical multiculturalism can be introduced, in practice, without effectively opening the classroom to a flood of moonshine.

36. Powell and Frankenstein, *Ethnomathematics*; Adams, *African-American Baseline Essays*.

37. D. Jonassen, T. Mayse, and R. McAleese, "A Manifesto for a Constructivist Approach to Technology in Higher Education."

38. M. Osborne, "Call for Papers: 1997 Special Edition of *Journal of Research on Science Teaching*."

39. The phrase "authoritarian form of knowledge" is taken from A. Bettencourt, "They Don't Come Easy to Us," an essay which, leaking resentment at every pore, loses no opportunity to denounce science as odiously "hegemonic." Bettencourt's knowledge of science seems to derive solely from the musings of postmodern philosophers and their handmaidens in "constructivist" sociology of scientific knowledge—that their wisdom might be "hegemonic" in some circles seems not to trouble him. In any event, the references for his paper contain no actual, concrete works of scientific exposition, whether on the research or on the textbook level. All the science is at second or third hand, sometimes through the clouded perceptions of "constructivist" sociologists.

40. Ibid.

41. W. W. Cobern, "A Cultural Constructivist Approach to the Teaching of Evolution." (Essentially the same paper appears as "Belief, Understanding, and the Teaching of Evolution"; I quote from the electronically published version.) See B. J. Alters, "Should Student Belief of Evolution Be a Goal," for a contrasting view.

42. Ibid.

43. For instance, S. E. Anderson ("Worldmath Curriculum," 303) includes the following in his account of his instructional methods: "I further point out that calculus was created to facilitate the study of ballistics and the wars of land consolidation waged by England and Germany. I also show how the needs of the military . . . have inspired and continue to inspire many mathematicians and scientists to pursue the War Machine." Elsewhere in *Ethnomathematics*, M. Fasheh, a Palestinian radical, informs us that, "In short, I came to believe that the teaching of math, like the teaching of any other subject in schools, is a 'political' activity. It either helps to create attitudes and intellectual models that will in their turn help students grow, develop, be critical, more aware, and more involved, and thus more confident and able to go beyond the existing structures; or it produces students who are passive, rigid, timid, and alienated" ("Mathematics, Culture, and Authority," 286–287).

44. See J. Kronholz, "Numbers Racket," and J. Steinberg, "California Goes to War over Math Instruction," for an overview, with special emphasis on the California situation, of the controversy concerning the new "standards" for elementary mathematics education. See also R. L. Colvin, "Spurned Nobelists Appeal Science Standards Rejection," and P. R. Gross, "Science without Scientists," for an ironic tale of how a consortium led by Nobel Prize-winning chemists (including Glenn Seaborg) met with frustration when their offer to help draft standards for California science education (at no fee) were rejected, and the job was offered (at the stipulated price) to a group comprised largely of education school professors.

45. See E. Rothstein, "The Subjectivce Hardbelly of Hardheaded Math," for some observations on the ways in which multiculturalism and the ethnomathematics movement are linked to some of the California reforms in mathematics education.

46. J. Steinberg, "In New Math Standards, California Stresses Skills over Analysis" and "Clashing over Education's One True Faith." Note, however, that Raimi and Braden (*State Mathematics Standards*, 25) give the new California Standards their highest rating. See the California State Board of Education's 1997 "Mathematics Framework."

47. J. Price, interview, the *Roger Hedgecock Show*, April 24, 1996.

48. This criticism comes from a document, "The California Mathematics Framework," produced by the Palo Alto parents' group H.O.L.D., which fought the 1992 Framework.

49. Ibid.

50. A. Thompson, personal communication, August 13, 1997.

51. D. L. Kroll, J. O. Masingila, and S. T. Mau, "Cooperative Problem Solving: But What about Grading?"

52. Hostility to the very notion of tracking still pervades even generally well-thought-out proposals for math education reform. See J. Kronholz, "Low X-pectations."

53. J. Garofalo and D. K. Mtetwa, "Mathematics as Reasoning."

54. Here is another example (R. B. Corwin, "Doing Mathematics Together: Creating a Mathematical Culture"):

The Professional Standards for Teaching Mathematics (1991) recommend a shift in classroom environments—

- toward classrooms as mathematical communities—away from classrooms as simply collections of individuals;
- toward logic and mathematical evidence as verification—away from the teacher as the sole authority for the right answers;
- toward mathematical reasoning—away from merely memorizing procedures.

Note that the ideological element is here a little more visible than in the first example through the insistence on the communal nature of the classroom and the derogation of the teacher's role as authority.

55. Raimi and Braden, *State Mathematics Standards*, 11.

56. B. Evers, "Keep the Academic Bar High."

57. R. Askey, "More on Addison-Wesley *Focus on Algebra.*"

58. The formal designation for the committee is American Mathematical Society Association Resource Group. Its membership, in addition to Askey, includes a number of other mathematicians of the highest distinction: Roger Howe, Hyman Bass, and Bill Thurston.

59. R. Howe et al., "Reports of the AMS Association Resource Group," 271. See also R. Howe, "The AMS and Mathematics Education." The tone of these comments is diplomatic, but clearly they are based on serious misgivings about some of the more exotic aspects of the NCTM Standards.

60. See E. C. Scott, "The Struggle for the Schools," for a sobering account of the continuing strength of the movement to teach biblical creationism in public schools. D. U. Wise, "Creationism's Geologic Time Scale," provides an interesting example of the kind of work now necessary to refute a creationist movement that is more voluble than ever and which is now able to create a simulacrum of scientific expertise.

61. Much of the impetus for "Whole Math" and similar efforts came from the NSF, in particular the Directorate of Education and Human Resources, which issued the "Guiding Principle" cited previously. Says Abby Thompson (personal communication): "Looking up grants from the NSF-EHR division is very instructive. They have poured millions of dollars into math reform in the last 5–10 years. A lot of this is their fault." The triumph of the traditionalist reaction in California evoked instant fury from Luther S. Williams, one of the high NSF officials advocating the "Systemic Initiatives" (see n. 16) in which the "Guiding Principle" is to be found. (See J. Steinberg, "In New Math Standards, California Stresses Skills over Analysis.")

A quick search of the NSF Website reveals that at least forty-four recent awards in mathematics and science education invoke the term "constructivism." Here is the abstract of one successful grant proposal (A. Cuocco and W. Harvey, Investigators):

This project is building [*sic*] on previous research and curriculum development work to develop flexible understandings of topics in precalculus, calculus, and linear algebra. Examplary course materials will be produced using a modular approach to package the concepts and activities. Based on the broad mathematical themes, these materials will make use of constructivist pedagogies, involving cooperative learning, computer technology, and alternatives to traditional lecturing. Field testing of the curriculum will provide sites for research into the way students learn the topics and environments for teacher enhancement.

The amount of this award is $782,184. Whether any of it will contribute to any student's ability to learn about the Intermediate Value Theorem or Jordan canonical form for matrices is, to say the least, doubtful. That this sort of faddish educational theory gets funded at a time when NSF has put serious mathematical research on a starvation diet is infuriating beyond words.

62. See M. M. Jennings, "'Rain Forest' Algebra Course Teaches Everything but Algebra," and D. J. Saunders, "Math Gadfly Calls Math Faddists' Bluff."

63. Astute readers will have noted that since Hill's problem reduces to a quadratic equation, it has two solutions: one of these yields two hundred survivors, but the other gives zero, that is, the wretched Pequods were slaughtered to the last man, woman, and child on the Mystic. Two other problems from Hill's algebra book, whose true polemical intent is probably more transparent, read as follows:

A gentleman in Richmond expressed a willingness to liberate his slave, valued at $1000, upon the receipt of that sum from charitable persons. He received contributions from 24 persons; and of these there were 14/19ths fewer from the North than from the South, and the average donation of the former 4/5ths smaller than that of the latter. What was the entire amount given by the latter? (Answer: $950 came from Southerners.)

In the year 1692, the people of Massachusetts executed, imprisoned, or privately persecuted 469 persons, of both sexes, and all ages, for the alleged crime of witchcraft. Of these, twice as many were privately persecuted as were imprisoned, and 7 17/19ths as many more were imprisoned than were executed. Required the number of sufferers of each kind. (Answer: 19 executed, 150 imprisoned, 300 "persecuted privately.")

Those who are instinctively inclined to denounce Hill for his exploitation of Indian misery on behalf of Southern slaveholders might want to reflect on the fact that, following Appomattox, the most intransigent Confederate general—the last one to give up the fight—was a full-blooded Cherokee, Brigadier General Stand Waitie (June 23, 1865). See S. Foote, *The Civil War*, III, 1022.

64. See S. A. Garfunkel and G. S. Young, "The Sky Is Falling," for a brief account of the parlous state of K–12 mathematics education in this country.

65. The NCTM has been feeling considerable pressure from the professional mathematics community because of the questionable aspects of its "Standards." The forthcoming revision may reflect that pressure—at least it is to be hoped it will. S. A. Jackson's "Interview with Gail Burril." (Burril is the NCTM president and closely identified with the "Standards.") See also M. Gardner, "The New New Math."

66. It is sobering to reflect how much effort has gone into discouraging young people with mathematical talent—all in the name of "inclusiveness" (and, of course, anti-elitism) and how little effort it would take to encourage them. I note, by way of nostalgia, that my own intense involvement with mathematics arose from an encounter with George Gamow's *One, Two, Three . . . Infinity*, unmediated by teachers or lectures. I'm sure many mathematicians recall a comparable experience with this or another similarly worthy book. Gamow was a Nobel laureate in physics and one of the inventors of what came to be called the Big Bang theory in cosmology. I'm delighted

to note that his book is still available in a reprint. Also still available is Lancelot Hogben's *Mathematics for the Million*, another work that did heroic service in initiating thousands of youngsters into serious mathematical thought. Hogben, as well as being a mathematician and educator, was a principled leftist, and his book, as its title indicates, was meant to debunk the image of mathematics as an "elitist" mystery and to make it available to all classes and conditions of people. This did not, however, induce him to fill his book with either sloppy mathematics or ethnic chauvinism, and one may reasonably surmise that he would have been horrified at much that has been done in the name of "democratizing" mathematics.

Both books are discussed in L. Horvitz, "Science for the People."

67. Some recent works of this kind include M. Ruse, ed., *But Is It Science?*; P. Kitcher, *Abusing Science*; R. Dawkins, *Climbing Mount Improbable*. Local education departments, as well as science education associations, turn out a continual stream of manifestos explaining why evolution must be taught in science classrooms and why creationism, overt or disguised, may not be. The following statement from the Wisconsin Department of Public Instruction is typical.

Evolution, Creation, and the Science Curriculum.
1. Alternate scientific theories may be compared in the science classroom, but only those that best explain evidence which has been validated by repeated scientific testing should be accepted, and that only tentatively.
2. Years of intensive geological, biological, and other scientific studies have provided the most acceptable explanations of the origin and development of the earth and life on the earth. The theory of evolution has the general consensus of the scientific community because it integrates and clarifies many otherwise isolated scientific facts, principles and concepts in a manner which is consistent with known evidence.
3. Like any scientific theory, evolution remains subject to modification and revision as new evidence is discovered. Therefore, evolution should never be presented to students as absolute fact. Good teaching dictates that students be reminded of the tentative nature of conclusions resulting from scientific enquiry.

68. See T. Gitlin, *The Twilight of Common Dreams*, and M. Tomasky, *Left for Dead*.

69. I note the recent rebuffs to the field at the Institute for Advanced Study, where the proposed appointments of science studies gurus like Bruno Latour and M. Norton Wise met a harsh fate (see L. McMillen, "The Science Wars Flare at the Institute for Advanced Study").

70. This is done with particular crudity by ethnic chauvinists like Vine Deloria (*Red Earth, White Lies*) and Molefi Asante (*Afrocentricity*), but it also does duty for radical feminists, "queer theorists," and even so-called post-Marxists.

71. One can view this process at work in the fatally "Sokalized" volume of *Social Text*. See, again, works by S. Harding, H. Rose, S. Traweek, E. Martin, and A. Ross.

72. See P. E. Johnson, "What, If Anything, Hath God Wrought?"

73. I once enraged a prominent postmodern literary critic by remarking, "We can read your papers, but you can't read ours." No doubt this was arrogant and impolite, but it also contains a large measure of truth. Sokal's famous parody validates it, as does a new book, *Impostures intellectuelles*, by Sokal and J. Bricmont (another skilled physicist). This work, not at all parodic, remorselessly demolishes the pretensions (especially as regards science and mathematics) of a flock of postmodernist grandmasters. It is highly doubtful that their admirers will respond with a detailed critique of Sokal and Bricmont's achievements in mathematical physics.

74. Some prominent postmodernists have, in fact, crossed that narrow gap. The late philosopher Paul Feyerabend notoriously recommended the inclusion of biblical myth alongside science in public school curricula. More recently, the "deconstructionist" sociologist of science Sheila Jasanoff has endorsed the right of fundamentalists to impose "creation science" on the schools in areas where they predominate (S. Jasanoff, plenary lecture, AAAS, 1996).

75. Again, see T. Edis, "Relativist Apologetics," on this point.

76. In contrast to this view, see S. J. Gould, "Drink Deep, or Taste Not the Pierian Spring." Gould's piece claims that if we do it right, instilling scientific literacy in the adult population will be swift and easy. I think this perspective is more admirable for its generous spirit than for its insight.

77. M. H. Shamos, *The Myth of Scientific Literacy*, 229.

78. See, for instance, N. Augustine, "What We Don't Know Does Hurt Us." See also J. Trefil, "Scientific Literacy."

79. R. M. Hazen and J. Trefil, *Science Matters*, xi.

80. M. H. Shamos, *The Myth of Scientific Literacy*, 201.

81. R. M. Hazen and J. Trefil, *Science Matters*, xii.

82. The list of the supposedly real-life drama *Lorenzo's Oil* alongside science fiction films is not an inadvertence; *Lorenzo's Oil is* a science fiction film.

83. Consider, for instance, the well-known T. Pinch and H. Collins, *The Golem*, a celebrated essay in "science studies" whose subtitle is at one time false, fatuous, and enormously self-revealing.

84. N. Koertge, "Postmodern Transformations of the Problem of Scientific Literacy," comments on the degree to which postmodern fads prospectively sabotage the prospects for scientific literacy programs. See also the curious piece by P. Forman ("Assailing the Seasons"), which slanders James Trefil as a purveyor of "snake oil." Trefil's crime? To devise a college-level syllabus for the general science education of nonscience majors. Here I am obliged to declare an interest. Forman's comments occur in the context of a review (in *Science* of all places!) of *The Flight from Science and Reason*, which contains the offending Trefil article, and which I edited with Paul R. Gross and Martin W. Lewis. My rather severe response, along with Forman's rejoinder, appears in a subsequent issue of *Science*. The special fondness of some of the editors of *Science* for the idiosyncrasies of Dr. Forman is itself a curious instance of the ideological penetration of the whimsicalities of postmodernism into seemingly unlikely quarters. Forman is elsewhere ("Independence, Not Transcendence, for the Historian of Science") on record as opining that the duty of historians and sociologists of science is to act as hanging judges in condemnation of the arrogance of scientists and the evil effects of science.

85. J. D. McInerney, "Animals in Education," 277.

NINE Health

1. *Newsweek*, October 29, 1997. The cover announces a feature story by J. Leland and C. Power, which is supplemented by W. Kaminer's insightful essay, "Why We Love Gurus." See also P. Molé, "Deepak's Dangerous Dogmas."

2. Cynics might want to argue that a truly comprehensive ranking would have to take into account such luminaries of conventional religion as Billy Graham and the Pope. I am cynic enough not to be inclined to dispute the point too strongly.

3. One of Chopra's most successful books is a confection called *Quantum Healing*.

4. William Jarvis, of the National Council against Health Fraud, is apparently one such skeptic, according to *Newsweek*. But, as Chopra seems to be an energetically

litigious sort of fellow, it might be best to withhold explicit evaluation of the justice of this charge.

5. See W. Morse, "On the Inevitability of Miracles."

6. G. Kolata, "Taking Charge."

7. A. Allen, "Injection Rejection," 22. Allen also notes the appeal of anti-vaccination theories to ethnic chauvinists.

8. A list of alternative conferences, recently come to hand from the Internet, lists (among others): American Association of Naturopathic Physicians; Frontiers in Consciousness and Healing; "Miracles of Transformation: Transforming Ourselves and Transforming the World"; 25th Annual Cancer Convention—Alternative Therapies; Dynamics of Spiritual Healing; Body and Soul Conference; 3rd International Conference on Spirituality in Business; Insight and Opening—The Power of Breath and Meditation. These, of course, appeared after only the most superficial of Web searches, and presumably betoken hundreds more such events.

9. This includes Oxford, the outfit which currently covers me and my family. A year or so ago Oxford scored a considerable coup in the highly competitive managed-care business by trumpeting its willingness to reimburse alternative practitioners for its clients in the New York City area. See E. Gilbert, "Medical Advances": "While alternative care has always had a solid core of followers, its popularity has skyrocketed in recent years as health costs shot up and comprehensive insurance coverage has become more difficult to obtain. The holistic approach is usually less expensive and intrusive than mainstream medical treatments." See also T. Dajer, "Hearts and Minds," for a wry anecdotal account of the problems facing an orthodox physician with a seriously ill patient who insists on coddling a preference for "alternative" therapy. As well, see B. L. Beyerstein, "Why Bogus Therapies Seem to Work."

10. Political support for alternative medicine and other nostrums unsupported by scientific evidence easily spans the supposedly wide gulf between "left" and "right." See V. Stenger, "Senator Harkin Sees that OAM Gets a Fat Increase in Its Budget . . . While Senator Hatch Proposes to Limit FDA." Harkin, one of the Senate's most emphatic liberals, and Hatch, a staunch cultural as well as political conservative, seem to be on the same page here. See, as well, W. Sampson, "Antiscience Trends in the Rise of the 'Alternative Medicine' Movement," and also E. Gilbert, "Good Vibrations."

11. L. Jaroff, "The Solution Is Not in the Solution."

12. See J. E. Brody, "U.S. Panel on Acupuncture Calls for Wider Acceptance," as well as follow-up pieces, "Acupuncture: An Expensive Placebo or a Legitimate Alternative," and "Weighing Pros and Cons While the Jury Is Out." But see also V. Stenger, "NIH Sticks It to the Public":

> This latest outburst of pseudoscience is the product of the disreputable NIH Office of Alternative Medicine. Despite explicitly claiming that it was prepared by a "non-advocate, non-Federal panel of experts," alternative medicine advocates with a direct interest were included on the panel. On the other hand, activists who work against health fraud and other alternative medicine critics were excluded. . . . According to Bob Park of the American Physical Society, "When panelists criticized the Western medical practice of relying on randomized, controlled clinical trials, the audience burst into applause."

This is seconded by Brody in "Weighing Pros and Cons": "Their [acupuncture researchers'] reports . . . more than anything begged for longer, larger, and better controlled studies to document carefully what acupuncture can and cannot do." But such caution plays only a minor role in Brody's largely favorable reportage. See, as well, V. Stenger, "NIH 'Consensus' Statement on Acupuncture Condemned."

13. R. A. Schweder, "Ancient Cures for Open Minds."

14. See D. L. Wheeler, "From Homeopathy to Herbal Therapy, Researchers Focus on Alternative Medicine." See also B. Furlow, "Medical School Faith Healing."

15. For a profile of a prominent physician who has leaped into the "alternative" camp (with New Age overtones), see W. Roush, "Herbert Benson." For a sharp critique of Benson's book *Timeless Healing*, see I. Tessman and J. Tessman, "Mind and Body."

16. R. M. Poses and A. M. Isen, "Qualitative Research in Medicine and Health Care."

17. R. M. Poses, personal communication. Indeed, a pair (D. A. Stone and J. A. Rich, "Letter to the Editor") of respondents to the article of Poses and Isen (ibid.) underscores this point:

> Furthermore, the ideological consequences of such an argument for the hegemony of a single way of knowing are most disturbing. Their stance, as they [Poses and Isen] have stated it, fails to appreciate the importance of the move toward a more human science. The implications of the "naturalistic paradigm" are significant. It has allowed for an approach to complex human phenomena from the perspective of the research "subject," seeking understanding of phenomena rather than a singular "cause." Furthermore, by acknowledging the value-laden nature of inquiry and moving away from the notion of a single truth defined by a privileged few, it has opened the door to a diversity of perspectives including those of women and people of color, previously marginalized by the hegemony of positivism.

Obviously, this patch of postmodern boilerplate could easily have been borrowed from the cliché closet of a thousand trendy English departments, where childish political self-righteousness has been posing as philosophy for more than a decade. The unfortunate authors of this screed also perpetrate a number of other standard-issue solecisms in their few brief paragraphs. Clearly, they have no idea of what "positivism" means; they merely follow the practice of their tribe in using it as an all-purpose cuss word. (As M. Bunge ["What is Science?" 262] observes, "In short, positivism is not materialist but phenomenologist, and, as such it is closely linked to Berkeley's immaterialism.") They also follow long-established postmodernist practices in misappropriating Thomas Kuhn, grotesquely misappropriating W.V.O. Quine, and worst of all, blithely accepting the "findings" of the tendentious "sociology of science" practiced by such as Bruno Latour and Steve Woolgar.

See B. Held, *Back to Reality*, "Constructivism in Psychotherapy," and "The Real Meaning of Constructivism," for an account of how deeply postmodernist, constructivist, and antirealist ideas have already affected psychotherapy and clinical psychology.

18. L. Rosa et al., "A Close Look at Therapeutic Touch." See also news accounts: G. Kolata, "A Child's Paper Poses a Medical Challenge," and D. L. Wheeler, "Top Medical Journal Publishes Paper by 6th Grader Debunking Therapeutic Touch."

19. G. Kolata, "A Child's Paper Poses a Medical Challenge." The uneasy professor is Dr. Henry Claman of the University of Colorado.

20. One aspect of the embrace of therapeutic touch by nurses and nursing schools that has not drawn sufficient attention is the underlying psychological motivation. By setting up as independent "therapists," even under a crackpot theory, these nurses are attempting to redress the disparity between the exalted status of physicians as "healers" and their own traditional role as anonymous drudges and technicians, competent only to provide palliative care at best. Clearly, sexual politics has a role in this.

21. It should be noted, however, that certain nineteenth-century quasi-medical cults, like chiropractic and osteopathy, had entrenched themselves in many states with condign legislative permission.

22. The term "intern" has been universally replaced by "first-year resident."

23. A. Zuger, "After a Bad Year for A.M.A., Doctors Debate Its Prognosis," discusses the recent steep decline in A.M.A. membership and influence.

24. See R. J. Glasser, "The Doctor Is Not In."

25. See P. Klass, "Managing Managed Care," for a physician's articulate analysis of the situation. See, as well, P. T. Kilborn, "For Managed Care, Dial the Keepers of the Cures," for an account of how the "gatekeeping" process works in practice, and I. Oransky, "Losing Patients," for a view of how managed care is affecting the training of young physicians. S. G. Stolberg, "As Doctors Trade Shingles for Marquees, Cries of Woe," examines the subjective and financial consequences of physicians' "proletarianization."

26. J. E. Brody, "In Vitamin Mania, Millions Take a Gamble on Health," 1.

27. I don't want to overstate my sympathy for drug companies, which are quite obscenely profitable in any event.

28. See T. Kuhn, *The Structure of Scientific Revolutions*, for the strict meaning of Kuhnian "paradigm shift."

29. This question is raised in a recent *London Review of Books* essay (P. Campbell, "How to See inside a French Milkman"). It is answered in a subsequent letter to the editor (P. Taylor, "Inside a French Milkman"), but in a way that will not, I'm afraid, do much to illuminate the issue for the mathematically unsophisticated.

30. The basic idea comes from a theorem proved by the notable Austrian mathematician, Radon. I will not bother the reader with details but the general idea is that if f is a function on the disk, then it is possible to specify what f is if one knows the values of the line integrals of f for chords through the disk. Apropos of nothing in particular, I might mention that one of the most delightful jokes in Alan Sokal's famous hoax article, "Transgressing the Boundaries" (242, n. 53), exploited the coincidence that Radon's name is also that of a radioactive gas.

31. For whatever it is worth, let me intrude my own experience in a few situations where a member of my family faced a serious health problem. One of the first things I did in each instance was to consult the American Medical Association's most recent consensus symposium on the disease in question. These outline what is known about the condition in terms of treatment and prognosis. In no case did I find the material especially difficult to read. (Presumably, it helped that I was comfortable with statistical language.)

32. W. Gratzer, "In the Hands of Any Fool."

33. But see R. Wurtman, "Cure All," 16, where it is argued that the uncritical acceptance of the ethos of "curiosity-driven" basic science has actually retarded the development of practical modes of treatment for a variety of diseases. Says Wurtman: "During the last few decades, fewer and fewer scientists have received the kind of training in clinical research, involving live human subjects, that treatment discovery invariably requires. In particular, many young clinicians have decided to forego studies involving human subjects in favor of much safer ones in basic science laboratories."

34. Of course, physicians work under strict scrutiny until fairly well advanced in their careers, and even senior physicians may have to undergo review of their judgment when they work in hospitals or clinics. Even when they practice on their own, their right to admit patients to hospitals (as "attending physicians") can depend on the good opinion of their peers. The point, however, is that independent practitioners are under no obligation to submit to such oversight, which in any case leaves enormous latitude to the individual doctor.

35. At times, the supposedly scientific assumptions that pervade a culture are so distorted as to be an almost comical source of misunderstanding. One example,

America's obsession with "germs" and its endless quest for batericidal perfection, is amusingly analyzed in H. Rosin, "Don't Touch This."

36. Consider, for instance, the late philosopher Paul Feyerabend, a charming gadfly whose celebrity arose from his pugnacious challenge to the epistemological monopoly of Western science. Nonetheless, as John Horgan relates in *The End of Science*, when gravely ill at the end of his life, Feyerabend effectively abjured the relativism that had made him famous and clung unquestioningly to conventional treatment.

37. R. A. Schweder, "Ancient Cures for Open Minds."

38. M. Angell, *Science on Trial*, 95–96.

39. Ibid., 27.

40. Ibid., 208.

41. See J. A. Paulos, *Innumeracy*.

TEN Law

1. It will be interesting to see how long the Simpson affair lingers in public memory compared, say, to the trial of the accused Lindbergh kidnapper. My own opinion (for the little it's worth) is that the affair will fade rather quickly from the common mythology, simply because of the endless cascade of "high profile" cases with which the public is endlessly inundated, a torrent which brutally erodes recollection of what was notorious a few brief months before. (Of course, as a confirmed Civil War buff, I tend to believe that the Sickles-Key love triangle murder, the O. J. Simpson case of its day [1859], will be remembered forever, if only in certain specialized circles.)

2. My (utterly valueless) personal opinion is that the evidence, scientific or otherwise, had almost no bearing on the outcome of the case. The prosecution was doomed once the trial was transferred to central Los Angeles, and the jury was essentially packed with blacks susceptible to racially charged conspiracy mongering. Of course, this was a mirror image of the process that resulted in the acquittals in the Rodney King beating case by a suburban jury. See M. Schoofs, "Genetic Justice," for a brief but useful discussion of the forensic use of DNA evidence, with particular attention to its exculpatory power in rape cases, as well as the civil libertarian concerns it evokes.

3. Basic scientific and statistical principles certainly seem to be completely opaque to the eminent academic theorists who contributed to T. Morrison and C. B. Lacour, eds., *Birth of a Nation'hood*, a collection of essays united by their determination to demonstrate, through the sophistical magic of postmodernism, that Simpson was really the victim of a racist frame-up. It is particularly depressing that the responsibility for this parade of imbecilities falls chiefly upon Nobel laureate Toni Morrison, one of our most gifted fiction writers.

4. Independence of two variables, in the statistical sense, means that measurement of one does not affect the probable value of the other. Height and weight, for instance, are not independent; knowing that a man is taller than six foot two is sufficient grounds for concluding that he is more likely than the "average" fellow to weigh more than two hundred pounds. In the case of DNA evidence, the key question is whether observing a certain allele (i.e., variant of a gene) at a certain locus affects the probability of observing another allele at another locus. If, for instance, allele A occurs at locus X in only 2 percent of the population, and allele B occurs at locus Y for only 1 percent of the population, and if the two variables are independent, then only .02 percent of the population has A at X and B at Y. On the other hand, if they are strongly dependent, the probability of having both conditions is much greater. For instance, if B occurs at Y only when A occurs at X, then there is a 1 percent chance that a random individual will exhibit both traits.

5. On the other hand, courts dealing with scientific issues are also wont to tolerate the endless pursuit of trivia by attorneys who are eager for any opportunity to concoct a smoke screen of irrelevancies. This can turn the supposed effort to unearth the scientific truth into a barrage of pointless ad hominems against scientists. See P. M. Fischer, "Science and Subpoenas."

6. N. Augustine, "What We Don't Know Does Hurt Us," 1640. Augustine makes this point in the context of a protest against large civil judgments against corporations in defiance of the weight of scientific evidence. Since Augustine is a corporate executive (chairman of Lockheed Martin), as well as an engineering professor at Princeton, it might be objected that his remark is a form of special pleading. Nonetheless, I think it is essentially sound.

7. C. Haberman, "The Court Would Have Been Pleased." See H. Rothwax, *Guilty*, for the judge's own account of the plea-bargaining system and its abuses. Whatever his excesses, Rothwax was passionately devoted to the cause of bringing actual courtroom justice much closer to its theoretical ideal.

8. Again, see Rothwax, ibid., for some strongly worded and unsettling arguments for abolishing some of the restrictive evidentiary and procedural rules that have accumulated in the criminal justice system since the 1960s. The knee-jerk reaction of many civil libertarians will be to dismiss Rothwax as an embittered dinosaur, but his arguments (and the experiences that underlie them) deserve careful scrutiny and, if they are to be rejected, counterarguments based on more than unexamined truisms.

9. See "Bush to Pardon Byrd" (AP wire story), and, concerning the similar Salazar case, "Bush Pardons Man Convicted of Rape" (UPI wire story).

10. Another index of the subscientific nature of courtroom evidentiary standards is the enormous credit given to eyewitness testimony, despite decades of psychological research that shows that such testimony is often wildly erroneous, especially when it comes to the subsequent identification of strangers. The victim's famous witness-stand cry, "I'll never forget that face as long as I live!" is not merely a cliché but probably a fair indication that one ought to regard an identification unsupported by independent evidence as quite unreliable.

11. See F. Crews et al., *Memory Wars*. See also E. Loftus, "Remembering Dangerously," especially in regard to the reluctance of psychotherapists to conform their practices—and their "expert" opinions—to reasonable scientific standards of veracity. Fortunately, recent court cases penalizing "recovered memory" enthusiasts have made the tribe considerably more cautious. (See V. Stenger, "Courts Hitting Hard at Recovered Memory Therapists," and E. Loftus, "The Price of Bad Memories.")

12. J. Malcolm, *The Journalist and the Murderer*. This book was a scathing denunciation of the supposed manipulativeness and dishonesty of journalist J. McGinniss in his coverage (*Fatal Vision*) of the notorious "Green Beret Doctor" murder case where Jeffrey MacDonald was convicted of butchering his wife and young daughters. Sharp-eyed readers will note that Malcolm is a subtle but emphatic adherent of MacDonald's innocence, not least because her own Freudian training has instilled considerable confidence in her own ability to divine truthfulness.

It's noteworthy that nowhere in this book did Malcolm reveal her own ethical problems with apostate Freudian Geoffrey Mousaieff Masson, who was, at the time, suing her for misrepresenting him in a *New Yorker* profile.

13. M. Angell, *Science on Trial*.

14. The defendants in the various breast-implant lawsuits have failed to reap the benefits of the Supreme Court decision in *Daubert v. Merrell-Dow Pharmaceuticals*, which enjoined judges to decide on the admissibility of scientific evidence according to the

standards employed by the scientific community. The decision, however, contained many ambiguities, which have been worsened, rather than eliminated, by subsequent litigation.

Dow-Corning's decision to concede billions of dollars to women claiming that implants caused them serious health problems is evidence of the present difficulty of employing the Daubert principles. At best, the weighty evidence attesting to the implausibility of the claims will be incorporated into a kind of legal postmortem on the matter. See G. Kolata, "In Implant Case, Science and the Law Have Different Agendas," J. Kaiser, "Breast Implant Ruling Sends a Message," and S. G. Stolberg, "Neutral Experts Begin Studying Dispute over Breast Implants."

15. For the record, the term "electromagnetic radiation" comprehends all sorts of phenomena, from the feeble emissions generated by the low-frequency oscillations of household currents, through TV and radio broadcasts, the visible light that comes from flames and lightbulbs (as well as the sun), X rays, and incredibly energetic "cosmic rays." The known health dangers come from radiation energetic enough to have ionizing effects.

16. See chapter 7.

17. R. L. Park, "Power Line Paranoia," and G. Kolata, "Big Study Sees No Evidence Power Lines Cause Leukemia."

18. The microwave radiation used to broadcast phone signals is, of course, far more energetic than that associated with 60-cycle power lines. ("Energy" here means the energy of the individual chemical reactions that can be triggered by impinging radiation, not the total radiative output; this is the most appropriate way of gauging the possibility of bioactivity.) Microwave radiation (like FM radio and TV broadcasts) is millions of times more energetic than the 60-cycle variety, but, given the low flux of cellular phone transmitters, it is highly unlikely to cause damage in humans. Nonetheless, I can attest from personal experience that it is an article of faith among many intelligent and educated people that cellular phone relay antennas pose a deadly menace.

19. See L. Greenhouse, "Trial Judges Are Backed on Rulings on Scientists." See also B. Black, "Science and Beyond," for an extensive survey of case law in which the principles of *Daubert* have been applied. Black argues, in an *obiter dictum*, that the principle of severe scrutiny of "expert" testimony should be extended even to ostensible experts in areas outside of science. By contrast, D. W. Holman, "Of Bumblebees and Street Gangs," advocates a very narrow construction of the *Daubert* doctrine, limiting it to matters strictly scientific. Holman, it must be noted, is hostile to *Daubert* as such; but then, his familiarity with philosophy of science may be surmised from such statements as: "This scientific method was standardized in the 17th century by Sir Isaac Newton."

20. Greenhouse, "Trial Judges Are Backed on Rulings on Scientists."

21. S. Breyer, "The Interdependence of Science and Law."

22. Ibid. See also "Justice Breyer Calls for Experts to Aid Courts in Complex Cases" (AP wire story).

23. See A. Mazur, "The Science Court," for a brief history of the idea. See also T. G. Field, "The Science Court Is Dead," and A. Kantrowitz, "The Separation of Facts and Values."

24. P. Huber, "Coping with Phantom Risks in the Courts."

25. J. D. Miller, R. Pardo, and F. Niwa, *Public Perceptions of Science and Technology*, 38, characterize the science court idea as "long discredited." This is a curious lapse in a book that is generally useful, objective, and free of resentment against science.

26. See C. F. Cranor, "Science Courts, Evidentiary Procedures, and Mixed Science-

Policy Decisions," for a statement of this skeptical view. Cranor's position is only lightly tinged with postmodernism, but it does reflect a deep distrust of professional scientists, especially as they might be seen as obstacles to freewheeling democracy.

27. S. Jasanoff, "Procedural Choices in Regulatory Science."

28. See D. A. Farber and S. Sherry, *Beyond All Reason*, for an account (by two liberals!) of how what they designate as "radical multiculturalism" has been enthusiastically taken up by advocates of feminism and minority rights in some of the nation's most prestigious law schools. "Radical multiculturalism" is, perhaps, an unfortunate term. "Identity politics" might have been a better choice. In fact, more might have been said about how the kind of postmodernism that has taken hold in departments of literature, and so forth, feeds into the theories Farber and Sherry describe and decry. For discussions of identity politics in general, see M. Tomasky, *Left for Dead*, and T. Gitlin, *The Twilight of Common Dreams*. See also R. A. Posner, "The Skin Trade," a review of the Farber-Sherry book by a generally conservative jurist, as well as E. Blumenson, "Mapping the Limits of Skepticism in Law and Morals," for an attempt to refute the anti-foundationalism of postmodern legal "theory."

29. The reader is once more referred to recent work scrutinizing the constructivist view of science and, more generally, postmodern denigrations of scientific epistemology: P. R. Gross and N. Levitt, *Higher Superstition*; N. Levitt and P. R. Gross, "Academic Anti-Science"; N. Levitt, "Vetenskapens pessimism, antivetenskapens nihilism" and "Knowledge, Knowingness, and Reality"; G. Holton, *Science and Antiscience*; P. R. Gross et al., eds., *The Flight from Science and Reason*; N. Koertge, ed., *A House Built on Sand*; P. Boghossian, "What the Sokal Hoax Ought to Teach Us"; A. Sokal and J. Bricmont, *Impostures intellectuelles*; S. Weinberg, *Dreams of a Final Theory*; M. F. Perutz, "The Pioneer Defended"; L. Wolpert, *The Unnatural Nature of Science*. It is pleasant to report that this genre seems to be growing as fast as necessary to set to rights the misapprehensions engendered by the social constructivists.

30. S. Jasanoff, "An Epidemic Challenges American Biomedicine," 477.

31. A. Mazur, "The Science Court."

32. S. Jasanoff, *Science at the Bar*; P. W. Huber, *Galileo's Revenge*.

33. D. Kevles, in a generally favorable review of *Science at the Bar*, nonetheless observes, "Jasanoff tends to press her social-constructionist commitments to questionable lengths. Broad areas of mainstream science do exist, and the expert testimony rooted in many of them . . . appears to be imported into the courtroom in numerous cases without much, if any, conflict."

34. S. Jasanoff, *Science at the Bar*, 207.

35. P. W. Huber, *Galileo's Revenge*, 222–223.

36. S. Jasanoff has just been offered a position in Harvard's Kennedy School of Government, after having been a central fixture of Cornell's science studies program for years.

37. P. W. Huber, *Galileo's Revenge*, 96.

38. Ibid., 98.

39. The relevant case seems to be *Sterling v. Velsicol Chemical Corp.*, where the court elected to exclude evidence from "clinical ecology" that was not vouched for by legitimate medicine.

40. Here, the reference is to the *Velsicol* case.

41. S. Jasanoff, *Science at the Bar*, 134.

42. X. Rarden and S. Jasanoff, "An Atypical Complaint."

43. Kuhn's *The Structure of Scientific Revolutions* seems to have scriptural status within the postmodern ambit. Like scripture generally, it tends to be interpreted according to the heart's desire of the interpreter.

44. S. Jasanoff, *Science at the Bar*, 52.

45. See R. A. Meserve, "Review of *Science at the Bar.*"

46. M. Shamos, *The Myth of Scientific Literacy*, 209.

47. S. Jasanoff, *Science at the Bar*, 222.

48. S. Jasanoff, plenary lecture, Annual Meeting of the AAAS, Baltimore, February 1996.

49. The question was aked by E. H. Manier, a philosopher at the University of Notre Dame.

50. S. Jasanoff, plenary lecture.

51. This, or something very much like it, is presumably the credo of Professor Warren Nord of the University of North Carolina's program in Humanities and Human Values. Nord allied himself with religious conservatives in protesting the monopoly of Darwinian evolutionary theory in schoolroom discussion of the history of life. (See chapter 4, n. 15.) Even more, W. W. Cobern's surly attitudes toward teaching evolution (chapter 9) seem explicitly rooted in the same constructivist eccentricities as Jasanoff's.

52. See, for instance, S. Jasanoff, "Research Subpoenas and the Sociology of Knowledge": "'Truth' [sneer-quotes in original] emerges not because nature, when interrogated by the scientific method, unambiguously reveals the answers, but because discipline-based scientists agree, through complex processes of negotiation and compromise, how they should choose among different possible readings of observations and experiments." Likewise, S. Jasanoff, "Procedural Choices in Regulatory Science": "Skepticism toward science, however, is not unwarranted and researchers should not be granted full immunity merely because their work has satisfied their own professional community's standards of criticism and peer review." The latter statement contrasts interestingly with Jasanoff's sturdy faith in the factual accuracy of her own field's "discoveries." In the same paper, while dismissing certain truisms about science as false assumptions, she notes, "each of these assumptions is at odds with well-established findings in the sociology of science." Apparently, sociology of science is allowed to have well-established findings, but not high-energy physics. A brief examination of the citations given to support these putative "well-established findings" reveals the inevitable Thomas Kuhn, grossly misconstrued as usual, plus a couple of equally inevitable "case studies" of science at work. These latter efforts are showcase examples of what philosopher Susan Haack has shrewdly described as "fake inquiry" where the conclusion was well in place long before the "research" was done (S. Haack, "Concern for Truth").

53. If this seems like hyperbole, consider the demise of Dow-Corning as a result of the breast implant lawsuits, and the peril still threatening its associated firms.

54. Here, consider not only the breast-implant cases, but the awards given to Vietnam War veterans who claimed—honestly but mistakenly—to be suffering from the supposed effects of Agent Orange. In the latter case, the claims were decisively refuted by epidemiological evidence. However, the desire of Congress and the Bush administration to placate veterans' organizations on the eve of the Gulf War overcame mere evidence. For an accurate account of the danger—or lack thereof—associated with dioxin residues, see M. Fumento, *Science under Siege*.

ELEVEN *Journalism*

1. J. H. Tanne, in an address to the New York Academy of Sciences. I thank Ms. Tanne for her kindness in providing me with a written text of her talk.

2. J. Hartz and R. Chappell, *Worlds Apart*, 7.

3. A recent item in the "Hot Type" column of the *Chronicle of Higher Education* announces that one W.J.T. Mitchell, professor at the University of Chicago and editor of *Critical Inquiry* (hence, a major presence in cultural studies), is at work on a book about the role of dinosaur imagery in our culture. Professor Mitchell personally finds dinosaurs boring. One pities him.

4. See any issue of the newsletter *SETIQuest* (available from Helmers Publishing, Peterborough, NH) to get an idea of what interests the community of SETI afficionados. It is very "hard" science indeed. Those on the lookout for imaginative images of ravenous aliens will be greatly disappointed. On the other hand, it is fascinating to note how hard people are willing to work to develop a field that might well turn out to be empty of content.

5. T. Ferris, "Not Rocket Science," takes ironic note of the inaccuracy of the "science"—both content and method—presented in most "science fiction" films, even those that try to be "realistic."

6. Fortunately, the amount of money needed to keep the various SETI projects going is so small that it has been relatively easy to find private funding to make up for the cutoff.

7. Hartz and Chappell, *Worlds Apart*, 107.

8. In fact, Einstein at times seriously pursued practical "applied physics" by inventing a number of very clever devices, such as refrigerators, from which he hoped to obtain substantial royalties. His failure to cash in has more to do with the political turmoil of the time—the late 1920s—than with the viability of the ideas. If he had chosen to be an engineer, Einstein would still have been one of the titans of the century. See P. Galison, "Three Laboratories," and G. Dannen, "The Einstein-Szilard Refrigerators."

9. The OSI's hand, in the Baltimore case, was greatly strengthened by the support of Congressman D. Riegle of Michigan, whose antagonism to what he regarded as the scientific "establishment" was virulent.

10. After having been compelled to resign as president of Rockefeller University at the height of the "scandal," Baltimore was ultimately vindicated by his selection as president of Caltech in 1997. See D. J. Kevles, *The Baltimore Case*.

11. W. Broad and N. Wade, *Betrayers of the Truth*.

12. See D. Goodstein, "Conduct and Misconduct in Science."

13. J. Horgan, "Science Set Free from Truth."

14. J. Horgan, *The End of Science*.

15. Ibid., 93.

16. See N. Levitt, "*The End of Science*, the Central Dogma of Science Studies, Monsieur Jourdain, and Uncle Vanya," for a detailed analysis of why Horgan's brand of pessimism is, at the least, premature. See also, for unfriendly reviews, J. Casti, "Lighter than Air," D. Goodstein, "The Age of Irony," B. Hayes, "The End of Science Writing?" and L. Horvitz, "The Enemies of Science." For a suitably quirky encounter between two eccentrics, see B. Latour's review ("La fin de Science?") of *The End of Science*.

17. The editor in question is Katherine Livingston. The hubbub which led to her resignation after a lengthy career at *Science* was precipitated by an outpouring of protests, including mine, against an unusually ferocious review of a book (*The Flight from Science and Reason*) which I edited, along with P. R. Gross and M. W. Lewis. (See P. Forman, "Assailing the Seasons"). Suffice it to say that this review was but one incident in a string of editorial eccentricities reflecting a deep sympathy for certain strands of academic anti-science.

18. See, again, P. R. Gross and N. Levitt, *Higher Superstition*. See also P. Boghossian,

"What the Sokal Hoax Ought to Teach Us" and M. L. Perutz, "The Pioneer Defended," for interesting analyses of the growth of anti-scientism among certain academics.

19. R. Bell, *Impure Science*, 114. The author is, in fact, an academic but has written widely in newspapers and mass-circulation magazines.

20. D. Kevles, *The Baltimore Case.*

21. P. Brodeur, *The Great Power-Line Cover-Up.*

22. The rare journalist who conscientiously tracks down the exaggerations of environmental alarmists is not necessarily lionized by fellow journalists. See, for instance, M. Dowie, "What's Wrong with the *New York Times*'s Science Reporting?" a hammer-and-tongs assault on *Times* science writer Gina Bari Kolata for failing to keep faith with ecoradical doctrine on a laundry list of issues—silicone breast implants, low-frequency electromagnetic fields, food irradiation, and so forth.

23. T. Ferris, "Some Like It Hot," 16.

24. This provides some amusing moments for readers of D. Noble's *The Religion of Technology*. Noble is concerned to show, among other things, that the ethos of science is deeply linked to the Western religious viewpoint and its transcendental longings. He attempts to illustrate this premise by pointing to the fervently expressed religious sentiments of a number of astronauts. He has missed the point badly; not only are the astronauts he describes "scientists" in only the loosest and most undemanding sense, but their ostentatious religiosity is an index, not of the typical spiritual convictions of scientists, or even engineers, but rather of the effort government authorities expended to make sure that their highly publicized spacemen would not offend the sensibilities of the conventionally religious.

25. See S. Weinberg, *Dreams of a Final Theory*. See also E. Witten, "Matter Matters," for an account of the projects that have been devised to fill the gap by the demise of the supercollider, and a plea for ungrudging governmental support of these.

26. The program was predicated on commitments from a number of governments, including Russia, to fund and build part of the project. Russia's current economic prostration makes it likely that there will be a default in that quarter, which may provide a reason—or a long-overdue pretext—for canceling the monster outright.

27. "Once documentaries were the sole province of the network news departments. Today, this is no longer true. The new genre, labeled 'infotainment' or 'docudrama,' are most often produced by independent companies based in Hollywood. Needless to say, their values are not in concert with those of the network news departments" (Hartz and Chappell, *Worlds Apart*, 85). On Roswell, see Fox Network, "Alien Autopsy." On dinosaurs, see NBC, "The Mysterious Origins of Man."

28. ABC, "Nightline," February 7, 1996. The frightening thing is that the reporter chiefly responsible for this edition has a Ph.D. in physics, a fact which cannot but provoke strong suspicions of cynicism and mercenary motives.

29. C. Burr, "The Geophysics of God."

30. Methodological sins included arbitrarily changing the half-lives of radioactive elements and baselessly revising the thermal diffusivity of geological strata by a factor of 10,000. See M. Matsumura, "Miracles In, Creationism Out," for an exhaustive anatomization of the Burr piece and the actual claims of the geologist, John Baumgartner, which prompted it. It is clear that Burr's story was just as "cooked" as the data fed to Baumgartner's program.

31. Tanne, address to the New York Academy of Sciences.

32. W. Sampson, "Antiscience Trends in the Rise of the 'Alternative Medicine' Movement," 196.

33. An exception might be seen in the intense media coverage given Emily Rosa's

merciless demolition of "therapeutic touch" for her fourth-grade science fair project (subsequently published in the *Journal of the American Medical Association*). However, it is clear that the media were not drawn by the implications of the work itself, but rather by the novelty of a ten-year-old researcher.

34. I have already mentioned (chapter 5) the *New York Times*'s brief flirtation with psychic detectives.

35. The Moyers series exists in published form as a book of the same title. See G. Weissmann, "Sucking with Vampires," for a deservedly acerbic evaluation of Moyers's adventures in the realm of alternative medicine.

36. See D. Ruelle, *Chance and Chaos*, for a firsthand account from an authentic maverick who was largely responsible for a genuine "pardigm shift" (in this case, the adaptation of mathematical models from chaos theory in the study of turbulence in fluid flow). As Ruelle's account makes clear, although there was initial resistance from an established "orthodoxy," the resulting fireworks were rather muted, all in all, and the fuddy-duddies began to fade away with remarkable celerity after a few critical experiments endorsed the new ideas.

37. John Horgan, trained in literary studies, has been influenced to a modest extent by postmodern skepticism. (See J. Horgan, "Science Set Free from Truth.") Moreover, Katherine Livingston's heavy flirtation with postmodernism was involved in the controversies that led to her resignation as book review editor of *Science*.

38. One notes with some dismay that the recent rise of "science studies" raises the serious possibility that an academic background in history and philosophy of science (as those subjects are propounded in certain institutions) will be a flaw, rather than an asset, in a science reporter.

39. See P. R. Gross and N. Levitt, *Higher Superstition*, chapter 4.

40. The New York Academy of Sciences' publication *The Sciences* is a partial exception because of its artwork. Articles are often "illustrated" by contemporary paintings and sculptures chosen by editors who are au courant with the contemporary art world, and who have a sharp eye for resonances—and ironies.

41. M. Robinson, " Darwinism" (see chapter 4, note 10).

42. William Jennings Bryan seems to be Robinson's hero.

43. One might retort that such a colloquy was, indeed, at the heart of the recent Hollywood film *Contact*, which hypothesized a successful outcome of a SETI project—more than successful, in that it results in the visit of the scientist-heroine to an alien world. However, it may well be that the box-office failure of the film is traceable to its rather skittish and stagey treatment of the conflict between the heroine's atheism and the rather genial theism of the (theologically and politically liberal) clergyman with whom she becomes romantically involved. In any case, it seems unlikely that Hollywood will again go down this particular road anytime soon.

44. In one recent episode of *The Simpsons*, the precocious Lisa unearths what appears to be the skeleton of an angel during an archaeological dig. This provokes a nasty conflict between Lisa's own stubborn atheism, which seeks for a naturalistic explanation, and the rush to credulity which grips virtually all the other townsfolk of Springfield. It turns out that the "angel" was planted by a publicity-thirsty shopping mall. Stephen J. Gould, another noted atheist, makes a guest appearance in the episode. For me, at any rate, it was gratifying to see the sardonic vindication of unbelief, even in a cartoon world. Brief accounts may also be found in *Skeptic* 5, no. 4, and in M. Berman, "Skepticism in Action."

45. See F. Capra, *The Tao of Physics*, or M. Kafatos and R. Nadeau, *The Conscious Universe*, for examples of this kind of literature.

46. See J. D. Barrow and F. Tipler, *The Anthropic Cosmological Principle* or, at an even

more intense level of speculation, F. J. Tipler, *The Physics of Immortality*. See L. Smolin, *The Life of the Cosmos* (chapter 15) for a critique of the anthropic principle. See also J. Polkinghorne, *Belief in God in an Age of Science*, and G. Easterbrook, "Science Sees the Light."

47. See S. Goldstein, "The Flight from Reason in Science," for a brief but telling critique of how the attempt to endow quantum mechanics with spiritual resonances has led even some fine physicists to misunderstand the science itself. It is notable that the work of the mathematical physicist David Bohm, himself possessed of a mystical bent, decidedly undercuts the "subjectivist" interpretation of standard quantum mechanics.

48. M. Kafatos and R. Nadeau, *The Conscious Universe*, 10.

49. See S. Weinberg, *Dreams of a Final Theory*.

TWELVE Plutocracy

1. Here, in evaluating "quality" of scientific knowledge, I am applying what I take to be a standard based on the intrinsic interest and depth of what is learned, rather than its applicability or utility in the practical sense. Thus, in making this judgment, I am applying the standards of a rather old-fashioned idealism.

2. See A. Allen, "Mighty Mice," for a disconcerting example of this phenomenon.

3. See E. Bronner, "Voracious Computers Are Siphoning Talent from Academia."

4. See P. Galison, "Three Laboratories," and G. Dannen, "The Einstein-Szilard Refrigerators."

5. See J. Høyrup, *In Measure, Number, and Weight*, chapter 8, for an interesting historical and moral analysis of the relation between the military and mathematical research.

6. J. Ziman, *Prometheus Bound*, 142–143.

7. The full quote (cited in M. Cartmill, "Oppressed by Evolution," 81) derives from science historian Mario Biagioli: "Claims about the universality of science should be understood as a form of cognitive colonialism." Biagioli teaches at Harvard—a fact which this aging Harvard alum, at least, regards with considerable misgiving.

8. See A. Sokal and J. Bricmont, *Impostures intellectuelles*.

9. This, perhaps, should be qualified by noting the strongly anti-Darwinist tendency in left-populist ideology in the early part of the century, which derived from a facile equation of evolutionary theory with Social Darwinism and eugenics, and therefore with the ethos of dog-eat-dog capitalism.

Inasmuch as I have been accused in certain quarters (see, for instance, A. Ross, ed., *Science Wars*, a somewhat modified book version of *Social Text* 46/47) of laying the cutbacks in science funding at the door of academic science studies, let me make clear that as it now exists, this rather minuscule faction has very little power (thank goodness!) over science policy. In fact, I would not even allege that the anti-science mood within the liberal community is due to its having been tainted with ideological conceits diffusing out of science studies. On the contrary, the rather ambiguous and unfocused anti-science sentiments that waft through the large corpus of left-liberalism may be said to have crystallized in particularly sharp and truculent form in the venue of radical science studies.

10. Indeed, the "boom" of the Reagan years was driven by huge federal deficits; it was the consequence of an elephantine version of Keynsianism.

11. This needs a little qualification. Right-wing Texans had no trouble supporting the supercollider since the lion's share of the funding was to go to that state. Moreover, the traditional right-wing coalition came together to support the scientifically

worthless space station project against a serious attempt to abort it; here, the explanation is that the space station stands to benefit greatly those military-oriented corporations with which the right has always maintained cordial relations.

12. R. Sclove, "Science, Inc., versus Science-for-Everyone." It is depressing to note that one recommendation for altering government funding so that it supports "science-for-everyone" is to increase the appropriation for the egregious Office of Alternative Medicine!

13. T. Kealey, *The Economic Laws of Scientific Research*, 330. See R. R. Nelson, "A Science Funding Contrarian," for a perceptive review of Kealey's curious tome.

14. Kealey, *The Economic Laws of Scientific Research*, 265.

15. Ibid., 217.

16. It is noteworthy that *The Economic Laws of Scientific Research* makes no mention of either mathematics or quantum mechanics.

17. According to *Issues in Science and Technology* (Fall 1997), the planned increases for NSF in fiscal 1998 (above the preliminary proposals) come to 8.6 percent, or 6.6 percent above 1997 levels; for NIH, there is a 7.5 percent upward revision, 6 percent more than 1997. Figures for other agencies are roughly similar. (Unfortunately, part of the increase is for the space station.) See also C. Cordes, "Senators Back Increase in NSF Authorization." But squabbles over funding policy hold these increases hostage as of the spring of 1998; see P. W. Campbell, "Big Increases in Federal Budget for Science Appear Unlikely." See also J. H. Moore, "The Changing Face of University R&D Funding."

18. Among the favorites are Philip E. Johnson, Michael Behe, and the quondam mathematician, David Berlinsky. As I write this (literally), these gentlemen, along with William F. Buckley, are engaging in a debate with defenders of evolution (Eugenie Scott, Michael Ruse, Barry Lynn, and Kenneth Miller) on Buckley's *Firing Line* show. I must also note the praise bestowed on "intelligent design theory" by the doughty Robert Bork in *Slouching toward Gomorrah*. See comments of P. R Gross ("Downsizing Darwin"). That book also contains some measured praise for a book of ours (P. R. Gross and N. Levitt, *Higher Superstition*), which I am forced to regard as more embarrassing than gratifying.

19. See M. M. Scott, "Intellectual Property Rights," which considers how corporate sponsorship of university-based research might constrict the freedom of researchers and subvert the ethic of open publication. See also A. Allen, "Mighty Mice," and D. Zalewski, "Ties That Bind."

20. See S. S. Hall, "Success Is Like a Drug," for an account of one such biotech company.

21. See, e.g., J. Ellis, *Literature Lost*.

22. A few years ago, for instance, the University of Rochester proposed ending its graduate program in mathematics, despite, or even because of, having a distinguished faculty in the subject. The thinking may have been that, without graduate courses to teach, many of the more accomplished—and more expensive—professors would drift off to other institutions, leaving a "service-oriented" department composed of teachers without research commitments. In any case, a vigorous campaign from the professional mathematics community obliged the university to reverse its decision.

23. Earlier this decade, there was a serious government attempt to limit the publication of research in mathematics that might be useful for creating—or breaking—encryption schemes. By and large, these efforts failed in the face of the strong commitment to open research and publication.

24. S. G. Stolberg, "Gifts to Science Researchers Have Strings, Study Finds."

25. Ibid. The victim of corporate censorship in this episode was Dr. Betty Dong of

the University of California at San Francisco. The implications are all the more chilling when one realizes that UCSF is one of the most important and prestigious medical research institutions in the world.

26. See G. Blumenstyk, "Conflict-of-Interest Fears Rise as Universities Chase Industry Support."

27. See M. Lerner, "Researcher Investigating Toxin Becomes Subject of Investigation."

28. D. A. Swackhamer and R. A. Hites, "Reducing Uncertainty in Estimating Taxophene Loading to the Great Lakes," EPA Grant No. R825246.

29. On low-frequency radiation, see P. Brodeur, *The Great Power-Line Cover-Up*, where, as the title suggests, the author frequently questions the motives of those scientists who contradict his allegations, especially (see, e.g., p. 251) when they take research money from the electric power industry. On implants, see M. Angell, *Science on Trial*, especially pp. 143–146, which discusses the attempts of a plaintiff's attorneys to impeach a Mayo Clinic study exonerating silicone breast implants by claiming that a remote financial connection with Dow-Corning money had corrupted the researchers.

30. See B. Meier, "A Silent Witness in Cigarette Trials," for a portrait of one such miscreant.

THIRTEEN *Democracy*

1. See J. Harr, *A Civil Action*, for some fascinating examples of this dubious art. The larger point of this book, however, is that corrupt lawyers and crooked judges, rather than prevaricating scientists, present the most serious threat to justice.

2. Sometimes these misgivings are presented as actual philosophical doubts as to the validity of science as a way of knowing—this is the tactic of "social constructivists" and other critics of science influenced by postmodern fashions. However, when pressed to the wall, such people will usually concede through gritted teeth that modern science really does deserve most of the epistemological supremacy claimed through it. At that point, it becomes clear that their real concern is whether the unrestricted sway of scientific fact is really consistent with human happiness for the great majority.

3. The late Paul Feyerabend (*Against Method*) is famous—or, depending on one's point of view, notorious—for advancing just such an argument.

4. S. Traweek, "Unity, Dyads, Triads, Quads, and Complexity," 137.

5. See, for instance, K. L. Feder, "Indians and Archaeologists: Conflicting Views of Myth and Science."

6. See V. Deloria, *Red Earth, White Lies*.

7. See, e.g., H. H. Adams, *African-American Baseline Essays*, or I. Van Sertima, ed., *Blacks in Science, Ancient and Modern*.

8. See, again, A. Powell and M. Frankenstein, *Ethnomathematics*, for examples of such schemes, as well as a great deal of unfiltered Afrocentrism.

9. There are of course exceptions to this observation. As noted, S. Jasanoff (see chapter 12) seems to approve of the case I have just put on behalf of classroom creationism, although she is clearly personally identified with the academic left and appears to take seriously the epistemological position of the social constructivist theories of science. Also note Marilyn Robinson's left-wing "creationist" position (see chapter 4).

10. This view is argued intelligently and cogently in C. Taylor, *Multiculturalism*.

11. See T. Gitlin, *The Twilight of Common Dreams*, and M. Tomasky, *Left for Dead*, for critiques (from the left) of identity politics.

12. See Kealey, *Economic Laws of Scientific Research*, for a scientist's endorsement of this kind of latitudinarianism.

13. Traditionally, of course, the left has been an ardent defender of unrestricted free speech rights, but the advent, mostly on university campuses, of "political correctness" has put that commitment into question for many.

14. Unfortunately, despite the cheerleading of the film, the supposed treatment discovered by the real-life counterparts of the film's heroes turns out to be of little use for the condition.

15. See F. Lentricchia, "Last Will and Testament of an Ex-Literary Critic," or T. Eagleton, *The Illusions of Postmodernism*, for symptoms of the crack-up in the postmodernist consensus among leftist literary critics.

16. For a grimly comical account of the mare's nest of fantasies that can ensnare a more or less typical rural community, see M. A. Churchill, "A Skeptic Living in Roswell."

17. W. Lunch, "Contemporary Attacks on Science." For historical background, see J. C. Burnham, *How Superstition Won and Science Lost*.

18. It might be conjectured that the clotted rhetoric and sodden density of postmodernist scholarly prose is in part a compensatory mechanism for the guilt of having abandoned depth and complexity in the thought and art where they have traditionally been located. The apotheosis of the postmodern mood is a cultural-studies monograph that celebrates utter fluff in impenetrable terminology.

19. But see K. Zoeteman, "Middle Earth in the Balance," for a defense of an alternative epistemology in dealing with environmental problems. In this interview, Zoeteman, deputy director-general for the environment in the government of the Netherlands, declares his belief in gnomes, elves, and various other species of fairy-folk: "Gnomes are a few decimeters tall. Elves are very small, a few centimeters. A landscape angel is meters tall. It's a glowing being of various colors and it uses energy to maintain nature's growth processes."

20. "Objectivity," especially as the term is applied to scientific findings, is a term that maddens many current theorists of the "social" nature of science. In the face of all that, it is vital to insist that the term is meaningful and often, in the context of actual scientific findings, accurate.

21. Obviously, this general principle excludes cults or movements that behave in a flagrantly illegal or dangerous manner, or which terrorize their own members. But this is a side issue.

22. M. Walzer, "Pluralism and Social Democracy," 48.

23. Ibid., 52.

24. D. H. Guston and K. Kenniston, "Updating the Social Contract for Science," 65.

25. See J. D. Miller, R. Pardo, and F. Niwa, *Public Perceptions of Science and Technology*, for a useful survey of the actual degree of "scientific literacy" (under various definitions) to be found in industrial societies. See M. Shamos, *The Myth of Scientific Literacy*, for a sharp critique of the notion. See also M. Bunge, "The Popular Perception of Science in North America."

26. J. M. Bishop, "Enemies of Promise," 21.

27. The fate of the proposal to ban or cripple much genetic research in Switzerland (it was voted down two to one) is more encouraging, but not without grim implications. Another way of looking at it is that a swing of just 16 or 17 percent of the vote would have reversed the outcome. See M. Shields, "Swiss Vote Down Bid to Curb Genetic Research."

28. It is interesting to note that in Europe, paranoia over genetically modified food crops seems stronger and better organized than in the U.S. (See M. Specter, "Europe,

Bucking Trend in U.S., Blocks Genetically Altered Food.") Of course, the situation is exactly opposite with respect to teaching of evolutionary theory.

29. A documentary broadcast in the Public Broadcasting System's *Frontline* series did a fine job of illuminating the degree to which continuing opposition to nuclear power has escaped the compass of serious scientific and technological debate.

30. S. Harding, *The Science Question in Feminism* and *Whose Science? Whose Knowledge*, as well as "After the Neutrality Ideal" and "Science is Good to Think With." See critiques of Harding in P. R. Gross and N. Levitt, *Higher Superstition*, and C. Pinnick, "Feminist Epistemology." H. Longino, *Science as Social Knowledge*.

31. See, e.g., S. Aronowitz, *Science as Power*, or A. Ross, *Strange Weather* and "Introduction to *Science Wars*."

32. R. Sclove, "Science, Inc., versus Science-for-Everyone."

33. It is interesting that part of Sclove's justification is that the supporters of alternative medicine include many people with college educations. Unfortunately, this says more about the questionable nature of science education at the college level than about the validity of alternative medicine. No mention is made of this aspect of democratization in Sclove's guest editorial for *Science*, "Better Approaches to Science Policy."

34. See N. Levitt and P. R. Gross, "The Perils of Democratizing Science," for additional observations on this point.

FOURTEEN *Authority*

1. E. O. Wilson, "Science and Ideology."

2. Whether and to what degree this respect is eroding is still rather unclear. Poll data (National Science Foundation, Science and Engineering Indicators 1996, 7–16) indicate that enthusiasm for science among Americans has held steady at a strong but not overwhelming 40 percent for two decades. On the other hand, many scholars, taking account of more qualitative social indicators, trace a definite decline (with varying degrees of alarm or enthusiasm). (See, for example, D. Nelkin, "Science, Technology, and Political Conflict," or L. Marx, *The Pilot and the Passenger*, chapter 16. For contrast, see also J. Berman and D. J. Weitzner, "Technology and Democracy.") A recent survey of European attitudes (INRA [Europe] and Report International, *Europeans, Science and Technology*, chapter 3) reports largely positive views of science among respondents, outdoing the figures for the USA. Europeans and Americans seem comparably knowledgeable about science—except for questions involving evolution! (Compare chapter 2 of the INRA survey with an exactly comparable National Science Foundation survey summarized in K. A. McDonald, "Public's Interest in Science Is at an All-Time High, but Knowledge Still Poor, Survey Shows."

3. S. J. Gould, "The Great Asymmetry," 812.

4. See A. Stanley, "Power to Fascinate Intact, Shroud of Turin is Unfurled." The shroud has been revealed as a medieval artifact not only by carbon-14 dating (by several labs, working independently under well-blinded conditions, which came up with the same figure), but by the presence of a number of substances in common use as artists' pigments in the Middle Ages. The difficulty of our "newspaper of record" in accepting these simple facts without hedging is grimly revealing.

5. See R. Dawkins, "Science, Delusion, and the Appetite for Wonder," for a sobering account of the effects of this resentment on a number of well-known writers and journalists.

6. The recent film *Good Will Hunting* plays amusingly with these archetypes. In its depiction of a mathematical prodigy working as a janitor in the halls of MIT's math department, it presents Prospero quite convincingly disguised as Caliban.

7. This may strike some readers as a cruel and unfair exaggeration; but see D. Patai, "Why Not a Feminist Overhaul of Higher Education?" which describes a feminist master (mistress?) plan for the lustration of our familiar educational and scholarly practices, the central theme of which is that power to set up and cast down is to be placed in the hands of Women's Studies departments. Some major institutions in the Northeast appear to have given at least preliminary approval to this idea (although its ultimate fate is quite uncertain). Of course, if this scheme is put into effect, even in modified form, the sciences on the affected campuses are likely to be in a tough spot, since they are notoriously the strongholds of objectivist resistance to the touchy-feely epistemology so prized in the feminist academy. (I point out that despite the title of her piece, which might create the opposite impression, Patai is a resolute opponent of the proposed "overhaul.")

See also J. P. Diggins, *The Rise and Fall of the American Left*, J. M. Ellis, *Literature Lost*, and A. Kernan, ed., *What's Happened to the Humanities?* for useful discussions of how the current clerisy of self-ascribed political virtue came to be a fixture of the academy. A short, and very amusing, anatomization of the phenomenon may be found in F. Kermode, "Toe-Lining." See N. Levitt, "What's Post Is Quaalude," for my own brief views on how and why the cult of postmodernism, with its inane view of science, came to play so large a role in academic life.

8. An example, recently come to hand, is "Postmodernity's War with Science," chapter 8 of J. Natoli, *A Primer to Postmodernism*. Likewise, J. Dean's absurd book *Aliens in America* (p. 8) praises public distrust of science in strident pseudo-leftist tones: "Given the political and politicized position of science today, funded by corporations and by the military, itself discriminatory and elitist, this attitude towards scientific authority makes sense. Its impact, moreover, is potentially democratic."

9. For the record, I here use the term "science studies" to refer to its epistemologically (and politically) radical wing, which may be somewhat unfair to scholars who take a much more moderate, pro-science view yet still regard themselves as members of the "science studies" community. Yet I think that the fact that the radical version has been able to dominate the conversation to such a degree justifies my usage, which has become commonplace. See S. Cole, "Voodoo Sociology," for an account of how "social constructivism" came to occupy its dominant, indeed, monopolistic, position in the field.

10. See C. Geertz, "The Legacy of Thomas Kuhn," for a typical example of the antipathy of social science (in its postmodern mode) toward the hard sciences.

11. Beyond mere discontent, Goethe was led to make extravagant claims for his own "scientific discoveries" and, indeed, for his own radically original methodology. Most pathetically, he claimed to have disproved the composite nature of white light—a notion singularly odious to his philosophy. This was part of his general disdain for Newtonian science, which, in Goethe's view, had to be wrong merely because it was so unpleasantly mathematical.

12. C. James, "Sex and Reason," 104.

13. E. O. Wilson argues in "Resuming the Enlightenment Quest" (excerpted from his new book, *Consilience*) that the increasing commonality of the natural sciences and the growing philosophical and methodological unanimity of their practitioners presages the unification of knowledge in an even broader sphere, including the social sciences. His philosophy here is unabashedly reductionist, which may seem a little strange to those who remember the bitter wars between the "naturalist" Wilson and the "reductionist" hotshots of the new molecular biology (but time has a way of repairing such breaches). In his response, Richard Rorty, in "Against Unity," argues, as did Paul Feyerabend before him, for epistemic and ontological pluralism and for an

end to science's insistence on sovereign exclusivity in the realms where it has entered a judgment. But behind Rorty's pleas, one discerns not a body of new insight that might qualify the primacy of science, but increasing dismay at the growing power, sweep, and consistency of the scientific worldview. In other words, Rorty is alarmed because science has advanced so far in its program, not because it has stalled. See also the comments by P. R. Gross, "The Icarian Impulse," as well as S. Boxer, "Science Confronts the Unknowable."

14. Trendy denigration of science on the part of the "postmodern left" has, of course, been heavily criticized by many ardent leftists as a betrayal of progressive ideals and as a self-defeating retreat into campus hermeticism. See, for instance, M. Albert, "Science, Post-Modernism, and the Left," or A. Sokal, "A Plea for Reason, Evidence, and Logic."

15. I speak from a certain amount of experience as a scientist who has been involved in a number of these face-offs—and also, I confess, with a certain amount of braggadocio. I hope the reader will forgive me for the latter.

16. A current favorite is that one need not know science well at all, merely the theories of the sociology of scientific knowledge, in order to participate in public debate about science and technology issues. "I believe," says one prominent proponent of this view, "that most of what non-scientists need to know in order to make informed public judgments about science falls under the rubric of history, philosophy, and sociology of science, rather than the technical content of scientific subjects" (S. Fuller, *Science*, 10). This, it seems to me, is rather like saying that in order to drive a car, one need only read the business pages to see how the manufacturer is doing; the functions of gas pedal, brakes, and steering wheel are irrelevant. This is also the point of view taken in T. Pinch and H. Collins, *The Golem*. Perhaps the attitude advocated by these thinkers has already penetrated the educational establishment. At any rate, this is suggested by a cartoon reprinted in J. Hartz and R. Chappell, *Worlds Apart* (p. 64):

Schoolkid: Our new science curriculum is teaching us the philosophical ramifications of science. We're learning the limitations of science and the appropriate circumstances for its use, along with the interdependence of science and cultural institutions.

Dad: Wow! I guess you won't need to come home and ask *me* any questions . . .

Schoolkid: Actually, there was one thing I've kinda been wondering . . . what's science?

17. An instance of this kind of pseudosophisticated raving is found in another remark of S. Fuller (*Science*, 18): "Finally, we must ask whether the collusion in the various branches of a science, such as physics, somehow corrupts the entire scientific enterprise. At the very least, scientists can be charged with hyperbole in the import they assign to their agreements (which is no doubt useful in securing continuing funding)." Fuller, it must be pointed out, is not a fringe figure in the science studies community, but a voluble and well-known leader in the field. He cranks out thousands of pages along much the same lines as illustrated. No doubt, he finds it useful in securing continuing funding, which comes his way with remarkable munificence.

18. For an example, see the discussion of S. Jasanoff in chapter 11.

19. See N. Koertge, "Are Feminists Alienating Women from the Sciences?"

20. See, for instance, B. Ortiz de Montellano, "Afrocentric Pseudoscience," or M. Nanda, "The Science Question in Postcolonial Feminism," "Is Modern Science a Western, Patriarchal Myth?" and "Modern Science and the Progressive Agenda."

21. See chapter 9.

22. J. B. Elshtain, "Authority Figures," 11–12.

23. A premonition of the consequences of abandoning any notion of social authority

is spelled out, somewhat humorously, in a description of the fringe of the British radical ecology movement (J. Griffiths, "Diary"): "They have unlearnt centuries of Western history, and learnt an aboriginal bewilderment at the very concept of owning land or owning anything." No doubt this vision has a certain amount of appeal—not too much, one hopes—in the contemporary United States as well.

24. E. Rothstein, "Where a Democracy and Its Money Have No Place."

25. G. Holton, "Address, AAAS Fellows' Forum."

26. See F. Dyson, "Science in Trouble," for a prominent scientist's meditation on these issues. See also M. A. Hammerton, *Science under Siege*.

27. A. Phillips, "You Have to Be Educated to Be Educated," 24.

28. C. T. Rubin, "The Troubled Relationship between Science and Policy."

29. For a very mildly "Comtean" or "scientistic" political program, see M. Bunge, "Technodemocracy."

30. Or perhaps for a not-so-imaginary present. E. O. Wilson, *Consilience*, seems to flirt with the idea, though in a thoroughly thoughtful and undogmatic manner, more interested in suggesting that social science might become much more scientific than in urging a takeover of that discipline by natural scientists.

31. K. Popper, *The Open Society and Its Enemies*, ii, 284, n. 6.

32. M. Holub, *Shedding Life*, 190.

33. K. Popper, *The Myth of the Framework*, 110.

34. One must reckon with the possibility that "tort reform" may take shape as a symbolic fight between Everyman and the megacorporation. Opposition will probably be orchestrated by a lobby of legal opportunists for whom any gimmick used in pursuit of a jury's sympathy is legitimate science. But it might also turn out that proposed "reform" gives short shrift to good science as well as bad.

References

ABC. "Nightline." Broadcast, February 7, 1996.

Adams, Hunter Havelin III. *African-American Baseline Essays—Science Baseline Essay: African and African-American Contributions to Science and Technology*. Portland, Ore.: Multnomah School District IJ, Portland Public Schools, 1990.

Admiraal, Pieter, et al. "Declaration in Defense of Cloning and the Integrity of Scientific Research." *Free Inquiry*, 17, no. 3 (1997), 11–12.

Adorno, Theodor W., et al. *The Positivist Controversy in German Sociology* (trans. G. Adley and D. Frisby). New York: Harper Torchbooks, 1976.

Albert, David Z. *Quantum Mechanics and Experience*. Cambridge, Mass.: Harvard University Press, 1992.

Albert, David Z. "Bohm's Alternative to Quantum Mechanics." *Scientific American*, May 1994, 58–63.

Albert, Michael. "Science, Post-Modernism, and the Left." *Z Magazine*, July/August 1996, 64–69.

Alcock, James E. "The Propensity to Believe." In Gross, Levitt, and Lewis, eds., *The Flight from Science and Reason*, 64–78.

Allen, Arthur. "Injection Rejection." *New Republic*, March 23, 1998, 20–23.

Allen, Arthur. "Mighty Mice." *New Republic*, August 10, 1998, 16–18.

Alters, Brian J. "Should Student Belief of Evolution Be a Goal?" *Reports of the National Center for Science Education*, January/February 1997, 15–16.

Ames, Bruce N. "Does Current Cancer Risk Assessment Harm Health?" Lecture, Washington Roundtable on Science and Public Policy, May 2, 1995. Electronically published, 1995.

Anderson, S. E. "Worldmath Curriculum: Fighting Eurocentrism in Mathematics." In Powell and Frankenstein, eds., *Ethnomathematics*, 291–306.

Angell, Marcia. *Science on Trial*. New York: W. W. Norton, 1996.

Anonymous. "Grouping Practices." *Pathways Critical Issues in Mathematics*. Electronically published, 1997.

Anonymous. "The Simpsons Debunks Angels." *Skeptic* 5, no. 4 (1997), 34.

Anonymous. "Bush to Pardon Byrd: Accused Rapist Cleared by DNA." AP wire story, October 8, 1997.

Anonymous. "Bush Pardons Man Convicted of Rape." UPI wire story, November 20, 1997.

Anonymous. "Hot Type." *Chronicle of Higher Education*, November 28, 1997, A17.

Anonymous. "Irradiate That Barbecue." *Economist*, December 13, 1997, 23.

Anonymous. "Justice Breyer Calls for Experts to Aid Courts in Complex Cases." AP wire story. *New York Times*, February 17, 1998, A17.

Appiah, K. Anthony. "The Multiculturalist Misunderstanding." *New York Review of Books*, October 9, 1997, 30–36.

Arnold, Matthew. *General Grant by Matthew Arnold; with a Rejoinder by Mark Twain.* Ed. John Y. Simon. Kent, Ohio: Kent State University Press, 1995.

Aronowitz, Stanley. *Science as Power: Discourse and Ideology in Modern Society.* Minneapolis: University of Minnesota Press, 1988.

Asante, Molefi K. *Afrocentricity.* Trenton, N.J.: African World Press, 1988.

Askey, Richard. "More on Addison-Wesley Focus on Algebra." Electronically published, 1997.

Atwater, Mary M. "Social Constructivism: Infusion into the Multicultural Science Education Research Agenda." *Journal of Research in Science Teaching*, 33, no. 8 (1996), 821–837.

Augustine, Norman R. "A New Business Agenda for Improving U.S. Schools." *Issues in Science and Technology*, 13, no. 3 (1997), 57– 62.

Augustine, Norman R. "What We Don't Know Does Hurt Us. How Scientific Illiteracy Hobbles Society." *Science*, 279 (March 13, 1998), 1640–1641.

Auletta, Ken. "The Microsoft Provocateur." *New Yorker*, May 12, 1997, 66–72.

Ausubel, Jesse H. *Rails and Snails and the Debate over Goals for Science.* Jerusalem: Israel Academy of Sciences and Humanities, 1994.

Ausubel, Jesse H. "The Liberation of the Environment." *Daedalus* 125, no. 3 (1996), 1–17.

Ausubel, Jesse H. "Environmental Trends." *Issues in Science and Technology*, 13, no. 2 (Winter 1996/97), 78–81.

Barad, Karen. "A Feminist Approach to Teaching Quantum Physics." In Rosser, ed., *Teaching the Majority*, 43–75.

Barr, Stephen M. "Debating Darwin." Review of M. J. Behe, *Darwin's Black Box. Public Interest*, 126 (Winter 1997), 108–112.

Barrow, J. D., and Frank J. Tipler. *The Anthropic Cosmological Principle.* Oxford: Clarendon Press, 1986.

Barry, Dan. "Using Advice of Psychic, Police Hunt for a Body." *New York Times*, July 4, 1997, B3.

Barry, Dan. "The NYPD's Psychic Friend: When Technology Fails, Detectives Call on a New Jersey Woman's 'Visions.'" *New York Times*, July 21, 1997, B1 ff.

Barzun, Jacques. *Science: The Glorious Entertainment.* New York: Harper and Row, 1964.

Bass, Gary. "Geek Chic." *New Republic*, September 8 and 15, 1997, 11–12.

Behe, Michael J. *Darwin's Black Box: The Biochemical Challenge to Evolution.* New York: Free Press, 1996.

Bell, Daniel. *The Cultural Contradictions of Capitalism.* New York: Basic Books, 1976.

Bell, Daniel. "Of 'Trotskyites' and 'Trotskyists.'" Letter to the editor. *Forward*, January 2, 1998, 6.

Bell, Diane. "Desperately Seeking Redemption." *Natural History*, March 1997, 52–53.

Bell, Robert. *Impure Science: Fraud, Compromise, and Political Influence in Scientific Research.* New York: John Wiley & Sons, 1992.

Bennett, Jane, and William Chaloupka, eds. *In the Nature of Things: Language, Politics, and the Environment.* Minneapolis: University of Minnesota Press, 1993.

Benson, Herbert. *Timeless Healing: The Power and Biology of Belief.* New York: Scribner (Simon and Schuster), 1996.

Bereano, Phil. "Don't Take Liberties with Our Genes." *Seattle Times*, July 17, 1997. (Also electronically published.)

Berman, Jerry, and Daniel J. Weitzner. "Technology and Democracy." *Social Research*, 64, no. 3 (1997), 1313–1319.

Berman, Marshall. "Skepticism in Action: Simpsons Religion vs. Science Episode." *Skeptical Inquirer*, March/April 1998, 19.

Bernal, Martin. "Animadversions on the Origins of Western Science." In Powell and Frankenstein, eds., *Ethnomathematics*, 83–99; also, *Isis* 83, no. 4 (1992), 596–607.

Bernstein, Jeremy. "The Merely Very Good." *American Scholar*, 66, no. 1 (1997), 31–39.

Bettencourt, Antonio. "They Don't Come Easy to Us: On the 'Alienness' of the Natural Sciences and Their Appropriation." Electronically published.

Beyer, Landon E., and Daniel P. Liston. *Curriculum Conflict: Social Visions, Educational Agendas, and Progressive School Reform*. New York: Teachers College Press, 1996.

Beyerstein, Barry L. "Why Bogus Therapies Seem to Work." *Skeptical Inquirer*, September/October 1997, 29–34.

Bilger, Burkhard. "Cell Block." *The Sciences*, September/October 1997, 17–19.

Bishop, J. Michael. "Paradoxical Strife: Science and Society in 1993." *Ethics, Values, and the Promise of Science*. Forum proceedings, February 25–26, 1993. Sigma Xi, 95–113.

Bishop, J. Michael. "Enemies of Promise." *Academe*, January/February 1996, 19–21.

Black, Bert. "Science and Beyond: Applying Daubert and Robinson to All Experts." *Advocate* (Texas Center for Legal Ethics and Professionalism), 1997. Electronically published.

Blank, Lisa M., and Hans O. Andersen. "Teaching Evolution: Coming to a Classroom Near You?" *Reports of the National Center for Science Education*, May/June 1997, 10–13.

Bleifuss, Joel. "Recipe for Disaster." *In These Times*, November 11, 1996, 12–13.

Blinder, Alan S., and Richard E. Quant. "The Computer and the Economy." *Atlantic Monthly*, December 1997, 26–32.

Blumenson, Eric. "Mapping the Limits of Skepticism in Law and Morals." *Texas Law Review*, 74, no. 3 (1996), 523–576.

Blumenstyk, Goldie. "Conflict-of-Interest Fears Rise as Universities Chase Industry Support." *Chronicle of Higher Education*, May 22, 1998, A41–A43.

Bodmer, Walter, and Robin McKie. *The Book of Man: The Human Genome Project and the Quest to Discover Our Genetic Heritage*. New York: Oxford University Press, 1994.

Boghossian, Paul A. "What the Sokal Hoax Ought to Teach Us." *Times Literary Supplement*, December 13, 1996, 14–15. (Also in Koertge, ed., *A House Built on Sand*, 23–31.)

Bohm, David, and Basil Hiley. *The Undivided Universe*. London: Routledge, 1995.

Borba, Marcelo C. "Ethnomathematics and Education." In Powell and Frankenstein, eds., *Ethnomathematics*, 261–272.

Bouchard, Thomas J., Jr. "Whenever the Twain Shall Meet." *The Sciences*, September/October 1997, 52–57.

Boudourides, Moses A. "Constructivism and Education: A Shopper's Guide." International Conference on the Teaching of Mathematics, Samos, July 3–6, 1998.

Boxer, Sarah. "Science Confronts the Unknowable." *New York Times*, January 24, 1997, B7–B9.

Brace, C. Loring, et al. "Clines and Clusters versus 'Race': A Test in Ancient Egypt, and a Case of Death on the Nile." In Lefkowitz and Rogers, eds., *Black Athena Revisited*, 129–164.

Brenna, Susan. "Hard Lessons." *New York* magazine, April 13, 1998, 20–24.

Breyer, Stephen G. "The Interdependence of Science and Law." Address, Annual Meeting of the American Association for the Advancement of Science, February 16, 1998. Electronically published, 1998.

Bricmont, Jean. "Science Studies—What's Wrong?" *Physics World,* December 1997, 15–16.

Broad, William J. "Air Force Details New Theory in U.F.O. Case." *New York Times,* June 25, 1997, B7.

Broad, William J. "Another Possible Climate Culprit: The Sun." *New York Times,* September 23, 1997, F1 ff, F8.

Broad, William, and Nicholas Wade. *Betrayers of the Truth.* New York: Simon and Schuster, 1983.

Brodeur, Paul. *The Great Power-Line Cover-Up: How the Utilities and the Government Are Trying to Hide the Cancer Hazard Posed by Electromagnetic Fields.* New York: Little, Brown, 1993.

Brody, Jane E. "In Vitamin Mania, Millions Take a Gamble on Health." *New York Times,* October 26, 1997, 1 ff, 28.

Brody, Jane E. "U.S. Panel on Acupuncture Calls for Wider Acceptance." *New York Times,* November 6, 1997, A12.

Brody, Jane. "Acupuncture: An Expensive Placebo or a Legitmate Alternative?" *New York Times,* November 18, 1997.

Brody, Jane. "Weighing Pros and Cons While the Jury Is Out." *New York Times,* November 18, 1997.

Bromwich, David. *Politics by Other Means: Higher Education and Group Thinking.* New Haven: Yale University Press, 1992.

Bromwich, David. "Democracy, Merit, and Presumptive Virtue." *Dissent,* Fall 1997, 63–68.

Bronner, Ethan. "Voracious Computers Are Siphoning Talent from Academia." *New York Times,* June 25, 1998, A1 ff, A14.

Brown, Peter G. "What Hath Wilmut Wrought?" *The Sciences,* September/October 1997, 4–5.

Brown, Richard Harvey. *Toward a Democratic Science: Scientific Narration and Civic Communication.* New Haven: Yale University Press, 1998.

Browne, Malcolm W. "International Language of Physics Ties Physicists' Tongues." *New York Times,* June 16, 1998, F4.

Bruce, Robert V. *Lincoln and the Tools of War.* Urbana: University of Illinois Press, 1989.

Bunge, Mario. "The Popular Perception of Science in North America." *Transactions of the Royal Society of Canada,* series 5, 4 (1989), 269–280.

Bunge, Mario. "What Is Science? Does It Matter to Distinguish It from Pseudoscience? A Reply to My Commentators." *New Ideas in Psychology,* 9, no. 2 (1991), 245–283.

Bunge, Mario. "Realism and Antirealism in Social Science." *Theory and Decision,* 35 (1993), 207–235.

Bunge, Mario. "Technodemocracy: An Alternative to Capitalism and Socialism." *Concordia,* no. 25 (1994), 93–99.

Burnham, John C. *How Superstition Won and Science Lost: Popularizing Science and Health in the United States.* New Brunswick, N.J.: Rutgers University Press, 1987.

Burr, Chandler. "The Geophysics of God." *US News & World Report,* June 16, 1997, 55–58.

Byrd, Alexander. "Squaring the Circle: Hobbes on Philosophy and Geometry." *Journal of the History of Ideas,* 57, no. 2 (1996), 217–232.

Byrne, Frank L., and Andrew T. Weaver, eds. *Haskell of Gettysburg: His Life and Civil War Papers.* Kent, Ohio: Kent State University Press, 1989.

California State Board of Education. "Mathematics Framework for California Public Schools, Kindergarten through Grade 12." Draft, September 1997. Electronically published.

Campbell, George, Jr. "Raising Expectations." Review of D. E. Drew, *Aptitude Revisited. Issues in Science and Technology*, 13, no. 3 (1997), 90–95.

Campbell, Mary Anne, and Randall K. Campbell-Wright. "Toward a Feminist Algebra." In Rosser, ed., *Teaching the Majority*, 127–144.

Campbell, Paulette Walker. "Big Increases in Federal Budget for Science Appear Unlikely." *Chronicle of Higher Education*, May 29, 1998, A34.

Campbell, Peter. "How to See Inside a French Milkman." Review of B.H. Kevles, *Naked to the Bone. London Review of Books*, July 31, 1997, 30–31.

Capra, Fritjof. *The Tao of Physics: An Exploration of the Parallels between Modern Physics and Eastern Mysticism*. New York: Random House, 1983.

Cardwell, Donald. *The Norton History of Technology*. New York: W. W. Norton, 1994.

Cartmill, Frank. "Reinventing Anthropology." *Yearbook of Physical Anthropology*, 37 (1994), 1–9.

Cartmill, Frank. "Oppressed by Evolution." *Discover*, March 1998, 78–83.

Casti, John. "Lighter than Air." Review of J. Horgan, *The End of Science. Nature*, 382 (August 29, 1996), 769–770.

Chalmers, Alan. *Science and Its Fabrication*. Minneapolis: University of Minnesota Press, 1991.

Chopra, Deepak. *Quantum Healing*. New York: Bantam Books, 1990.

Churchill, Martha A. "A Skeptic Living in Roswell." *Skeptical Inquirer*, May/June 1998, 40–43 ff, 60.

Cobern, W. W. "Belief, Understanding, and the Teaching of Evolution." *Journal of Research in Science Teaching*, 31, no. 5 (1994), 583–590.

Cobern, W. W. "A Cultural Constructivist Approach to the Teaching of Evolution." Working paper no. 12, Scientific Literacy and Cultural Studies Project, Western Michigan University. Electronically published, 1994.

Cobern, William W. "Defining 'Science' in a Multicultural World: Implications for Science Education." Conference of the National Association for Research in Science Teaching, 1998. Electronically published.

Cohen, I. Bernard. *Science and the Founding Fathers: Science in the Political Thought of Jefferson, Franklin, Adams, and Madison*. New York: W. W. Norton, 1997.

Cole, Stephen. "Voodoo Sociology: Recent Developments in the Sociology of Science." In Gross, Levitt, and Lewis, eds., *The Flight from Science and Reason*, 274–287.

Coleman, Simon, and Leslie Carlin. "No Contest: The Non-Debate between Creationism and Evolutionary Theory in Britain." *Creation/Evolution*, 38 (1995), 1–9.

Colvin, Richard Lee. "Spurned Nobelists Appeal Science Standards Rejection." *Los Angeles Times*, November 17, 1997.

Cordes, Colleen. "Senators Back Increase in NSF Authorization." *Chronicle of Higher Education*, May 22, 1998, A40.

Corwin, Rebecca B. "Doing Mathematics Together: Creating a Mathematical Culture." *Arithmetic Teacher*, 40, no. 6 (1993), 338–341.

Crandall, B. C. *Nanotechnology: Molecular Speculations on Global Abundance*. Cambridge, Mass.: MIT Press, 1996.

Cranor, Carl F. "Science Courts, Evidentiary Procedures, and Mixed Science-Policy Decisions." Conference, "Which Scientist Do You Believe? Process Alternatives in Technological Controversies," Concord, N.H., October 6–7, 1994. Electronically published, Franklin Pierce Law Center Homepage.

Crews, Frederick. "Freudian Suspicion versus Suspicion of Freud." In Gross, Levitt, and Lewis, eds., *The Flight from Science and Reason*, 470–482.

Crews, Frederick. "The Mindsnatchers." *New York Review of Books*, June 25, 1998, 14–19.

Crews, Frederick, et al. *Memory Wars: Freud's Legacy in Dispute*. New York: New York Review of Books, 1997.

Cringely, Bob. *Triumph of the Nerds*. Videotape. New York: Ambrose Video Publishing, 1996.

Cromer, Alan. *Connected Knowledge: Science, Philosophy, and Education*. New York: Oxford University Press, 1997.

Cuocco, Albert, and Wayne Harvey. "Gateways to Advanced Mathematical Thinking: Linear Algebra and Precalculus." Abstract, NSF Award 9450731, 1994. Electronically published.

Cushing, James T. *Quantum Mechanics: Historical Contingency and the Copenhagen Hegemony*. Chicago: University of Chicago Press, 1994.

Cushman, John H. "Industrial Group Plans to Battle Climate Treaty." *New York Times*, April 26, 1998, 1 ff, 24.

Dajer, Tony. "Hearts and Minds." *Discover*, December 1997, 49–53.

D'Andrade, Roy. "Moral Models in Anthropology." *Current Anthropology*, 36 (1995), 399–408.

Dannen, Gene. "The Einstein-Szilard Refrigerators." *Scientific American*, January 1997, 90–95.

Dawkins, Richard. *Climbing Mount Improbable*. New York: W. W. Norton, 1996.

Dawkins, Richard. *Humanist*, January/February 1997, 26–29. (Quoted in "An Apostle of Science," *Wilson Quarterly*, 21, no. 2 [1997], 134.)

Dawkins, Richard. "Thinking Clearly about Clones: How Dogma and Ignorance Get in the Way." *Free Inquiry*, 17, no. 3 (1997), 13–14.

Dawkins, Richard. "Science, Delusion, and the Appetite for Wonder." *Reports of the National Center for Science Education*, January/February 1997, 8–14.

Dean, Jodi. *Aliens in America: Conspiracy Cultures from Outerspace to Cyberspace*. Ithaca, N.Y.: Cornell University Press, 1998.

Degler, Carl N. *In Search of Human Nature: The Decline and Revival of Darwinism in American Social Thought*. New York: Oxford University Press, 1991.

Deloria, Vine, Jr. *Red Earth, White Lies*. New York: Scribner's, 1995.

Denby, David. "In Darwin's Wake." *New Yorker*, July 21, 1997, 50–62.

Denfeld, Renee. "Old Messages: Ecofeminism and the Alienation of Young People from Environmental Activism." In Gross, Levitt, and Lewis, eds., *The Flight from Science and Reason*, 246–255.

Dennett, Daniel. *Darwin's Dangerous Idea*. New York: Simon and Schuster, 1996.

Dennis, Carl. "On the Bus to Utica" (poem). *The New Republic*, July 6, 1998, 44.

DePalma, Anthony. "Ford Joins in a Global Alliance to Develop Fuel-Cell Auto Engines." *New York Times*, December 16, 1997, D1 ff, D6.

Di Berardino, Marie A., and Robert G. McKinnell. "Backward Compatible." *The Sciences*, September/October 1997, 32–37.

Dickison, Mike. "Maori Science." *New Zealand Science Monthly*, 5, no. 4 (May 1994), 6–7. (Also electronically published.)

Diggins, John Patrick. *The Rise and Fall of the American Left*. New York: W. W. Norton, 1992.

Doron, Shalom, and R. B. Bernstein. "Exploring the Age of Experiment in Government. Review of I. B. Cohen, *Science and the Founding Fathers*. H-Law, H-Net Reviews, 1998. Electronically published.

Douglas, Mary. *Purity and Danger: An Analysis of Concepts of Pollution and Taboo*. Harmondsworth, U.K.: Penguin Books, 1970.

Douglas, Mary, and Aaron B. Wildavsky. *Risk and Blame: An Essay on the Selection of Technical and Environmental Dangers*. Berkeley: University of California Press, 1982.

Dowie, Mark. "What's Wrong with the *New York Times*'s Science Reporting." *Nation*, June 28, 1998, 13–14.

Dowling, William C. *Jameson, Althusser, Marx: An Introduction to the Political Unconscious*. Ithaca, N.Y.: Cornell University Press, 1984.

Drew, David E. *Aptitude Revisited: Rethinking Math and Science Education for America's Next Century*. Baltimore: Johns Hopkins University Press, 1996.

Drosnin, Michael. *The Bible Code*. New York: Simon and Schuster, 1997.

D'Souza, Dinesh. *Illiberal Education: The Politics of Race and Sex on Campus*. New York: Free Press, 1991.

Duffy, Thomas, Jost Lowyck, and David Jonassen, eds. *Designing Constructivist Learning Environments*. Heidelberg: Springer-Verlag, 1993.

Dyson, Freeman J. "Science in Trouble." *American Scholar*, 62, no. 4 (1993), 513–525.

Dyson, Freeman J. "Is God in the Lab?" *New York Review of Books*, May 28, 1998, 8–10.

Eagleton, Terry. *The Illusions of Postmodernism*. Oxford: Blackwell, 1996.

Eakin, Emily. "Paper Trail." *Lingua Franca*, August, 1997, 12.

Easterbrook, Gregg. "Hot and Not Bothered." *New Republic*, May 4, 1998, 20–25.

Easterbrook, Gregg. "Science Sees the Light." *New Republic*, October 12, 1998, 24–29.

Edis, Taner. "Relativist Apologetics: The Future of Creationism." *Reports of the National Center for Science Education*, 17, no. 1 (1997), 17–19.

Edmundson, Mark. "Save Sigmund Freud." *New York Times Magazine*, July 13, 1997, 34–35.

Egan, Timothy. "Old Skull Gets White Looks, Stirring Debate." *New York Times*, April 2, 1998, A12.

Ehrenreich, Barbara, and Janet MacIntosh. "The New Creationism." *Nation*, June 9, 1997, 11–16.

Ehrlich, Paul R., and Anne H. Ehrlich. *The Betrayal of Science and Reason: How Anti-Environmental Rhetoric Threatens Our Future*. Washington, D.C.: Island Press, 1996.

Ehrlich, Paul R., et al. "No Middle Way on the Environment." *Atlantic Monthly*, December 1997, 98–104.

Eichman, Eric. "The End of the Affair." *New Criterion*, December 1996, 77–80.

Ellis, John M. *Against Deconstruction*. Princeton N.J.: Princeton University Press, 1989.

Ellis, John M. *Literature Lost: Social Agendas and the Corruption of the Humanities*. New Haven: Yale University Press, 1997.

Elshtain, Jean Bethke. "Ewegenics." *New Republic*, March 31, 1997, 25.

Elshtain, Jean Bethke. "Our Bodies, Our Clones." *New Republic*, August 4, 1997, 25.

Elshtain, Jean Bethke. "Authority Figures." *New Republic*, December 22, 1997, 11–12.

Evans, Lawrence. "Should We Care about Science 'Studies?'" *Duke Faculty Forum*, 8, no. 1 (1996), 1.

Evers, Bill. "Keep the Academic Bar High." *New York Times*, September 23, 1997, A27.

Eysenck, Hans. *Decline and Fall of the Freudian Empire*. New York: Viking Press, 1985.

Ezrahi, Yaron. *The Descent of Icarus: Science and the Transformation of Contemporary Democracy*. Cambridge, Mass.: Harvard University Press, 1990.

Ezrahi, Yaron, Everett Mendelsohn, and Howard P. Segal, eds. *Technology, Pessimism, and Postmodernism*. Amherst: University of Massachusetts Press, 1994.

Farber, Daniel A., and Suzanna Sherry. *Beyond All Reason: The Radical Assault on Truth in American Law*. New York: Oxford University Press, 1997.

Fasheh, Munir. "Mathematics, Culture, and Authority." In Powell and Frankenstein, eds., *Ethnomathematics*, 273–290.

Feder, Kenneth L. "Indians and Archaeologists: Conflicting Views of Myth and Science." *Skeptic*, 5, no. 3 (1997), 74–80.

Ferris, Timothy. "The Risks and Rewards of Popularizing Science." *Chronicle of Higher Education*, April 4, 1997, B6.

Ferris, Timothy. "The Wrong Stuff." *New Yorker*, April 14, 1997, 33.

Ferris, Timothy. "Some Like It Hot." *New York Review of Books*, September 25, 1997, 16–20.

Ferris, Timothy. "Not Rocket Science." *New Yorker*, July 20, 1998, 4–5.

Feyerabend, Paul K. *Against Method*. London: Verso, 1993.

Feynman, Richard P. *The Meaning of It All: Thoughts of a Citizen Scientist*. New York: Addison-Wesley, 1998.

Fezer, Karl D. "Is Science's Naturalism Metaphysical or Methodological?" Review of P. E. Johnson, *Reason in the Balance*. *Creation/Evolution* 39 (Winter 1996), 31–35.

Field, Thomas G., Jr. "The Science Court Is Dead: Long Live the Science Court." Conference, "Which Scientist Do You Believe? Process Alternatives in Technological Controversies," Concord, N.H., October 6–7, 1994. Electronically published, Franklin Pierce Law Center Homepage.

Firenze, Richard. "Lamarck vs. Darwin: Dueling Theories." *Reports of the National Center for Science Education*, July/August 1997, 9–11.

Fischer, Paul M. "Science and Subpoenas: When Do the Courts Become Instruments of Manipulation?" *Law and Contemporary Problems*, 59, no. 3 (1996), 59–68.

Foote, Shelby. *The Civil War: A Narrative*. New York: Random House, 1974.

Forman, Paul. "Independence, Not Transcendence, for the Historian of Science." *Isis*, 82 (1991), 71–86.

Forman, Paul. "Assailing the Seasons." Review of Gross, Levitt, and Lewis, eds., *The Flight from Science and Reason*. *Science*, May 2, 1997, 750–752.

Forman, Paul. "Recent Science: Late-Modern and Post-Modern." In Soderqvist, ed., *The Historiography of Contemporary Science and Technology*, 179–213.

Fox Network. "Alien Autopsy: Fact or Fiction?" Broadcast, August 28, 1995.

Fox Network. *The Simpsons*. Broadcast, November 22, 1997.

Fox, Robin. "Anthropology and the 'Teddy Bear Picnic.'" *Society*, November/December 1992, 47–55.

Fox, Robin. "State of the Art/Science in Anthropology." In Gross, Levitt, and Lewis, eds., *The Flight from Science and Reason*, 327–345.

Frankel, Mark S. "Multicultural Science." *Chronicle of Higher Education*, November 10, 1993, B1–B2.

Friedman, Alan J. "Exhibits and Expectations." *Public Understanding of Science* 4 (1995), 305–313.

Fromm, Harold. "Ecology and Ideology." *Hudson Review*, 45, no. 1 (1992), 23–36.

Fromm, Harold. "My Science Wars." *Hudson Review*, 49, no. 4 (1997), 599–609.

Frosch, Robert A. "Reductionism and the Unity of Science." *American Scientist*, 83, no. 1 (January/February 1997), 2.

Fuller, Steve. *Science*. Buckingham, U.K.: Open University Press, 1997.

Fuller, Steve. "Does Science Put an End to History, or History to Science? Or, Why Being Pro-science Is Harder than You Think." In Ross, ed., *Science Wars*, 29–60.

Fumento, Michael. *Science under Siege: How the Environmental Misinformation Campaign Is Affecting Our Laws, Taxes, and Our Daily Life*. New York: Quill (William Morrow), 1993.

Fumento, Michael. "Pesticides Are Not the Main Problem." *New York Times*, June 30, 1998, A23.

Furlow, Bryant. "Medical School Faith Healing: Report on a Spirituality Conference." *Skeptical Inquirer*, September/October 1997, 49–50.

Galison, Peter. "Three Laboratories." *Social Research*, 64, no. 3 (1997), 1127–1161.

Gamow, George. *One, Two, Three . . . Infinity*. New York: Dover Books, 1988.

Gardner, Martin. "Intelligent Design and Philip Johnson." *Skeptical Inquirer*, November/December 1997, 17–20.

Gardner, Martin. "The New New Math." *New York Review of Books*, September 24, 1998, 9–12.

Garfunkel, Solomon A., and Gail S. Young. "The Sky Is Falling." *Notices of the American Mathematical Society*, 45, no. 2 (1998), 256–257.

Garofalo, Joe, and David Kfakwami Mtetwa. "Mathematics as Reasoning." National Council of Teachers of Mathematics. Electronically published. (Reprinted from *Arithmetic Teacher*.)

Geertz, Clifford. "The Legacy of Thomas Kuhn: The Right Text at the Right Time." *Common Knowledge*, 6, no. 1 (1997), 1–5.

Gerlich, Nick. "A Skeptic Crashes in Roswell." *Skeptic*, 5, no. 2 (1997), 19–22.

Gessen, Masha. "The Anti-Sakharov." *Lingua Franca*, September/October 1995, 68–73.

Gilbert, Evelyn. "Medical Advances: Insurance Companies Are Taking Notice as Alternative Health Care Becomes More Popular." *Village Voice*, December 10, 1996, 52–53.

Gilbert, Evelyn. "Good Vibrations: Energy Healing Goes Mainstream, but Not without Static." *Village Voice*, July 29, 1997, 37.

Gitlin, Todd. *The Twilight of Common Dreams: Why America Is Wracked by Culture Wars*. New York: Henry Holt (Metropolitan Books), 1995.

Gitlin, Todd. "The Dumb-Down." *Nation*, March 17, 1997, 28.

Gladwell, Malcolm. "A Matter of Gravity." *New Yorker*, November 11, 1996, 36–38.

Glasser, Ronald J. "The Doctor Is Not In." *Harper's Magazine*, March 1998, 35–41.

Gleick, James. "Legal Eagles." *New York Times Magazine*, September 14, 1997, 48–50.

Goldberg, Steven. *The Inevitability of Patriarchy*. New York: William Morrow, 1973.

Goldberg, Steven. *When Wish Replaces Thought: Why So Much of What You Believe Is False*. Buffalo, N.Y.: Prometheus Books, 1991.

Goldberg, Steven. *Why Men Rule: A Theory of Male Dominance*. Chicago: Open Court, 1993.

Goldberg, Vicki. "Art and Science, the Yin and Yang of Culture." *New York Times*, April 27, 1997, sect. 2, 35 ff.

Goldstein, Sheldon. "Review Essay: Bohmian Mechanics and the Quantum Revolution." *Synthese*, 107, no. 1 (1996), 145–165.

Goldstein, Sheldon. "Quantum Philosophy: The Flight from Reason in Science." In Gross, Levitt, and Lewis, eds., *The Flight from Science and Reason*, 119–125.

Gonzales, Dave. "The Origins of Indigenous Americans." *Chronicle of Higher Education*, April 24, 1998, B12.

Good, Byron. "Comment on 'A Methodology for Cross-cultural Ethnomedical Research.'" *Current Anthropology*, 29, no. 5 (1988), 693.

Goodstein, David. "Scientific Elites and Scientific Illiterates." Address, Sigma Xi Forum on Ethics, Values, and the Prospects of Science, February 1993. Electronically published.

Goodstein, David. "The Age of Irony." Review of J. Horgan, *The End of Science*. *Science*, 272, June 14, 1996, 1594.

Goodstein, David. "On the Shoulders of Giants." Review of B. L. Silver, *The Ascent of Science. New York Times Book Review*, March 22, 1998.

Goodstein, David. "Conduct and Misconduct in Science." In Gross, Levitt, and Lewis, eds., *The Flight from Science and Reason*, 31–38.

Gottfried, Kurt, and Kenneth G. Wilson. "Science as a Social Construct." *Nature*, 386 (April 10, 1997), 645–647.

Gould, Stephen J. *The Flamingo's Smile*. New York: W. W. Norton, 1985.

Gould, Stephen J. "Nonoverlapping Magisteria." *Natural History*, March 1997, 16 ff.

Gould, Stephen J. "Drink Deep, or Taste Not the Pierian Spring." *Natural History*, September 1997, 24–25.

Gould, Stephen J. "Individuality." *The Sciences*, September/October 1997, 14–16.

Gould, Stephen J. "The Great Asymmetry." *Science*, 279 (February 6, 1998), 812–813.

Grant, Ulysses S. *The Personal Memoirs of Ulysses S. Grant*. New York: Konecky and Konecky, 1992.

Gratzer, Walter. "In the Hands of Any Fool." Review of P. Fleming, *A Short History of Cardiology. London Review of Books*, July 3, 1997, 24.

Greenhouse, Linda. "Trial Judges Are Backed on Rulings on Scientists." *New York Times*, December 16, 1997, A25.

Griffiths, Jay. "Diary." *London Review of Books*, May 8, 1997, 33.

Griffiths, Paul E., Hamish Spencer, and John Stenhouse. "Onward and Upward." Reviews of M. Ruse, *Monad to Man. Metascience*, 7, no. 1 (1998), 52–64.

Gross, Paul R. "Science without Scientists." *New York Times*, December 1, 1997, A21.

Gross, Paul R. "Bête Noire of the Science Worshippers." *History of the Human Sciences*, 10, no. 1 (1997), 125–128.

Gross, Paul R. "Downsizing Darwin." *Boston Globe*, May 17, 1998, E7.

Gross, Paul R. "Author's Response." *Metascience*, 7, no. 1 (1998), 45–50.

Gross, Paul R. "The Icarian Impulse." *Wilson Quarterly*, 22, no. 1 (1998), 39–49.

Gross, Paul R. "Bashful Eggs, Macho Sperm, and Tonypandy." In Koertge, ed., *A House Built on Sand*, 59–70.

Gross, Paul R., and Norman Levitt. "The Natural Sciences: Trouble Ahead? Yes." *Academic Questions*, Spring 1994, 13–29.

Gross, Paul R., and Norman Levitt. *Higher Superstition: The Academic Left and Its Quarrels with Science*. Baltimore: Johns Hopkins University Press, 1994.

Gross, Paul R., and Norman Levitt. "Knocking Science for Fun and Profit." *Skeptical Inquirer*, March/April 1995, 38–42.

Gross, Paul R., Norman Levitt, and Martin W. Lewis, eds. *The Flight from Science and Reason. Annals of the New York Academy of Sciences*, 775 (1995); also, Baltimore: Johns Hopkins University Press, 1997.

Grudin, Robert. *Book*. New York: Penguin Books, 1993.

Grünbaum, Adolf. *The Foundations of Psychoanalysis: A Philosophical Critique*. Berkeley: University of California Press, 1984.

Gurdon, J. B. "The Birth of Cloning." *The Sciences*, September/October 1997, 26–31.

Guston, David H., and Kenneth Kenniston. "Updating the Social Contract for Science." *Technology Review*, November/December 1994, 60–68.

Haack, Susan. "Concern for Truth: What It Means, Why It Matters." In Gross, Levitt, and Lewis, eds., *The Flight from Science and Reason*, 57–63.

Haack, Susan. "Towards a Sober Sociology of Science." In Gross, Levitt, and Lewis, eds., *The Flight from Science and Reason*, 259–265.

Haack, Susan. "Misinterpretation and the Rhetoric of Science: or, What Was the Color of the Horse?" *Proceedings of the American Catholic Philosophical Association*. In press.

Haberman, Clyde. "The Court Would Have Been Pleased." *New York Times*, October 28, 1997, B1.

Habermas, Jürgen. "The Analytical Theory of Science and Dialectics: A Postscript to the Controversy between Popper and Adorno." In Adorno et al., *The Positivist Controversy in German Sociology*, 131–162.

Hackerman, Norman, and Kenneth Ashworth. *Conversations on the Uses of Science and Technology*. Denton: University of North Texas Press, 1996.

Hacking, Ian. "Taking Bad Arguments Seriously: Ian Hacking on Psychopathology and Social Construction." *London Review of Books*, August 21, 1997, 14–16.

Hale, Alan. "Astronomer Hale Speaks Out about Poor Prospects for Science Careers." *Hawaii Rational Inquirer*, 2, no. 24 (1997). Electronically published.

Hall, Stephen S. "Success Is Like a Drug." *New York Times Magazine*, November 23, 1997, 64–70.

Hammerton, M. A. *Science under Siege*. Newcastle, U.K.: University of Newcastle Upon Tyne, 1974.

Harding, Sandra. *The Science Question in Feminism*. Ithaca, N.Y.: Cornell University Press, 1986.

Harding, Sandra. *Whose Science? Whose Knowledge: Thinking from Women's Lives*. Ithaca, N.Y.: Cornell University Press, 1991.

Harding, Sandra. "After the Neutrality Ideal: Science, Politics, and 'Strong Objectivity.'" *Social Research*, 59, no. 3 (1992), 567–587.

Harding, Sandra. "Science Is 'Good to Think With.'" In Ross, ed., *Science Wars*, 16–28.

Harmon, Amy. "NASA Flew to Mars for Rocks? Sure." *New York Times*, July 20, 1997, sec. 4, 4.

Harr, Jonathan. *A Civil Action*. New York: Vintage, 1996.

Harris, Marvin. "Post-Modern Anti-Scientism." Lecture, American Association for the Advancement of Science, Boston, February 12, 1993.

Hart, Ronald, Angelo Turturro, and Julian Leakey. "Born Again." *The Sciences*, September/October 1997, 47–51.

Hartz, Jim, and Rick Chappell. *Worlds Apart: How the Distance between Science and Journalism Threatens America's Future*. Nashville: First Amendment Center, 1998.

Hayes, Brian. "The End of Science Writing?" Review of J. Horgan, *The End of Science*. *American Scientist*, 84, no. 5 (September/October 1996), 495–496.

Hazen, Robert M., and James Trefil. *Science Matters: Achieving Scientific Literacy*. New York: Anchor Books, 1992.

Heginbotham, Stanley J. "The Power of HIV-Positive Thinking." Review of S. Epstein, *Impure Science*. *The Sciences*, May/June 1997, 38–42.

Heilemann, John. "The Perceptionist: How Steve Jobs Took Back Apple." *New Yorker*, September 8, 1997, 34–41.

Held, Barbara. *Back to Reality*. New York: W. W. Norton, 1995.

Held, Barbara. "The Real Meaning of Constructivism." *Journal of Constructivist Psychology*, 8, no. 2 (1995), 305–315.

Held, Barbara. "Constructivism in Psychotherapy: Truth and Consequences." In Gross, Levitt, and Lewis, eds., *The Flight from Science and Reason*, 198–206.

Herschbach, Dudley. "Imaginary Gardens with Real Toads." In Gross, Levitt, and Lewis, eds., *The Flight from Science and Reason*, 11–30.

Hill, Daniel Harvey. *Elements of Algebra*. Philadelphia: Lippincott, 1857.

Hirsch, E. D. *The Schools We Need and Why We Don't Have Them*. New York: Doubleday, 1996.

Hirsch, Robert L., Gerald Kulcinski, and Ramy Shanny. "Fusion Research with a Future." *Issues in Science and Technology*, 13, no. 4 (1997), 60–64.

Hively, Will. "The Incredible Shrinking Finger Factory." *Discover*, March 1998, 84–93.

Hoffman, David. Review of J. Horgan, *The End of Science*. *Notices of the American Mathematical Society*, 45, no. 2 (1998), 260–266.

Hoffmann, Roald. "Movement of the People." *Science*, 280 (April 17, 1998), 386–387.

Hogben, Lancelot. *Mathematics for the Million*. New York: W. W. Norton, 1993.

H.O.L.D. (Palo Alto, Calif.) "The California Mathematics Framework." Newsletter, 1995. Electronically published.

Hollinger, David A. *Jews, Science, and Secular Culture*. Princeton, N.J.: Princeton University Press, 1996.

Holman, David W. "Of Bumblebees and Street Gangs: The Limited Scope of Daubert/Robinson." *Advocate* (Texas Center for Legal Ethics and Professionalism), 1997. Electronically published.

Holt, Jim. "Their Days Are Numbered." *Lingua Franca*, September/October 1994, 46–53.

Holt, Jim. "Let's Make a Deal." Review of M. Piatelli-Palmarini, *Inevitable Illusions*. *American Scholar*, 65, no. 2 (1996), 306–309.

Holton, Gerald. *Science and Anti-Science*. Cambridge, Mass.: Harvard University Press, 1993.

Holton, Gerald. "Address." AAAS Fellows' Forum. San Francisco, February 20, 1994.

Holton, Gerald. *Einstein, History, and Other Passions: The Rebellion against Science at the End of the Twentieth Century*. New York: Addison-Wesley, 1996.

Holton, Gerald. "Einstein and the Cultural Roots of Modern Science." *Daedalus*, 127, no. 1 (1998), 1–44.

Holton, Gerald. "Science Education and the Sense of Self." In Gross, Levitt, and Lewis, eds., *The Flight from Science and Reason*, 551–560.

Holub, Miroslav. *Shedding Life: Disease, Politics, and Other Human Conditions*. Minneapolis: Milkwood Editions, 1998.

Horgan, John. "Science Set Free from Truth." *New York Times*, July 16, 1996, A17.

Horgan, John. *The End of Science: Facing the Limits of Knowledge in the Twilight of the Scientific Age*. New York: Addison-Wesley, 1996.

Horgan, John. "Darwin on His Mind." *Lingua Franca*, November 1997, 40–48.

Horvitz, Lucy. "The Enemies of Science." *Boston Book Review*, 4, issue 1 (January 1997), 28–29.

Horvitz, Lucy. "Science for the People." *Boston Book Review*, 4, issue 8 (October 1997), 20–21.

Howe, Roger. "The AMS and Mathematics Education: The Revision of the 'NCTM Standards.'" *Notices of the American Mathematical Society*, 45, no. 2 (1998), 243–247.

Howe, Roger, et al. "Reports of AMS Association Resource Group." *Notices of the American Mathematical Society*, 45, no. 2 (1998), 270–276.

Høyrup, Jens. *In Measure, Number, and Weight: Studies in Mathematics and Culture*. Albany: State University of New York Press, 1994.

Huber, Peter W. *Galileo's Revenge: Junk Science in the Courtroom*. New York: Basic Books, 1993.

Huber, Peter W. "Coping with Phantom Risks in the Courts." Conference, "Which Scientist Do You Believe? Process Alternatives in Technological Controversies," Concord, N.H., October 6–7, 1994. Summary of presentation; electronically published, Franklin Pierce Law Center Homepage.

Hughes, Robert. *Culture of Complaint: The Fraying of America*. New York: Oxford University Press, 1993.

Hull, Richard T. "No Fear: How a Humanist Faces Science's New Creation." *Free Inquiry*, 17, no. 3 (1997), 18–20.

Hunt, James B., Jr., et al. *Mathematics and Science Achievement for the 21st Century*. Washington, D.C.: National Education Goals Panel, 1997.

Hynes, James. *Publish and Perish: Three Tales of Tenure and Terror*. New York: Picador, 1997.

INRA (Europe) and Report International. *Europeans, Science and Technology: Public Understanding and Attitudes*. Brussels: Commission of the European Communities, 1993.

Irigaray, Luce. *The Sex Which Is Not One*. Trans. Catherine Porter with Carolyn Burke. Ithaca, N.Y.: Cornell University Press, 1985.

Jackson, Allyn. "Interview with Gail Burill." *Notices of the American Mathematical Society*, 45, no. 1 (1998), 87–90.

James, Clive. "Sex and Reason." *New Yorker*, December 9, 1996, 104–112.

Jardine, Nick, and Marina Frasca-Spada. "Splendours and Miseries of the Science Wars." *Studies in the History and Philosophy of Science*, 28, no. 2 (1997), 219–235.

Jaroff, Leon. "The Solution Is Not in the Solution." *Skeptic*, 5, no. 3 (1997), 51.

Jasanoff, Sheila. "Procedural Choices in Regulatory Science." Conference, "Which Scientist Do You Believe? Process Alternatives in Technological Controversies," Concord, N.H., October 6–7, 1994. Electronically published, Franklin Pierce Law Center Homepage.

Jasanoff, Sheila. *Science at the Bar: Law, Science, and Technology in America*. Cambridge, Mass.: Harvard University Press, 1995.

Jasanoff, Sheila. Plenary Lecture, Annual Meeting of the American Association for the Advancement of Science, Baltimore, February 1996.

Jasanoff, Sheila. "Research Subpoenas and the Sociology of Knowledge." *Law and Contemporary Problems*, 59, no. 3 (1996), 95–118.

Jasanoff, Sheila. "An Epidemic Challenges American Biomedicine." Review of S. Epstein, *Impure Science. American Scientist*, 85, no. 5 (September/October 1997), 476–477.

Jasanoff, Sheila, et al. "Conversations with the Community: AAAS at the Millennium." *Science*, December 19, 1997, 2066–2067.

Jastrow, Robert. *God and the Astronomers*. New York: W. W. Norton, 1978.

Jennings, Marianne M. "'Rain Forest' Algebra Course Teaches Everything but Algebra." Electronically published; reprinted from the *Christian Science Monitor*, April 2, 1996.

Johnson, George. "Don't Worry. A Brain Still Can't Be Cloned." *New York Times, Week in Review*, March 2, 1997, 1 ff.

Johnson, George. "The Unspeakable Things That Particles Do." *New York Times, Week in Review*, July 27, 1997, 5.

Johnson, George. "Ethical Fears Aside, Science Plunges On." *New York Times, Week in Review*, December 7, 1997, 6.

Johnson, Philip E. *Darwin on Trial*. Washington, D.C.: Regnery Gateway, 1991.

Johnson, Philip E. *Reason in the Balance: The Case against Naturalism in Science, Law, and Education*. Downer's Grove, Ill.: InterVarsity Press, 1995.

Johnson, Philip E. "What (If Anything) Hath God Wrought?" *Academe*, September/October 1996.

Jonassen, David, Terry Mayse, and Ray McAleese. "A Manifesto for a Constructivist Approach to Technology in Higher Education." In T. Duffy, J. Lowyck, and D. Jonassen, eds., *Designing Constructivist Learning Environments*.

Jones, Steve. "The Set within the Skull." Review of S. Pinker, *How the Mind Works. New York Review of Books*, November 6, 1997, 13–16.

Jones, Steve. "In the Genetic Toyshop." *New York Review of Books*, April 23, 1998, 14–16.

Kafatos, Menas, and Robert Nadeau. *The Conscious Universe*. New York: Springer-Verlag, 1990.

Kaiser, Jocelyn. "Breast-Implant Ruling Sends a Message." *Science*, January 3, 1997, 21.

Kaminer, Wendy. "The Latest Fashion in Irrationality." *Atlantic Monthly*, July 1996, 103–106.

Kaminer, Wendy. "Why We Love Gurus." *Newsweek*, October 29, 1997, 60.

Kantrowitz, Arthur. "The Separation of Facts and Values." Conference, "Which Scientist Do You Believe? Process Alternatives in Technological Controversies," Concord, N.H., October 6–7, 1994. Electronically published, Franklin Pierce Law Center Homepage.

Kaplan, Jonathan Michael. "Problematizing Reifications and Naturalizations." *Stanford Humanities Review*, 5 (suppl.) (1996), 126–131.

Kaplan, Robert D. "Was Democracy Just a Moment?" *Atlantic Monthly*, December 1997, 55–80.

Karl, Thomas R., Neville Nicholls, and Jonathan Gregory. "The Coming Climate." *Scientific American*, May 1997, 78–83.

Kass, Leon R. "The Wisdom of Repugnance." *New Republic*, June 2, 1997, 17–26.

Kealey, Terence. *The Economic Laws of Scientific Research*. New York: St. Martin's Press, 1996.

Kellert, Stephen R. *Kinship to Mastery: Biophilia in Human Evolution and Development*. Washington, D.C.: Island Press, 1997.

Kenshur, Oscar. "The Rhetoric of Incommensurability." *Journal of Aesthetics and Art Criticism*, 42/4 (1994), 375–381.

Kermode, Frank. "The Academy vs. the Humanities." Review of J. M. Ellis, *Literature Lost*. *Atlantic Monthly*, August 1997, 93–96.

Kermode, Frank. "Toe-Lining." *London Review of Books*, January 22, 1998, 9–10.

Kernan, Alvin, ed. *What's Happened to the Humanities?* Princeton, N.J.: Princeton University Press, 1997.

Kevles, Bettyann H., and Daniel J. Kevles. "Scapegoat Biology." *Discover*, October 1997, 58–62.

Kevles, Daniel J. Review of S. Jasanoff, *Science at the Bar*. *American Scientist*, 85, no. 4 (July/August 1997), 393.

Kevles, Daniel J. "Grounds for Breeding: The Amazing Persistence of Eugenics in Europe and North America." *Times Literary Supplement*, January 2, 1998, 3–4.

Kevles, Daniel J. "The New Enlightenment." Review of E. O. Wilson, *Consilience*. *New York Times Book Review*, April 26, 1998, 11–12.

Kevles, Daniel J. *The Baltimore Case: A Trial of Politics, Science, and Character*. New York: W. W. Norton, 1998.

Kilborn, Peter T. "For Managed Care, Dial the Keepers of the Cures." *New York Times*, November 3, 1997, A14.

Kimball, Roger. *Tenured Radicals: How Politics Has Corrupted Our Higher Education*. New York: Harper and Row, 1990.

Kitcher, Philip. *Abusing Science: The Case against Creationism*. Cambridge, Mass.: MIT Press, 1982.

Kitcher, Philip. "Whose Self Is It Anyway?" *The Sciences*, September/October 1997, 58–62.

Kitcher, Philip. "Tall, Slender, Straight and Intelligent." *London Review of Books*, March 5, 1998, 15–16.

Klass, Perri. "Managing Managed Care." *New York Times Magazine*, October 5, 1997.

Klinkenborg, Verlyn. "Imagining the Consequences of Cassini." *New York Times*, October 16, 1997, A28.

Klinkenborg, Verlyn. "Biotechnology and the Future of Agriculture." *New York Times*, December 8, 1997, A24.

Koertge, Noretta. "Are Feminists Alienating Women from the Sciences?" *Chronicle of Higher Education*, September 14, 1994, A80.

Koertge, Noretta. "Feminist Epistemology: Stalking an Un-Dead Horse." In Gross, Levitt, and Lewis, eds., *The Flight from Science and Reason*, 413–419.

Koertge, Noretta. "Postmodern Transformations of the Problem of Scientific Literacy." In Koertge, ed., *A House Built on Sand.*

Kohn, Alphie. "Only for My Kid: How Privileged Parents Are Undermining School Reform." *Phi Delta Kappan*, 79, no. 8 (1998), 568–577.

Kolata, Gina. "Big Study Sees No Evidence Power Lines Cause Leukemia." *New York Times*, July 3, 1997, 1 ff.

Kolata, Gina. "Scientists Face New Ethical Quandaries in Baby-Making." *New York Times*, August 19, 1997, C1 ff, C8.

Kolata, Gina. "Taking Charge." Review of G. W. Smith and S. Naifeh, *Making Miracles Happen. New York Times Book Review*, September 7, 1997, 19.

Kolata, Gina. "On Cloning Humans, 'Never' Turns Swiftly into 'Why Not.'" *New York Times*, December 2, 1997, A1 ff, A24.

Kolata, Gina. "Proposal for Human Cloning Draws Dismay and Disbelief." *New York Times*, January 8, 1998, A22.

Kolata, Gina. "A Child's Paper Poses a Medical Challenge." *New York Times*, April 1, 1998, A1 ff, A20.

Kolata, Gina. "Cows Cloned in First Case Like Dolly's, Japanese Say." *New York Times*, July 7, 1998, A8.

Kolata, Gina. "In Implant Case, Science and the Law Have Different Agendas." *New York Times*, July 11, 1998, A8.

Konig, Hans. "Notes on the Twentieth Century." *Atlantic Monthly*, September 1997, 90–100.

Krause, Bernie. "What Does Western Music Have to Do with Nature?" *Terra Nova*, 2, no. 2 (Summer 1997), 108–114.

Kroll, Diana Lambdin, Joanna O. Masingila, and Sue Tinsley Mau. "Cooperative Problem Solving: But What about Grading?" *Arithmetic Teacher*, 39, no. 6 (1992), 17–23.

Kronholz, June. "Numbers Racket: 'Standards' Math Is Creating a Big Division in Education Circles." *Wall Street Journal*, November 5, 1997.

Kronholz, June. "Low X-pectations." *Wall Street Journal*, June 16, 1998, 1 ff.

Kuznar, Lawrence A. *Reclaiming a Scientific Anthropology.* Walnut Creek, Calif.: Altamira Press, 1997.

Labaton, Stephen. "Gates Helped Draft Microsoft's Response to Judge." *New York Times*, January 15, 1998, D1 ff, D6.

Larson, E. J., and L. Witham. "Leading Scientists Still Reject God." *Nature*, 394 (July 23, 1998), 313.

Latour, Bruno. "La fin de science?" *La Recherche*, 294, January 1997, 97.

Lazurus, Neil, Steven Evans, Anthony Arnove, and Anne Menke. "The Necessity of Universalism." *differences: A Journal of Feminist Cultural Studies*, 7, no. 1 (1997), 75–145.

Leacock, Eleanor Burke. *Myths of Male Dominance.* New York: Monthly Review Press, 1981.

Lee, Keekok. "Designer Mountains: The Ethics of Nanotechnology." *Terra Nova*, 2, no. 2 (Spring 1997), 129–134.

Lefkowitz, Mary. *Not Out of Africa: How Afrocentrism Became an Excuse to Teach Myth as History.* New York: Basic Books, 1996.

Lefkowitz, Mary R., and Guy Maclean Rogers, eds. *Black Athena Revisited*. Chapel Hill, N.C.: University of North Carolina Press, 1996.

Leland, John, and Carla Power. "Deepak's Instant Karma." *Newsweek*, October 29, 1997, 52–58.

Lemann, Nicholas. "Rewarding the Best, Forgetting the Rest." *New York Times, Week in Review*, April 26, 1998, 15.

Lentricchia, Frank. "Last Will and Testament of an Ex-Literary Critic." *Lingua Franca*, September/October 1996, 59–67.

Leonard, John. "Culture Watch: Alien Nation." *Nation*, June 15–22, 1998, 23–28.

Lerner, Maura. "Researcher Investigating Toxin Becomes Subject of Investigation." *Minneapolis Star-Tribune*, May 17, 1998, 1.

Levitt, Norman. "Letter to the Editor." *New Republic*, August 16, 1993, 4.

Levitt, Norman. "Sniper Attack." *Lingua Franca*, November/December 1994, 72.

Levitt, Norman. Review of J. T. Cushing, *Quantum Mechanics*. *Physics Today*, November 1995, 84.

Levitt, Norman. "Vetenskapens pessimism, antivetenskapens nihilism." *Tvärsnitt*, 4 (1995), 62–69.

Levitt, Norman. "Knowledge, Knowingness, and Reality." *Skeptic* 4, no. 4 (1996), 78–82.

Levitt, Norman. "Author's Response: Three Sonnets Contra-Fullerine." *Metascience*, 7, no. 1 (1998), 50–51.

Levitt, Norman. "*The End of Science*, the Central Dogma of Science Studies, Monsieur Jordain, and Uncle Vanya." In Koertge, ed., *A House Built on Sand*, 272–285.

Levitt, Norman. "Mathematics as the Stepchild of Contemporary Culture." In Gross, Levitt, and Lewis, eds., *The Flight from Science and Reason*, 39–53.

Levitt, Norman. "What's Post Is Quaalude." *2B*. In press.

Levitt, Norman, and Paul R. Gross. "The Perils of Democratizing Science." *Chronicle of Higher Education*, October 5, 1994, B1–B2.

Levitt, Norman, and Paul R. Gross. "Academic Anti-Science." *Academe*, November/December 1996, 38–42.

Lewis, Martin W. *Green Delusions*. Durham, N.C.: Duke University Press, 1992.

Lewis, Martin W. Review of J. Bennett and W. Chaloupka, eds., *In the Nature of Things*. *Annals of the Association of American Geographers*, 84 (1994), 514–516.

Lewis, Martin W. "In Defense of Environmentalism." Review of P. R. Ehrlich and A. H. Ehrlich, *The Betrayal of Science and Reason*. *Issues in Science and Technology*, 13, no. 2 (Winter 1996/97), 82–84.

Lewis, Martin W. "Radical Environmental Philosophy and the Assault on Reason." In Gross, Levitt, and Lewis, eds., *The Flight from Science and Reason*, 209–230.

Lilla, Mark. "A Tale of Two Reactions." *New York Review of Books*, May 14, 1998, 4–7.

Lindsay, Ronald A. "Taboos without a Clue: Sizing Up Religious Objections to Cloning." *Free Inquiry*, 17, no. 3 (1997), 15–17.

Loftus, Elizabeth. "Remembering Dangerously." *Skeptical Inquirer*, March/April 1995, 20–29.

Loftus, Elizabeth. "The Price of Bad Memories." *Skeptical Inquirer*, March/April 1998, 23–24.

Longino, Helen. *Science as Social Knowledge*. Princeton, N.J.: Princeton University Press, 1990.

Loving, Cathleen C. "Cortes' Multicultural Empowerment Model and Generative Teaching and Learning in Science." To appear in *Science and Education*. Electronically published, 1998.

Lumpkin, Beatrice. "Africa in the Mainstream of Mathematics History." In Powell and Frankenstein, eds., *Ethnomathematics*, 101–117.

Lunch, William. "Contemporary Attacks on Science." Lecture, International Political Science Association World Congress, Berlin, August 25, 1994.

Malcolm, Janet. *The Journalist and the Murderer.* New York: Knopf, 1990.

Marcuse, Herbert. *One-Dimensional Man.* Boston: Beacon Press, 1994.

Marx, Leo. *The Pilot and the Passenger.* New York: Oxford University Press, 1988.

Marx, Leo. "The Idea of 'Technology' and Postmodern Pessimism." In Ezrahi, Mendelsohn, and Segal, eds., *Technology, Pessimism, and Postmodernism*, 11–28.

Matsumura, Molleen. "Miracles In, Creationism Out: 'The Geophysics of God.'" *Reports of the National Center for Science Education*, 17, no. 3 (1997), 29–32.

Matthews, Michael R. *Science Teaching: The Role of History and Philosophy of Science.* New York: Routledge, 1994.

May, Robert M. "The Scientific Wealth of Nations." *Science*, February 7, 1997, 793–796.

Mazur, Allan. "The Science Court: Reminiscence and Retrospective." Conference, "Which Scientist Do You Believe? Process Alternatives in Technological Controversies," Concord, N.H., October 6–7, 1994. Electronically published, Franklin Pierce Law Center Homepage.

McDonald, Kim A. "Researchers Battle for Access to a 9,300-Year-Old Skeleton." *Chronicle of Higher Education*, May 22, 1998, A18–A22.

McDonald, Kim A. "Public's Interest in Science Is at All-Time High, but Knowledge Still Poor, Survey Shows." *Chronicle of Higher Education*, Daily News, July 1, 1998. Electronically published.

McElroy, Michael. "Clean Machines." *New Republic*, May 4, 1998, 24.

McGinn, Colin. "Reason the Need." Review of T. Nagel, *The Last Word. New Republic*, August 4, 1997, 38–41.

McGinniss, Joe. *Fatal Vision.* New York: Putnam, 1983.

McInerney, Joseph D. "Public Education." *Biotechnology Education*, 2, no. 3 (1991), 98–100.

McInerney, Joseph D. "Animals in Education: Are We the Prisoners of False Sentiment?" *American Biology Teacher*, 55, no. 5 (1993), 276–277.

McMillen, Liz. "The Science Wars Flare at the Institute for Advanced Study." *Chronicle of Higher Education*, May 16, 1997, A13.

Meade, Harry M. "Dairy Gene." *The Sciences*, September/October 1997, 20–25.

Meier, Barry. "A Silent Witness in Cigarette Trials." *New York Times*, March 30, 1998, A10.

Menand, Louis. "The Demise of Disciplinary Authority." In Kernan, ed., *What's Happened to the Humanities?* 201–219.

Mendelsohn, Everett. "The Politics of Pessimism: Science and Technology Circa 1968." In Ezrahi, Mendelsohn, and Segal, eds., *Technology, Pessimism, and Postmodernism*, 151–173.

Mermin, N. David. "The Science of Science: A Physicist Reads Barnes, Bloor, and Henry." *Social Studies of Science*, 28, no. 4, 603–623.

Meserve, Richard A. Review of S. Jasanoff, *Science at the Bar. Issues in Science and Technology*, 12, no. 3 (1996), 88–91.

Michod, Richard E. "What Good Is Sex." *The Sciences*, September/October 1997, 42–46.

Miele, Frank. "Living without Limits: An Interview with Julian Simon." *Skeptic*, 5, no. 1 (1997), 54–59.

Miele, Frank. "IQ in Review: Getting at the Hyphen in the Nature-Nurture Debate." *Skeptic*, 5, no. 4 (1997), 91–95.

Miller, Jon D., Rafael Pardo, and Fujio Niwa. *Public Perceptions of Science and Technology: A Comparative Study of the European Union, the United States, Japan, Canada.* Bilbao: Fundación BBV, 1997.

Minsky, Marvin. "Technology and Culture." *Social Research*, 64, no. 3 (1997), 1119–1126.

Mirsky, Steve. "The Emperor's New Toilet Paper." *Scientific American*, July 1997, 24.

Mishkin, Barbara. "The Needless Agony and Expense of Conflict among Scientists." *Chronicle of Higher Education*, February 23, 1994, B1–B2.

Molé, Phil. "Deepak's Dangerous Dogmas." *Skeptic*, 6, no. 2 (1998), 38–45.

Moore, John H. "The Changing Face of University R&D Funding." *American Scientist*, 86, no. 5 (September/October 1998), 402.

Morrison, Toni, and Claudia Brodsky Lacour. *Birth of a Nation'hood: Gaze, Script, and Spectacle in the O. J. Simpson Case.* New York: Pantheon Books, 1997.

Morse, Whitney (pseud.). "On the Inevitability of Miracles." *Topics in Geriatric Rehabilitation*, 10, no. 1 (1994), 1–6.

Moyers, Bill. *Healing and the Mind.* New York: Doubleday, 1993.

Myhrvold, Nathan. "The Dawn of Technomania." *New Yorker*, October 20 and 27, 1997, 236–237.

Naess, Arne. *Ecology, Community, and Lifestyle: Outline of an Ecosophy.* Cambridge, U.K.: Cambridge University Press, 1988.

Nakičenovič, Nebojša. "Freeing Energy from Carbon." *Daedalus*, 125, no. 3 (1996), 95–112.

Nanda, Meera. "Is Modern Science a Western, Patriarchal Myth? A Critique of the Populist Orthodoxy." *South Asia Bulletin*, 11, nos. 1 & 2 (1991), 32–61.

Nanda, Meera. "Reclaiming Modern Science for Third World Progressive Social Movements." *Economic and Political Weekly*, 33 (April 18, 1998), 915–922.

Nanda, Meera. "The Epistemic Charity of the Social Constructivist Critics of Science and Why the Third World Should Refuse the Offer." In Koertge, ed., *A House Built on Sand*, 286–311.

Nanda, Meera. "The Science Question in Postcolonial Feminism." In Gross, Levitt, and Lewis, eds., *The Flight from Science and Reason*, 420–436.

National Association of Biology Teachers. Statement on evolution versus creation. *Creation/Evolution*, 37 (1995), s3–s4.

National Science Foundation. *Science and Engineering Indicators 1996.* Washington, D.C.: National Science Foundation, 1996.

National Science Teachers Association. "Inclusion of Nonscience Tenets in Science Instruction." *Creation/Evolution*, 37 (1995), s8–s9.

Natoli, Joseph. *A Primer to Postmodernity.* Oxford: Blackwell, 1997.

Nattier, Jan. "Buddhism Comes to Main Street." *Wilson Quarterly*, 21, no. 2 (Spring 1997), 72–90.

NBC. "The Mysterious Origins of Man." Broadcast, February 25, 1996.

Nelkin, Dorothy. "Science, Technology and Political Conflict: Analyzing the Issues." In Nelkin, ed., *Controversy* (3d ed.), ix–xxv.

Nelkin, Dorothy. "The Science Wars: Responses to a Marriage Failed." In Ross, ed., *Science Wars*, 114–122.

Nelkin, Dorothy, ed. *Controversy: Politics of Technical Decisions.* 3d ed. Thousand Oaks, Calif.: Sage, 1992.

Nelson, Richard R. "A Science Funding Contrarian." Review of T. Kealey, *The Economic Laws of Scientific Research. Issues in Science and Technology*, 14, no. 1 (1997), 90–92.

Nemecek, Sasha. "Frankly, My Dear, I Don't Want a Dam." *Scientific American*, August 1997, 20–22.

Niebuhr, Gustav. "Lutherans Bridge Old Divisions, but Decide to Keep One in Place." *New York Times*, August 19, 1997, A1 ff, A12.

Niebuhr, Gustav. "Lutherans Reconsider Episcopal Concordat." *New York Times*, November 30, 1997, 25.

Noble, David F. *The Religion of Technology: The Divinity of Man and the Spirit of Invention*. New York: Alfred A. Knopf, 1997.

O'Meara, Tim. "Anthropology as Empirical Science." *American Anthropologist*, 91, no. 2 (June 1989), 354–369.

O'Neill, Gerard K. *The High Frontier: Human Colonies in Space*. New York: Morrow, 1977.

Oransky, Ivan. "Losing Patients." *New Republic*, November 17, 1997, 16–17.

Ortiz de Montellano, Bernard. "Afrocentric Pseudoscience: The Miseducation of African-Americans." In Gross, Levitt, and Lewis, eds., *The Flight from Science and Reason*, 561–572.

Ortiz de Montellano, Bernard, Gabriel Haslip-Viera, and Warren Barbour. "They Were NOT Here Before Columbus: Afrocentric Hyperdiffusionism in the 1990s." *Ethnohistory*, 44, no. 2 (1997), 199–254.

Osborne, Mary. "Call for Papers: 1997 Special Edition of *Journal of Research on Science Teaching*; 'Pedagogies and Science Education.'" E-mail communication, July 16, 1996.

Osborne, R. J., and P. Freyberg. *Learning in Science: The Implication of Children's Science*. London: Heinemann, 1985.

Park, Robert L. "Power Line Paranoia." *New York Times*, November 13, 1996, A23.

Park, Robert L. "Scientists and Their Political Passions." *New York Times*, May 2, 1998, A15.

Patai, Daphne. "Why Not a Feminist Overhaul of Higher Education?" *Chronicle of Higher Education*, January 23, 1998, A56.

Patai, Daphne, and Noretta Koertge. *Professing Feminism: Cautionary Tales from the Strange World of Women's Studies*. New York: Basic Books, 1994.

Paulos, John Allen. *Innumeracy: Mathematical Illiteracy and Its Consequences*. New York: Hill and Wang, 1988.

Paulos, John Allen. *A Mathematician Reads the Newspaper*. New York: Basic Books, 1995.

PBS. "Frontline: Nuclear Reaction." Broadcast, April 22, 1997.

Pearce, Fred. "Greenhouse Wars." *New Scientist*, July 19, 1997, 38–43.

Pennock, Robert T. "Naturalism, Creationism, and the Meaning of Life: The Case of Philip Johnson Revisited." *Creation/Evolution*, 39 (1996), 10–27.

Penzias, Arno. "Technology and the Rest of Culture." *Social Research*, 64, no. 3 (1997), 1021–1041.

Perutz, M. F. "The Pioneer Defended." *New York Review of Books*, December 21, 1995, 54–58.

Petroski, Henry. "Development and Research." *American Scientist*, 85, no. 3 (May/June 1997), 210–213.

Phillips, Adam. "You Have to Be Educated to Be Educated." Review of S. Shapin, *The Scientific Revolution. London Review of Books*, April 3, 1997, 24–25.

Piatelli-Palmarini, Massimo. *Inevitable Illusions: How Mistakes of Reason Rule Our Minds*. New York: John Wiley & Sons, 1996.

Pinch, Trevor, and Harry M. Collins. *The Golem: What Everyone Should Know about Science*. Cambridge, U.K.: Cambridge University Press, 1994.

Pinker, Steven. "Why They Kill Their Newborns." *New York Times Magazine*, November 2, 1997, 52–54.

Pinnick, Cassandra. "Feminist Epistemology: Implications for Philosophy of Science." *Philosophy of Science*, 61, no. 4 (1994), 646–657.

Pinxten, Rik. "Applications in the Teaching of Mathematics." In Powell and Frankenstein, eds., *Ethnomathematics*, 373–401.

Plumwood, Val. *Feminism and the Mastery of Nature*. London: Routledge, 1993.

Plumwood, Val. "Being Prey." *Terra Nova* 1, no. 3 (Summer 1996), 33–44.

Polkinghorne, John. *Belief in God in an Age of Science*. New Haven: Yale University Press, 1998.

Pomeroy, D. "Science across Cultures: Building Bridges between Traditional Western and Alaskan Native Cultures." In Hills, ed., *History and Philosophy of Science in Science Education*, vol. 2, 257–268.

Popper, Karl R. *The Open Society and Its Enemies*, vol. 2. Princeton, N.J.: Princeton University Press, 1966.

Popper, Karl R. *The Myth of the Framework: In Defense of Science and Rationality*. London: Routledge, 1994.

Porter, Robert S. "Inevitable Illusions." *American Scholar*, 65, no. 3 (Summer 1996), 480.

Poses, Roy M., and Alice M. Isen. "Qualitative Research in Medicine and Health Care: Questions and Controversy." *Journal of General Internal Medicine*, 13 (1998), 32–38.

Posner, Richard A. "The Skin Trade." Review of D. A. Farber and S. Sherry, *Beyond All Reason*. *New Republic*, October 13, 1997, 40–43.

Powell, Arthur B., and Marilyn Frankenstein. *Ethnomathematics: Challenging Eurocentrism in Mathematics Education*. Albany: State University of New York Press, 1997.

Powell, Arthur B., and Marilyn Frankenstein. "Uncovering Distorted and Hidden History of Mathematical Knowledge." In Powell and Frankenstein, eds., *Ethnomathematics*, 51–59.

Preston, Douglas. "Skin & Bones." *New Yorker*, February 9, 1998, 52–53.

Price, Jack. On the *Roger Hedgecock Show*, April 24, 1996. Transcript electronically published.

Quine, Willard Van Orman. *From a Logical Point of View: Logico-Philosophical Essays* (2d ed.). New York: Harper and Row, 1963.

Quine, Willard Van Orman. "Commensurability and the Alien Mind." *Common Knowledge*, 1, no. 3 (Winter 1992), 1–2.

Quittner, Joshua, and Michelle Slatalla. *Speeding the Net: The Inside Story of Netscape and How It Challenged Microsoft*. New York: Atlantic Monthly Press, 1998.

Rabinow, Paul. *Essays on the Anthropology of Reason*. Princeton, N.J.: Princeton University Press, 1996.

Radcliffe-Richards, Janet. "Why Feminist Epistemology Isn't." In Gross, Levitt, and Lewis, eds., *The Flight from Science and Reason*, 385–412.

Raimi, Ralph A., and Lawrence S. Braden. *State Mathematics Standards: An Appraisal of Math Standards in 46 States, the District of Columbia, and Japan*. Washington, D.C.: Thomas B. Fordham Foundation, 1998.

Rarden, Xandra, and Sheila Jasanoff. "An Atypical Complaint: Science v. Subjectivity in Breast Implant Litigation." Lecture abstract, Workshop: "Making People: The Normal and Abnormal in Constructions of Personhood." Cornell University, Ithaca, N.Y., April 1998. Electronically published.

Raymo, Chet. "Spiritually Homeless in the Cosmos." *Boston Globe*, March 25, 1996, 38.

Reardon, Carol. *Pickett's Charge in History and Memory*. Chapel Hill, N.C.: University of North Carolina Press, 1997.

Ribalow, M. Z. "Take Two." *The Sciences*, September/October 1997, 38–41.

Ridpath, Ian. "Interview with J. Allen Hynek." *Nature*, 251 (October 4, 1975), 369.

Robinson, Marilynne. "Consequences of Darwinism." *Salmagundi*, 114/115 (1997), 13–47.

Robinson, Marilynne. *The Death of Adam: Essays on Modern Thought*. New York: Houghton Mifflin, 1998.

Rochester, J. Martin. "What's It All About, Alphie? A Parent/Educator's Response to Alphie Kohn." *Phi Delta Kappan*, 80, no. 2 (1998), 165–169.

Rorty, Richard. "Against Unity." *Wilson Quarterly*, 22, no. 1 (1998), 28–38.

Rosa, Linda, Emily Rosa, Larry Sarner, and Stephen Barrett. "A Close Look at Therapeutic Touch." *Journal of the American Medical Association*, 279 (1998), 1005–1010.

Rosin, Hanna. "Don't Touch This." *New Republic*, November 10, 1997, 24–31.

Ross, Andrew. *Strange Weather: Culture, Science and Technology in the Age of Limits*. London: Verso, 1991.

Ross, Andrew. "Science Backlash on Technoskeptics." *Nation*, October 2, 1995, 346–350.

Ross, Andrew, ed. *Science Wars*. Durham, N.C.: Duke University Press, 1997.

Ross, Andrew. "Introduction." In Ross, ed., *Science Wars*, 1–15.

Rosser, Sue V., ed. *Teaching the Majority: Breaking the Gender Barriers in Science, Mathematics, and Engineering*. New York: Teachers College Press, 1995.

Rothstein, Edward. "Where a Democracy and Its Money Have No Place." *New York Times*, October 26, 1997, AR1 ff, 39.

Rothstein, Edward. "The Subjective Underbelly of Hardheaded Math." *New York Times*, December 20, 1997, B7–B8.

Rothstein, Edward. "Now a Warm Welcome Instead of a Cold Bath." *New York Times*, May 2, 1998, B9–B11.

Rothwax, Harold J. *Guilty: The Collapse of Criminal Justice*. New York: Warner Books, 1996.

Roush, Wade. "Herbert Benson: Mind Body Maverick Pushes the Envelope." *Science*, April 18, 1997, 357–359.

Rubin, Charles T. "The Troubled Relationship between Science and Policy." Address, Washington Roundtable on Science and Public Policy, January 15, 1997. Electronically published.

Ruelle, David. *Chance and Chaos*. Princeton, N.J.: Princeton University Press, 1991.

Ruse, Michael. *Monad to Man: The Concept of Progress in Evolutionary Biology*. Cambridge, Mass.: Harvard University Press, 1996.

Sampson, Wallace. "Inconsistencies and Errors in Alternative Medical Research." *Skeptical Inquirer*, September/October 1997, 35–38.

Sampson, Wallace. "Antiscience Trends in the Rise of the 'Alternative Medicine' Movement." In Gross, Levitt, and Lewis, eds., *The Flight from Science and Reason*, 188–197.

Sapolsky, Robert. "A Gene for Nothing." *Discover*, October 1997, 40–46.

Sarewitz, Daniel. *Frontiers of Illusion: Science, Technology, and the Politics of Progress*. Philadelphia: Temple University Press, 1996.

Sarich, Vincent. "In Defense of *The Bell Curve*: The Reality of Race and the Importance of Human Difference." *Skeptic*, 3, no. 3 (1995), 84–93.

Sartwell, Crispin. "Science and Race in W.E.B. Du Bois." Abstract of a talk presented to the Rensselaer Polytechnic Institute Science and Technology Studies Colloquium. Electronically published.

Saunders, Debra J. "Math Gadfly Calls Math Faddists' Bluff." *San Francisco Chronicle*, August 7, 1998, A23.

Scarry, Elaine. "The Fall of TWA 800: The Possibility of Electromagnetic Interference." *New York Review of Books*, April 9, 1998, 59–76.

Schneider, Stephen H. *Laboratory Earth: The Planetary Gamble We Can't Afford to Lose.* New York: Basic Books, 1997.

Schoofs, Mark. "Fear and Wonder." *Village Voice*, September 30, 1997, 36–43.

Schoofs, Mark. "The Myth of Race." *Village Voice*, October 28, 1997, 34–39.

Schoofs, Mark. "Genetic Justice." *Village Voice*, November 18, 1997, 44–50.

Schweder, Richard A. "Ancient Cures for Open Minds." *New York Times Week in Review*, October 26, 1997, 6.

Sclove, Richard. *Democracy and Technology*. New York: Guilford Press, 1995.

Sclove, Richard. "Science, Inc., versus Science-for-Everyone." *Loka Alert,* 4, no. 4 (July 24, 1997). Electronically published.

Sclove, Richard E. "Better Approaches to Science Policy." *Science*, February 27, 1998, 1283.

Scott, Eugenie C. "The Struggle for the Schools." *Natural History*, July 1994, 9–13.

Scott, Eugenie C., and Robert M. West. "Again, Johnson Gets It Wrong." *Creation/Evolution*, 37 (1995), 26–29.

Scott, Eugenie C. "Antievolution and Creationism in the United States." *Annual Review of Anthropology*, 26 (1997), 263–289.

Scott, Eugenie C., and Kevin Padian. "The New Antievolution—and What to Do about It." *Tree*, 12, no. 2 (1997), 84.

Scott, Eugenie C. "Creationism, Ideology, and Science." In Gross, Levitt, and Lewis, eds., *The Flight from Science and Reason*, 505–522.

Scott, M. M. "Intellectual Property Rights: A Ticking Time Bomb in Academia." *Academe*, May/June 1998, 22–26.

Shamos, Morris. *The Myth of Scientific Literacy*. New Brunswick, N.J.: Rutgers University Press, 1995.

Shapin, Steven, and Simon Schaffer. *Leviathan and the Air-Pump*. Princeton, N.J.: Princeton University Press, 1985.

Shattuck, Roger. *Forbidden Knowledge: From Prometheus to Pornography*. New York: St. Martin's Press, 1996.

Shattuck, Roger. "Education: Higher and Lower." *Salmagundi* 116/117 (1998), 41–54.

Shenk, David. "Biocapitalism: What Price the Genetic Revolution?" *Harper's Magazine*, December 1997, 37–45.

Shermer, Michael. "The Beautiful People Myth: Why the Grass Is Always Greener in the Other Century." *Skeptic*, 5, no. 1 (1997), 72–79.

Shermer, Michael. *Why People Believe Weird Things*. New York: W. H. Freeman, 1997.

Shields, Michael. "Swiss Vote Down Bid to Curb Genetic Research." Reuters News Service, June 7, 1998. Electronically published.

Shiva, Vandana. *Monocultures of the Mind: Perspectives on Biodiversity and Biotechnology*. London: Zed, 1993.

Silver, Brian L. *The Ascent of Science*. Oxford: Oxford University Press, 1998.

Smolin, Lee. *The Life of the Cosmos*. Oxford: Oxford University Press, 1997.

Soderqvist, Thomas, ed. *The Historiography of Contemporary Science and Technology*. Amsterdam: Harwood Academic Publishers, 1997.

Sokal, Alan. "A Physicist Experiments with Cultural Studies." *Lingua Franca*, May/June 1996, 62–64.

Sokal, Alan. "Transgressing the Boundaries—An Afterword." *Dissent*, Fall 1996, 93–99.

Sokal, Alan. "Transgressing the Boundaries: The Transformative Hermeneutics of Quantum Gravity." *Social Text*, 46/47 (1996), 217–252.

Sokal, Alan, and Jean Bricmont. *Impostures intellectuelles*. Paris: Editions Odile Jacob, 1997.

Sokal, Alan. "A Plea for Reason, Evidence, and Logic." *New Politics*, 6, no. 2 (1997), 126–129.

Sommers, Christina Hoff. *Who Stole Feminism? How Women Have Betrayed Women*. New York: Simon and Schuster, 1994.

Sonleitner, Frank J. "Philip Johnson and the Philosophy of Science." *National Center for Science Education Reports*, 15, no. 4 (1996), 18–19.

Sonnert, Gerhard, and Gerald Holton. *Gender Differences in Science Careers*. New Brunswick, N.J.: Rutgers University Press, 1995.

Sonnert, Gerhard, and Gerald Holton. *Who Succeeds in Science?* New Brunswick, N.J.: Rutgers University Press, 1995.

Specter, Michael. "Europe, Bucking Trend in U.S., Blocks Genetically Altered Food." *New York Times*, July 20, 1998, A1 ff, A8.

Stacey, Weston M. "The ITER Decision and U.S. Fusion R&D." *Science and Technology*, 13, no. 4 (1997), 53–59.

Stanley, Alessandra. "Power to Fascinate Intact, Turin Shroud Is Unfurled." *New York Times*, April 19, 1998, A1.

Steinberg, Jacques. "California Goes to War over Math Instruction." *New York Times*, November 27, 1997, A1 ff, A34.

Steinberg, Jacques. "Clashing over Education's One True Faith." *New York Times, Week in Review*, December 14, 1997, 1 ff, 14.

Steinberg, Jacques. "In New Math Standards, California Stresses Skills over Analysis." *New York Times*, December 14, 1997, 43.

Stenger, Vic. "NIH Sticks It to the Public." *Hawaii Rational Inquirer*, 3, no. 12 (1997). Electronically published.

Stenger, Vic. "Courts Hitting Hard at Recovered Memory Therapists." *Hawaii Rational Inquirer*, 3, no. 13 (1997). Electronically published.

Stenger, Vic. "NIH 'Consensus' Statement on Acupuncture Condemned." *Hawaii Rational Inquirer*, 3, no. 13 (1997). Electronically published.

Stenger, Vic. "Senator Harkin Sees That OAM Gets a Fat Increase in Its Budget . . . while Senator Hatch Proposes to Limit FDA." *Hawaii Rational Inquirer*, 3, no. 14 (1997). Electronically published.

Stevens, William K. "Experts on Climate Change Ponder: How Urgent Is It?" *New York Times*, September 9, 1997, C1–C2.

Stevens, William K. "Computers Model World's Climate, but How Well?" *New York Times*, November 4, 1997, F1 ff, F6.

Stevens, William K., et al. "Global Warming: A Preview to the Kyoto Conference." *New York Times*, December 1, 1997, F1–F12.

Stevenson, Leslie, and Henry Byerly. *The Many Faces of Science*. Boulder, Colo.: Westview Press, 1995.

Stix, Gary. "Profile: Jeremy Rifkin." *Scientific American*, August 1997, 28–30.

Stolberg, Sheryl Gay. "Small Spark Ignites Debate on Cloning Humans." *New York Times*, January 20, 1998, A11.

Stolberg, Sheryl Gay. "Gifts to Science Researchers Have Strings, Study Finds." *New York Times*, April 1, 1998, A17.

Stolberg, Sheryl Gay. "Neutral Experts Begin Studying Dispute Over Breast Implants." *New York Times*, July 23, 1998, A21.

Stolberg, Sheryl Gay. "As Doctors Trade Shingles for Marquees, Cries of Woe." *New York Times*, August 3, 1998, A1 ff, A14.

Stone, David A., and John A. Rich. "Letter to the Editor: In Defense of Qualitative Research." *Journal of General Internal Medicine*, 13 (1998), 68–69.

Stork, David G., ed. *HAL's Legacy: 2001's Computer as Dream and Reality*. Cambridge, Mass.: MIT Press, 1997.

Sulloway, Frank. "Darwinian Virtues." Review of M. Ridley, *The Origin of Virtue*. *New York Review of Books*, April 9, 1998, 34–40.

Surowiecki, James. "Flame Wars." *New York Magazine*, June 15, 1998, 30–35.

Sutherland, John. "Unplug the Car and Let's Go!" *London Review of Books*, August 21, 1997, 17–18.

Swackhamer, Deborah L., and Ronald A. Hites. "Reducing Uncertainty in Estimating Toxaphene Loading to the Great Lakes." EPA Grant Number R825246. Electronically published.

Talbot, Margaret. "America Image Disorder." Review of M. Williamson, *The Healing of America*. *New Republic*, December 8, 1997, 31–38.

Talbot, Margaret. "The Perfectionist." Review of B. McKibben, *Maybe One*. *New Republic*, July 20 and 27, 1998, 32–38.

Tanne, Janice Hopkins. Address, New York Academy of Sciences, October 2, 1997.

Taylor, Bron. "Earth First! Fights Back." *Terra Nova*, 2, no. 2 (Spring 1997), 27–41.

Taylor, Charles. *Multiculturalism: Examining the Politics of Recognition*. Princeton, N.J.: Princeton University Press, 1994.

Taylor, Paul. "Inside a French Milkman." *London Review of Books*, August 21, 1997, 4.

Tenner, Edward. *Why Things Bite Back: Technology and the Revenge of Unintended Consequences*. New York: Knopf, 1996.

Tessman, Irvin, and Jack Tessman. "Mind and Body." Review of H. Benson, *Timeless Healing*. *Science*, April 18, 1997, 369–379.

Thurtle, Philip. "The Creation of Genetic Identity." *Stanford Humanities Review*, 5 (suppl.) (1996), 80–98.

Tipler, Frank J. *The Physics of Immortality: Modern Cosmology, God and the Resurrection of the Dead*. New York: Anchor Books, 1997.

Todorov, Tzvetan. "The Surrender to Nature." Review of E. O. Wilson, *Consilience*. *New Republic*, April 27, 1998, 29–33.

Tomasky, Michael. *Left for Dead: The Life, Death, and Possible Resurrection of Progressive Politics in America*. New York: Free Press, 1996.

Toumey, Christopher P. *Conjuring Science: Scientific Symbols and Cultural Meanings in American Life*. New Brunswick, N.J.: Rutgers University Press, 1996.

Traweek, Sharon. "Unity, Dyads, Triads, Quads, and Complexity: Cultural Choreographies of Science." In Ross, ed., *Science Wars*, 139–150.

Trefil, James. "Scientific Literacy." In Gross, Levitt, and Lewis, eds., *The Flight from Science and Reason*, 543–550.

Tribe, Laurence H. "Second Thoughts on Cloning." *New York Times*, December 5, 1997, A31.

Trollope, Anthony. *North America*. London: Penguin Books, 1992.

Van Sertima, Ivan. *Blacks in Science, Ancient and Modern*. New Brunswick, N.J.: Transaction Books, 1983.

Verran, Helen Watson, and David Turnbull. "Talking about Science and Other Knowledge Traditions in Australia." *EASST Review*, 14, no. 1 (March 1995). Electronically published.

Verrengia, Joseph B. "Hawaii Scientists Clone 50 Mice." AP wire story, July 22, 1998.

Visvanathan, Shiv. "A Celebration of Difference: Science and Democracy in India." *Science*, April 3, 1988, 42–43.

Vos Savant, Marilyn. *Parade's Ask Marilyn*. New York: St. Martin's Press, 1992.

Vos Savant, Marilyn. *The World's Most Famous Math Problem*. New York: St. Martin's Press, 1993.

Vyse, Stuart A. *Believing in Magic: The Psychology of Superstition*. New York: Oxford University Press, 1997.

Wagner, Peter. "A Gentleman's Anger: John Ziman's Report on Science." Review of J. Ziman, *Prometheus Bound. EASST Review*, 14, no. 1 (March 1995). Electronically published.

Wagner, Wendeline L. "They Shoot Monkeys, Don't They?" *Harper's Magazine*, August 1997, 27–30.

Wald, Matthew L. "Finding a Formula to Light the World but Guard the Bomb." *New York Times*, June 2, 1998, F3.

Walters, Ronald G., ed. *Scientific Authority and Twentieth-Century America*. Baltimore: Johns Hopkins University Press, 1997.

Walzer, Michael. "Pluralism and Social Democracy." *Dissent*, Winter 1998, 47–53.

Wargo, John. *Our Children's Toxic Legacy: How Science and Law Fail to Protect Us from Pesticides*. New Haven: Yale University Press, 1996.

Webb, Sara L. "Female-Friendly Environmental Science: Building Connections and Life Skills." In Rosser, ed., *Teaching the Majority*, 193–210.

Weill, Nicholas. "La mystification pedagogique du professeur Sokal." *Le Monde*, December 20, 1996, 1 ff, 6.

Weinberg, Steven. *Dreams of a Final Theory*. New York: Pantheon Books, 1992.

Weinberg, Steven. "Reductionism Redux." *New York Review of Books*, October 5, 1995, 39–42.

Weinberg, Steven. "Sokal's Hoax." *New York Review of Books*, August 8, 1996, 11–12.

Weinberg, Steven. "Physics and History." *Daedalus*, 127, no. 1 (1998), 151–164.

Weissmann, Gerald. "'Sucking with Vampires': The Medicine of Unreason." In Gross, Levitt, and Lewis, eds., *The Flight from Science and Reason*, 179–187.

Wheeler, David L. "From Homeopathy to Herbal Therapy, Researchers Focus on Alternative Medicine." *Academe Today*, March 27, 1998. Electronically published.

Wheeler, David L. "Top Medical Journal Publishes Paper by 6th Grader Debunking Therapeutic Touch." *Academe Today*, April 2, 1998. Electronically published.

Wick, David. *The Infamous Boundary: Seven Decades of Controversy in Quantum Physics*. Boston: Birkhäuser, 1995.

Wildavsky, Aaron B. *But Is It True? A Citizen's Guide to Environmental, Health, and Safety Issues*. Cambridge, Mass.: Harvard University Press, 1995.

William Mitchell Law Review, 22, no. 2 (1996).

Williams, George C. *The Pony Fish's Glow and Other Clues to Plan and Purpose in Nature*. New York: Basic Books, 1997.

Williams, Joy. "The Inhumanity of the Animal People." *Harper's Magazine*, August 1997, 60–67.

Williams, Luther. "National Science Foundation's Systemic Initiatives," March 23, 1997. Electronically published, National Science Foundation website.

Wilson, Edward O. *The Diversity of Life*. New York: W. W. Norton, 1992.

Wilson, Edward O. "Science and Ideology." *Academic Questions*, 8, no. 3 (1995), 73–81.

Wilson, Edward O. "Scientists, Scholars, Knaves and Fools." *American Scientist*, 86, no. 1 (January/February 1998), 6–7.

Wilson, Edward O. "Back from Chaos." *Atlantic Monthly*, March 1998, 41–62.

Wilson, Edward O. "The Biological Basis of Morality." *Atlantic Monthly*, April 1998, 53–70.

Wilson, Edward O. "Resuming the Enlightenment Quest." *Wilson Quarterly*, 22, no. 1 (1998), 16–27.

Wilson, Edward O. *Consilience: The Unity of Knowledge.* New York: Knopf, 1998.

Wisconsin Department of Public Instruction. "Evolution, Creation, and the Science Curriculum." *Creation/Evolution*, 37 (1995), s20–s23.

Wise, Donald U. "Creationism's Geologic Time Scale." *American Scientist*, 86, no. 2 (March/April 1998), 169–173.

Witten, Edward. "Matter Matters." *New Republic*, December 19, 1997, 16–17.

Wolpert, Lewis. *The Unnatural Nature of Science.* Cambridge, Mass.: Harvard University Press, 1993.

Wood, James. "Twister." *New Republic*, June 8, 1998, 46.

Wurtman, Richard. "Cure All." *New Republic*, November 10, 1997, 16–18.

Yoon, Carol Kaesuk. "Evolutionary Biology Begins Tackling Public Doubts." *New York Times*, July 8, 1998, B8.

Zalewski, Daniel. "Ties That Bind: Do Corporate Dollars Strangle Scientific Research?" *Lingua Franca*, June/July 1997, 51–59.

Ziman, J. *Prometheus Bound: Science in a Dynamic Steady State.* Cambridge, U.K.: Cambridge University Press, 1994.

Zoeteman, Kees. "Middle Earth in the Balance." *Harper's Magazine*, February 1998, 27.

Zuger, Abigail. "After Bad Year for A.M.A., Doctors Debate Its Prognosis." *New York Times*, December 2, 1997, F3.

Index

About the Author

Norman Levitt was educated at the Bronx High School of Science, Harvard, and Princeton. He is a professor of mathematics at Rutgers University, and does research in geometric topology. His work on questions involving science, politics, and cultural values includes *Higher Superstition* (with Paul R. Gross) and, as editor (with Gross and Martin W. Lewis), *The Flight from Science and Reason*. He has also addressed these issues in articles for a number of journals, including *Academe*, *The Chronicle of Higher Education*, *Skeptic*, *Academic Questions*, and *Skeptical Inquirer*.